T0310142

Fractography of Glasses and Ceramics VI

Fractography of Glasses and Ceramics VI

Ceramic Transactions, Volume 230

*Proceedings of the Sixth Conference on the
Fractography of Glasses and Ceramics
June 5–8, 2011
Jacksonville, Florida*

Edited by
James R. Varner
Marlene Wightman

A John Wiley & Sons, Inc., Publication

Published by John Wiley & Sons, Inc., Hoboken, New Jersey.
Published simultaneously in Canada.

For general information on our other products and services or for technical support, please contact our Customer Care Department within the United States at (800) 762-2974, outside the United States at (317) 572-3993 or fax (317) 572-4002.

Wiley also publishes its books in a variety of electronic formats. Some content that appears in print may not be available in electronic formats. For more information about Wiley products, visit our web site at www.wiley.com.

Library of Congress Cataloging-in-Publication Data is available.

ISBN: 978-1-118-27373-9
ISSN: 1042-1122

Printed in the United States of America.

10 9 8 7 6 5 4 3 2 1

Contents

Preface

These are the proceedings of the Sixth Alfred Conference on the Fractography of Glasses and Ceramics, which was held in Jacksonville, Florida, 5-8 June 2011. Van Fréchette and Jim Varner of the New York State College of Ceramics at Alfred University started this series in 1986.

The Fifth Conference was held six years ago (9–13 July 2006, Rochester, New York), and we felt that it was time to get another snapshot of research and practice of fractography of brittle materials. We asked S. Jill Glass (Sandia National Laboratory, Albuquerque, New Mexico), John J. Mecholsky, Jr. (University of Florida, Gainesville, Florida), and Jeffrey J. Swab (U.S. Army Research Laboratory, Aberdeen Proving Ground, Maryland) to join us as members of the conference organizing committee. We are grateful for their guidance and assistance in attracting speakers and participants. They were especially helpful in identifying conference topics and potential invited speakers. Their active participation in the conference was also one key to its success.

Fractography of Glasses and Ceramics VI had the following objectives: (1) review progress in fundamentals of fractography of glasses and ceramics; (2) highlight applications of fractography in research, production, testing, and product failure analysis; (3) promote discussion among diverse users of fractography of glasses and ceramics; (4) document the state of fractography and its importance in the field of glass and ceramics.

We met these objectives through over 30 high-quality presentations (including four invited papers) and lively, in-depth discussions in the formal sessions and during the breaks. We were pleased to have a mix of experienced fractographers and folks who are newer to the discipline. There were presentations on a variety of ceramics (silicon nitride, alumina, nanocomposites, laminates, dental ceramics, ceria, and zirconia), and glasses. Topics included effects of laser etching of glass, fractography of ion-exchange-strengthened glass, mixed-mode failure, effects on crack evolution and propagation, quantitative fractography, ballistic damage, and effects of processing on strength and fracture behavior. The invited talks provided in-depth coverage of four very different subjects. George Quinn's keynote lecture was a thorough look at the history of the fractography of glasses and ceramics. His careful study of the literature and excellent choice of images made this a fascinating presentation and addition to these proceedings. Matteo Ciccotti provided an excellent overview of phenomena at the crack tip during slow crack growth in glasses. Jan Dusza highlighted failure and damage mechanisms in ceramic nanocomposites. Chuck Kurkjian brought his insights into the issues of inherent strength and damage mechanisms in glasses.

The Sixth Alfred Conference on the Fractography of Glasses and Ceramics met our goal of providing a snapshot of the state of fractography of these brittle materials. We thank our co-organizers and all of the presenters and participants for making this a lively and successful meeting.

JAMES R. VARNER
MARLENE WIGHTMAN

Dedication

Roy Rice

We dedicate the proceedings of the Sixth Alfred Conference on the Fractography of Glasses and Ceramics to Roy W. Rice. Roy passed away on April 29, 2011, at the age of 76.

Roy Rice had a long and distinguished career of research in ceramics. He started with Boeing, later headed the Ceramics Section and Branch of the Naval Research Laboratory, and was Director of Materials Research of W.R. Grace & Co before his retirement in 1994. He published over 300 papers, authored three books, and was granted 30 patents. His research topics include pressure sintering and other processing techniques, machining and its effect on properties, and the effect of porosity on mechanical and physical properties. His colleagues recognized his innovations of concepts, processes, and structure-property relationships. Whenever Roy presented a paper or made a comment at a technical meeting, people listened.

Among his many accomplishments, Roy was a pre-eminent fractographer of ceramics. He was closely associated with the Alfred Conferences on fractography. He gave the plenary lecture, the very first lecture, at the first conference in 1986. The published version of his talk ("Perspective on Fractography") is still used by researchers today who want to gain perspective on failure analysis of polycrystalline ceramics and single crystals. Roy attended every other conference in this series, except this most recent one. His contributions in the published proceedings, like the first one, remain valued references. For example, there is no better summary of fracture modes in polycrystalline ceramics than his leadoff article in the proceedings of the Third Alfred Conference ("Ceramic Fracture Mode—Intergranular vs. Transgranular Fracture").

These Proceedings will be the first without a paper from Roy Rice. His absence at the meeting was felt, and we sorely miss having another print example of his remarkable insight into the complexities of fracture in polycrystalline ceramics and single crystals. Most of all, we miss his smile, his unassuming nature, his abiding interest in ceramics, and his inherent kindness.

JIM VARNER

In Memoriam

Janet B. Quinn

Janet B. Quinn, 58, passed away on July 19, 2008 after a brief but valiant battle with lung cancer. At the time of her death, Dr. Quinn was a Project Leader for the American Dental Association Foundation Paffenbarger Research Center. Her research was funded by grants from the National Institutes of Health and the Rockefeller Brothers Fund. Janet began her professional career at the U. S. Army Materials Testing Laboratory in Watertown, Massachusetts, while also rearing her two children. Later, she worked as a consultant at the National Institute of Standards and Technology (NIST). She also became a student again, and received her Ph.D. in Materials Science and Engineering from the University of Maryland 25 years after getting her M.S. in Mechanical Engineering.

Janet's work included time-dependent failure of ceramics, development of testing methods and programs, determining fracture energies of single crystals, and tensile testing of ceramic fibers. More recently, she turned her considerable intellect and deep scientific curiosity to the field of dental ceramics. In just a short period of time, her studies on dental restoration failure analysis earned her international recognition. Janet worked hard to hone her own skills as a fractographer. As her experience grew, so did her desire to share her knowledge with others. She developed and led the first dental fractography hands-on course in May 2007 at NIST. Goals of this course included understanding failure modes of dental restorations, producing guidelines and recommendations for improvements in longevity of dental restorations, and making more people in the field aware of available information. Janet drew upon her own experience and her participation in the Alfred short course on Failure Analysis of Brittle Materials in the summer of 2003 in creating her course focused on fractography of dental restorations. She led the course again in 2008, just a few days before receiving the news that she had lung cancer. The course continues to be offered each year and to be well attended, with Janet's husband, George, carrying on the leadership.

Janet Quinn was also well known outside of science through her accomplishments in Middle Eastern dance. She was a professional instructor of Middle Eastern dance, and she trained hundreds of belly-dance students in Boston, Washington, D.C., Cologne, Germany, and Strasbourg, France.

We know that Janet Quinn would have contributed much more to fractography of ceramics, especially of dental restorations, had she been given more time. However, we celebrate the significant contributions that she made throughout her entire professional career, and we remember her joy of life, her ability to brighten up any room that she entered, her ever-present smile, her affection for people, and the inspiration she provided to everyone who spent even five minutes with her.

JIM VARNER
MARLENE WIGHTMAN

A HISTORY OF THE FRACTOGRAPHY OF GLASSES AND CERAMICS

George D. Quinn
Consultant
North Potomac, MD 20878, USA

ABSTRACT

The science of fractography of brittle materials evolved from failure analysis problems involving brittle metals such as cast iron and early steels. Early analyses focused on general patterns of fracture and how they correlated to the loading conditions. Scientific and engineering explanations gradually were developed for the observed patterns. Advances in microscopy and flaw-based theories of strength and fracture mechanics led to dramatic advances in the state of the art. The Griffith theory of flaw-controlled strength gradually became accepted, especially when the microscopic flaws themselves were finally detected by Ernsberger. Improvements in processing control in the 1970s led to stronger ceramics that were more amenable to fractographic analysis and even fracture origin determination. This history is a story of the people who were pioneers in the field, of theoretical developments on the strength of brittle materials, of advances in materials science including the fabrication of stronger materials, of developments in microscopy, of the publication of key books, and of standardization.

INTRODUCTION

Some deem fractography as the study of fracture surfaces, but I take a broader view. Fractography is the means and methods for characterizing fractured specimens or components. A simple examination of the fragments and how they fit together to study the overall breakage pattern is a genuine fractographic analysis, even if the fracture surfaces are not examined.

When I wrote my fractography guide book between 2003 and 2007,[1] my curiosity was aroused about how the field had evolved. Articles existed about how the fractographic analysis of metals evolved,[2,3,4,5] but there was no analogue for ceramics and glasses. I first wrote about this topic in 2008 for the Stara Lesna conference on Fractography of Advanced Ceramics.[6] At the time, I was unable to include any illustrations of photographs of key fracture features, or photographs of our forebears. I therefore have written this more comprehensive illustrated article to remedy these shortcomings and to expand some key points, particularly on the work of Mr. Roy Rice and standardization activities.

The key scientists, engineers, and analysts who contributed to our field are shown in Figure 1. This figure is a slightly revised version of my figure in the 2008 paper.[6] In the text below, I have used underlined italics to highlight the first documented instances of new terms such as: "hackle," "mirror," and "Wallner lines."

EARLY STUDIES

Derek Hull[2] credits Robert Hooke[7] with the first reported observation of a fracture surface, of any subject made using a microscope (Figure 2). Hull also gave a brief history of the use of fractography for minerals, metals, and lithic structures over the centuries.

1

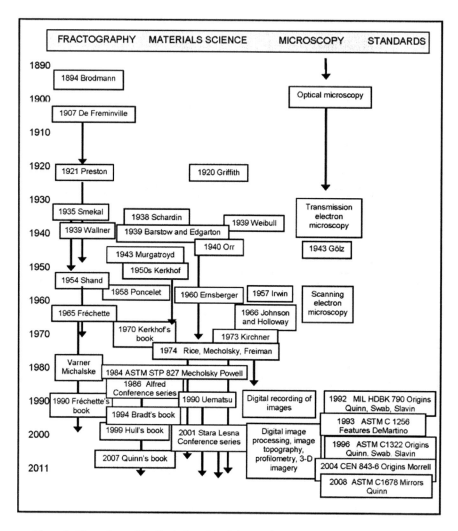

Figure 1. Chronology of the Evolution of the Science of Fractography of Brittle Materials

Figure 2. Illustration by Robert Hooke of the fracture surface of Ketton limestone, a shapeable stone with an interesting microstructure. It is formed by precipitation of calcite and is used for architectural projects. From *Micrographia,* 1665, (Ref. 7). The bonded spherical nodules are about 0.5 mm diameter according to Hull[2]. (Courtesy Project Gutenberg eBook.)

Although he showed no pictures, Brodmann[8] in 1894 made some of the earliest observations of overall fracture patterns and fracture surfaces of glass rods tested in tension, bending, and torsion. His paper "A few observations on the strength of glass articles" was published in Reports of the Scientific Society of Gottingen.[8] Most of the paper was about the testing procedure and the strength results. His excellent fractographic observations were in a single paragraph at the end of the paper. Brodmann used the word *mirrors* to describe the smooth area around an origin and noted that:

"In general, the radially measured size of these mirrors was larger, the smaller was the strength."

This was a very important observation, which eventually was quantified. Fracture mirror size analysis today is an important failure analysis tool.

Charles De Freminville (1856 to 1936) wrote the first comprehensive treatment of the subject. He was identified as a mechanical and industrial engineer in Paris, working as a consultant for Schneider works in 1919. He was concerned with brittle fractures in brittle steels and iron in which negligible deformation occurred. He observed that the fractures bore a good resemblance to those in glasses and bitumen. It seemed that some such fractures were so sudden and violent that they could be deemed "explosive." He wrote two long papers in *Revue de Metallurgie* in 1907[9] and 1914.[10] The first was fifty-one pages in length and was titled the "Character of Vibrations Accompanying Impact, Observations from the Examinations of Broken Pieces." Impact testing was proposed as one way of measuring the fracture resistance of brittle metals, a suggestion that was realized years later with the adoption of Charpy impact testing. Overall there are thirty-eight figures in the first paper, many with multiple parts, which show a variety of classical fracture patterns. The second paper was a major expansion and almost book length (eighty five pages). Each paper included superb combinations of schematics and photographs of both the overall breakage patterns and fracture surfaces. For example, Figure 3 shows several of his illustrations for bending fractures in a round axle as well as glass rods for comparison. This extraordinary sketch shows glass rod fracture patterns with multiple bending fractures. He wrote that once elastic waves reverberate, regions initially in tension could be exposed to transient compression stresses and vice versa. One whole section of his paper covers secondary fractures caused by reverberations of elastic waves once the primary fracture had occurred.

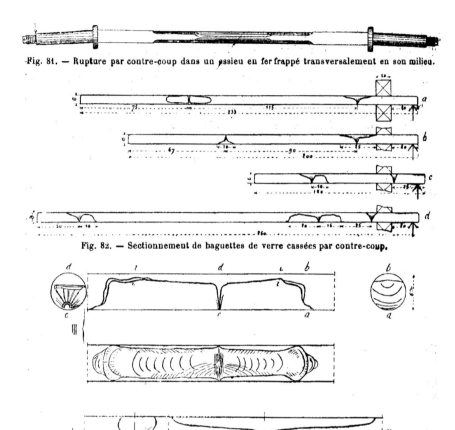

Fig. 81. — Rupture par contre-coup dans un essieu en fer frappé transversalement en son milieu.

Fig. 82. — Sectionnement de baguettes de verre cassées par contre-coup.

Fig. 83. — Types de cassures par contre-coup dans des baguettes de verre.

Figure 3. Illustrations by De Freminville in 1907. The top shows fracture in a brittle cast iron axle, and the lower figures show bending fractures in glass rods. (Reprinted with permission of EDP Sciences.)

Figures 4 and 5 show fracture types that are quite recognizable to the modern fractographer. De Freminville categorized fractures as "direct" or "indirect." "Indirect fractures," shown in Figure 4b were bending fractures by overloading or impact on the opposite side of the fracture origin. De Freminville observed that the fracture occurred opposite the struck surface and the crack propagated up

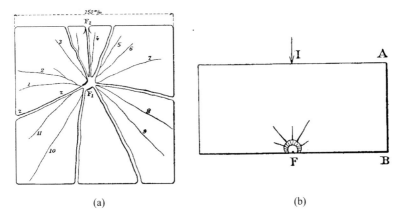

(a)   (b)

Figure 4. Illustrations by De Freminville in 1907 and 1914. (a) shows a glass plate. (b) shows a fracture surface of a beam broken in bending. De Freminville deemed this an "indirect fracture" since the initiation of fracture (as shown by the fracture mirror schematic on the bottom) occurred opposite to the impact site shown by the top arrow. (Reprinted with permission of EDP Sciences.)

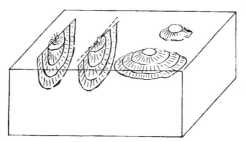

Fig. 74. — Schémas de commencements de cassure visibles par transparence dans u bloc de verre.

Figure 5. "Direct fractures" are those that occurred at an impact site, according to De Freminville. The two images on the left show sharp-contact-initiated fractures in a block of glass, and the two on the right show blunt contactor ("Hertzian") cone cracks. (Reprinted with permission of EDP Sciences.)

and joined with the impact site. "Direct fractures" were those where the origin was initiated at the impact site (Figure 5). Several illustrations such as Figure 5 illustrate classical Hertzian cone cracks in flat plates or in glass spheres dropped from a height. His carefully drawn breakage patterns reveal classic bending stress branching patterns in both square plates and long slabs. His paper included stress distributions from other sources that were relevant to the fracture. Little was said about the fracture origins themselves. De Freminville used the French word "le foyer" which may be translated as the "source." For the case of "direct fractures," the impact site itself was assumed to have been the origin.

Fig. 58. — Petit foyer d'éclatement suivi d'une dislocation importante. Pied d'un vase de cristal (13 diam.).

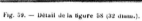

Fig. 59. — Détail de la figure 58 (32 diam.).

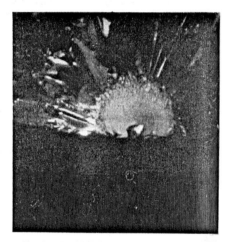

Fig. 60. — Détail de la figure 58, contre-partie de la figure 59 (30 diam.)

Figure 6. Fracture surface of a strong broken glass rod. The fracture mirror and even the initiating flaw are readily observable. De Freminville did not use the term "mirror," however, as Brodmann had before him. De Freminville also did not comment on the character of the flaw. (Reprinted with permission of EDP Sciences.)

Nevertheless, surface contact or abrasion damage flaws are in fact *easily seen* in several of De Freminville's figures, such as Figure 6. It is astonishing that his 1907 perceptive paper showed

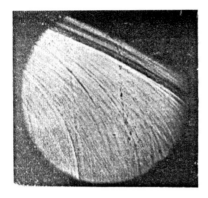
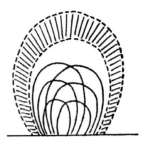

Fig. 40. — Schéma de la disposition des petites ondulations conjuguées dans un foyer d'éclatement.

Figure 7. De Freminville observed rib like undulations in the fracture surface. (Reprinted with permission of EDP Sciences.)

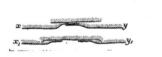

Fig. 37. — Cassure d'un pied de verre (13 diam.). Imbrication des surfaces de sectionnement.

Figure 8. De Freminville commented on lines that later became known as "hackle lines." In this illustration, he showed how overlapping crack segments create such lines. (Reprinted with permission of EDP Sciences.)

numerous examples of what later became known as "Wallner lines," the telltale gentle arc-shaped lines ("ribs") on fracture surfaces, as shown in Figure 7. He correctly interpreted these lines as undulations in the fracture surface as the crack radiated outwards from a fracture origin. The telltale lines created by an impact are concentric about the impact site and lead one's attention back to the origin. Hackle lines are also shown or depicted in the illustrations, such as Figure 8, although he described them as striae. He compared rib lines and hackle lines in glass and sandstone.

De Freminville set an example for all future fractographers by often showing matched drawings and photos such as Figure 8. He also showed reconstructed parts and illustrations of the fracture surfaces. A section of the 1914 paper also shows fascinating illustrations of broken diametrally-loaded 30 mm diameter glass balls, as shown in Figure 9.

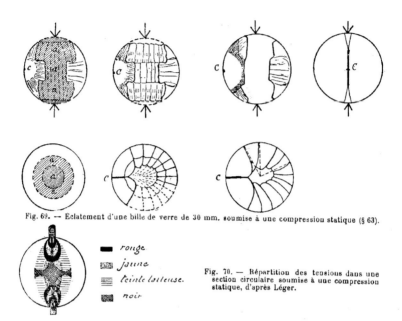

Fig. 69. — Eclatement d'une bille de verre de 30 mm. soumise à une compression statique (§ 63).

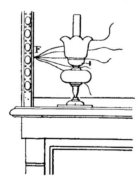

rouge.
jaune
teinte laiteuse.
noir

Fig. 70. — Répartition des tensions dans une section circulaire soumise à une compression statique, d'après Léger.

Figure 9. Fracture patterns in diametrically compressed glass balls. The top row shows side views where the fracture origins ("c") are on the outer rim at the equator. The figure on the right shows the fracture pattern from an interior origin site. The second row shows top views of the same pieces. The final figure on the bottom illustrates the stress distribution. (Reprinted with permission of EDP Sciences.)

Figure 10. Fracture of a glass mirror caused by center heating. The tensile stresses at the origin site F on the cooler rim were moderate to large since branching occurred. The waviness of the cracks once they propagated further into the warmer, compression-stressed portion of the plate, is a telltale characteristic of thermal stress fracture. (Reprinted with permission of EDP Sciences.)

One of his final illustrations in the 1914 paper, Figure 10, is a charming illustration of a glass mirror fracture due to center heating from an oil lamp placed too close to the mirror. Uneven heating creates tensile stresses on the cooler rim. Edge-initiated fractures in glass are a problem to this day!

De Freminville's 1914 paper was a major expansion of the first and included additional loading conditions such as thermal stresses. It is curious that the journal publishers allowed the repetition of so many illustrations in their journal only a few years later, but they are to be commended since the second paper can be used as a standalone document. Many of the figures are enlarged in the second paper. One addition was an explicit section on component reconstruction. Many more fracture examples were shown including some for brittle metals such as broken railroad tracks and wheel axles. Radiating hackle lines were termed striae. Some illustrations and schematics showed overlapping crack portions that formed hackle lines (lances). He described what we now call fracture mirrors although he did not use that term as Brodmann had done earlier. The smooth central region that was the focus of all the splintering lines seemed to surround an origin site. The smooth central region was also surrounded by a dull surface portion that he correctly attributed to surface roughening. He also ventured a discussion of the flaws located at the center of the mirror in cases of slab bending and impact bending fractures. Keeping in mind the photographic and microscopic limitations of the day, the flaws he showed were large surface contact damage, handling, or grinding flaws. He astutely observed that brittle materials are susceptible to surface flaws, but showed some examples of internal origins in brittle metallic tension specimens. In December 1919, De Freminville was invited to give a lecture at the Annual Meeting of the American Society of Mechanical Engineers.[11] A sixteen-page summary article of his fractographic work on the topic of the reliability of materials and the mechanisms of fracture was published in English after the meeting. De Freminville's three papers[9-11] constitute the first significant treatment of the fractography of brittle materials. A web search of his name indicates that he was a leader in the "scientific management" movement of the early 1900s, and was active in the American Society of Mechanical Engineers and used modern management techniques applied to auto manufacture. He lived from 1856 to 1936. I have not been able to locate a photograph of him.

PROGRESS IN THE 1920s – 1950s

Only a short time later in 1920, Alan Griffith (Figure 11) published his seminal paper[12] that identified flaws as the nuclei of fracture. There is no indication that he was aware of De Freminville's papers. Fine filaments of pristine drawn glass had tensile strengths approaching the theoretical strength, but strengths decreased with time and/or exposure to surface damage sources. He showed that the strength of a uniformly stressed plate containing an elliptical through-crack of size 2c in a uniform tensile stressed plate in plane stress is:

$$\sigma_f = \sqrt{\frac{2E\gamma_f}{\pi c}} \tag{1}$$

where σ_f is the fracture stress, E is the elastic modulus, and γ_f is the fracture surface energy to create unit surface area.[12,13] (Incidentally, the first paper had an extra Poisson's ratio in the denominator, that was corrected in his second paper without explanation.) The critical feature of this relationship is that strength is inversely proportional to the square root of flaw size. The larger the flaw, the weaker is the structure. Griffith stated:[13]

"the general conclusion may be drawn that the weakness of isotropic solids … is due to the presence of discontinuities, or flaws, as they may be more correctly called, whose ruling dimensions are large compared with molecular distances."

Figure 11. Alan A. Griffith (1893 – 1963).

Notwithstanding some confusion as to whether the surface energy was simply the thermodynamic surface energy or a larger effective fracture surface energy, researchers now had some guidance as to the size of the flaws they should look for with their microscopes. Griffith showed only one sketch of a hypothetical crack and no photomicrographs, but he estimated the crack size had to be 1.5 μm (Ref. 13) or 5 μm (Ref. 12) for his conventional tension strength tests. Griffith believed that flaws were molecular fault regions in glass that would act as fracture nuclei. The quest to find minute Griffith flaws took many years and was not completely settled until the advent of electron microscopy. De Freminville had already shown some relatively large strength-limiting flaws in glass in his photos and sketches (see Figure 6 of this paper.) The concept of Griffith flaws applies equally well to large, visibly observable flaws and to submicroscopic flaws in very high strength materials. Griffith's paper was not immediately accepted by many in the field. (For more on this see the fine review of the history of glass strength studies by Holloway.[14]) For many years researchers strove to find the submicroscopic flaws they could not see and argued over their true nature and whether they really existed. The expression "Griffith flaw" was typically used to describe submicroscopic sized flaws they could not readily detect. An outstanding biography on Griffith's life and his work[15] analyzes some of the minor mistakes in his equations, but these do not detract from the significance of the work.

Significant advances in understanding strength and flaws in glass were made by Frank Preston over a long and productive career (Figure 12). Born in Leicester, England, he began writing about his glass work in 1921 when he studied the flaws created in glass surfaces by grinding and polishing, contact with balls, and scoring with glazer's wheels.[16] Figure 13 shows some of his illustrations. He used the terms median and lateral to describe cracks created under a glazer's diamond, as shown in Figure 14, a nomenclature that has persisted to this day. He noted that:

"It is clear from these observations that … there are deep flaws extending far below the surface irregularities…"

This was one of the first observations that cracks can penetrate far deeper than the grinding surface roughness damage. It was recognized as "a rather startling conclusion" that was verified by additional work by a reviewer in the discussion section at the end of the paper. Another commenter said that:

"The present paper constituted a marked advance in the subject. Apart from its scientific value, it should be of great assistance to manufacturers."

Figure 12. Frank Preston (1896 – 1989). (Courtesy of the American Glass Research Company.)

(a)   (b)

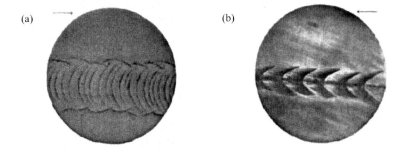

Figure 13. Preston's photos of damage caused by a hard ball.[15] (a) shows the top surface with ball motion from left to right (arrow). (b) shows a cross sectional view of the damage underneath as seen on the fracture surface. The top half is a mirror image of the actual fracture surface on the bottom. (Reprinted with permission of IOP publishing.)

Years later, Preston's colleagues published a paper[17] that showed how soft metals could damage glass surfaces by creating chatter sleeks that were a series of shallow partial cone cracks.

Preston emigrated to the United States in 1921 and soon founded Preston Laboratory in Butler, Pennsylvania. In 1926 Preston wrote the first of a series of perceptive papers on the strength of glass[18] in which he acknowledges Griffith's notion that flaws control strength. Preston discussed blunt contact cracks, stones from manufacture, and fractures produced by heating and cooling. It is curious that Preston, like De Freminville before him, referred to some virulent fractures (e.g., thermal shock) as "explosive." Figure 15 shows what we now refer to as a fracture mirror. Preston describes some of the fracture markings surrounding an "explosion center." Referring to the figure, he described X as a:

> "tiny semi-circular area of bright ('polished') fracture, surrounded by a dull fracture P, and it is succeeded by a coarser structure Q, . . . which may be recognized as hackly fracture with the unaided eye."

We now refer to these as the "mirror," "mist," and "hackle" regions.

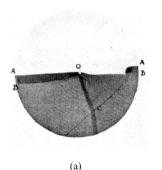

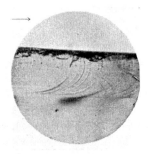

(a)   (b)

Figure 14. Preston's photos of damage caused by a glazer's diamond.[16] (a) is an end-on view where "O" shows the axis of scoring (in and out of the page). The original glass surface is A-A. B-B shows "lateral" cracks and C is the "median" crack. (b) shows the fracture surface of a scored plate broken in bending where the crack ran from left to right. The shallow scoring median cracks are visible at the top as well as the many curved Wallner lines typical of a bending fracture. (Reprinted with permission of IOP publishing.)

Fɪɢ. 12.

Diagram of an explosion centre and distribution of hackle-texture.

Figure 15. Preston's schematic of a fracture mirror (Reference 18). (Reprinted with permission of the Society of Glass Technology.)

Preston in 1926 said[18] that the "explosion center no doubt represents the site of some pre-existing minute 'flaw'" and "the fracture originates in a minute spot of possibly ultra-microscopic size, and spreads quietly over a tiny area-the semicircular spot." It is curious that he seemed reluctant to use the word "flaw" as Griffith had done earlier. Preston always showed the word in quotation marks.

Further into the paper, Preston described the arced or ribbed shape lines that we now call "Wallner lines." He called them "tide marks" or ripple marks, and he correctly deduced that they were formed by an oscillating stress system. He also correctly observed that for violent fractures the oscillations that produced the rib lines did not reach the crack front simultaneously, but at different times causing the curved arcs (Figure 16). This perceptive observation anticipates Wallner's analysis in 1939 and, had Preston been more mathematically inclined, the rib lines today may have been called "Preston Lines." The paper further discussed the effect of stress upon crack branching or forking. He even said:

"hackle is incipient forking (page 257), the first stage in the development of a radiant {crack}."

(a)

(b)

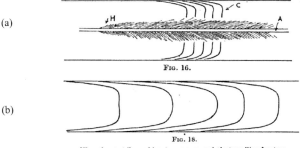

FIG. 16.

FIG. 18.

Wave-fronts (rib markings) on a more quietly travelling fracture.

Figure 16. Preston's schematics of the fracture surfaces of glass plates.[18] (a) shows a thick rolled plate which was not annealed, as was customary. The severe internal residual stresses cause a fracture surface similar to that of tempered glass. H is for "hackle" lines in a herringbone pattern. "C" denotes what we now refer to as Wallner lines, illustrating that the crack led in the interior due to stress gradients. (b) shows the fracture surface with no hackle and only curved Wallner lines in a location of the same plate with tensile stress of lower magnitude. (Reprinted with permission of the Society of Glass Technology.)

Preston felt that, once a hackle fissure parallel to the main crack surface grew broad enough such that it extended through the thickness of a plate, the crack would fork. Preston concluded that the object of his paper was to "correlate various fractures in glass and other brittle materials, …in a general system based on their genesis and on the stresses initiating and extending them." Most of his optical micrograph figures had magnifications of 44 to 120 power.

Preston's 1926 paper elicited a number of comments that were published as a group in the Journal of Glass Technology in 1927.[19] Preston replied with further comments on "flaws." In a brilliant anticipation of Weibull's 1939 analysis, he said:[19]

"A glass may contain a virtually infinite number of nuclei {flaws}, but rupture will not start in the absence of stress. On the other hand, glass would never break under the stresses to which it is commonly subjected if it had no nuclei. The particular condition that a nucleus should be present in, or very close to a particular point (the point where the tension is greatest) can only be expressed as a probability function. If the test is made on such a fashion that the tension is absolutely uniform over a large region, the probability that a nucleus will be located in the critical region amounts to certainty, ….. If on the other hand, very small specimens are used, or the stress is very uneven so that only a small region experiences the maximum tension, then the probability of a nucleus being in the critical region is by no means a certainty, and the stress throughout the mass may have to be raised much higher until a great enough tension is produced somewhere else where there is a nucleus."

"Under these conditions, a larger series of tests will show a high probable error {variability} for the individual measurements, and when we have tested ten thousand pieces, we shall still not be able to predict the strength of the next piece with any accuracy."

Preston continued to publish regularly with as many as 200 papers on glass and a like number on ornithology, geology, ecology, and even politics! One fascinating, but very short paper in 1935 showed a relationship between the branching angle and the stress state[20] as shown in Figure 17. For example, uniaxial stresses, such as in a tension or flexural strength test, cause 45 degree branches. Equibiaxial stresses, such as a ring-on-ring disk strength test, have 180 degree angles. Seventy six years later his trend curve of branching angle versus biaxial stress ratio needs further analytical and

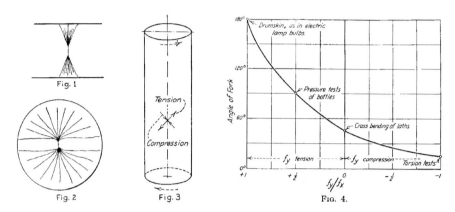

Figure 17. Preston's 1935 paper[20] showed that the angle of forking (branching) varied with stress state. His Fig. 1 is for uniaxial tension; 2 is for equibiaxial tension; 3 is torsion. The abscissa of his Figure 4 is the ratio of the two principal stresses.

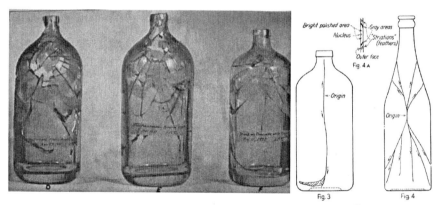

Figure 18. Figures from Preston's 1939 bottle-breakage paper.[21]

experimental verification, especially for ceramics. Practical problem solving led Preston to write papers in 1939[21] and 1942[22] on bottle breakages (Figure 18). In the former, he said that the principles of fractographic analysis and pattern analysis:

".. are often useful in determining whether a bottle was broken the way a complainant in a lawsuit says it was or whether the case contains elements of deceit."

Preston's 1939 bottle-breakage paper[21] and a second 1942 paper[23] on mechanical properties of glass treat the topic of static fatigue, or the time-dependent degradation of strength while under load. The later 1942 paper includes much colorful language and is entertaining to read:

"A lecture on the mechanical properties of glass a few years ago might have been regarded … on a par with the metaphysical significance of elephants or the moral significance of liquid air, or that mechanical properties of glass could be described in exactly the same word as the famous treatise on the snakes in Ireland, viz, 'There are none.'"

Later he wrote:

"No one will deny, however, that the behavior of cast iron is angelic, in respect to brittleness, as compared with the behavior of glass."

He also wrote simple declarative sentences:

"Under ordinary circumstances, glass breaks only in tension and the fracture is at right angles to the maximum tension." "The 'strength of glass' is a very elusive quantity."

The first sentence is evidently the first formal statement of what we now refer to as the law of normal crack propagation, which is:[1] "A crack propagates normal to the direction of local principal tension stress." It is very odd, but De Freminville, who was right in so many of his observations in his 1907 paper,[9] seemed to have gotten this wrong (on his page 878):

"Une Tenson normale au plan de la cassure n'est pas favorable à la propagation par fissilité. Elle ne peut produire qu'un arrangement."

which may be translated to:

"A tension normal to the fracture plane is not favorable to the propagation of the crack. This tension can only create a pull out (tear)."

In the conclusions in his 1942 Journal of Applied Physics paper,[23] Preston summarized the art of fractographic analysis:

"From the beginning of time to its end, no two cracked surfaces will fit each other, unless originally they were part of the same piece. Cracks are as distinctive as fingerprints. … Now the important thing about cracks is that they must be propagated. They do not originate all over the final fracture surface, but at one tiny spot, and from thence are propagated out to the rest of the area. This fact and the fact that the telltale marks or "fingerprints" of the process are left on the surface, provide the groundwork for the subject of Fracture Diagnosis, which in the last decade has become an important minor branch of physical science."

I have been asked many times over the years by forensic investigators whether fractures are as distinctive as fingerprints, and I have directed the inquirer to the quote above. Summaries of Preston's contributions and career have been published[24,25] and a web page about him is maintained by the American Glass Research Company,[26] the modern day descendant of Preston Labs.

In the meantime, considerable progress was being made in Europe in Professor Adolf Smekal's (Fig. 19) Institute of Theoretical Physics in the University of Halle in Germany between 1935 and 1945. Smekal was an Austrian who had a long-term interest in the practical aspects of brittle materials fracture in addition to his work on quantum physics. He published on glass fracture between 1935 and 1959.[27,28,29,30,31,32,33,34] He broadened the molecular physics theory of crack initiation and propagation beyond Griffith's approach. Some consider him an early pioneer of what would later become fracture mechanics theory.[35] Smekal focused on the arcs, ripples, fracture mirrors, and lances (hackle). Some of his figures are shown in Figures 20 and 21. He correctly deduced that the size of the fracture mirror was inversely related to the stress in the body at fracture, but unfortunately he used the mirror's area for the size and not its radius. He also analyzed the stress data in terms of the load divided by the remnant area outside the mirror region. Smekal came close to the fundamental relationship between stress and mirror size, and he even had tables of mirror constants for different materials, but he did not quite have it right. His later publications with images of crisscrossing Wallner lines are exquisite. He showed how crack velocities could be computed inside fracture mirrors and how terminal velocity was reached while the crack was still forming the smooth mirror. He used the term "mirror" ("der Spiegel") in these early publications as Brodmann had done in 1894. He also began to show similarities and differences between fracture markings in glass and Plexiglass.[33,34] The authors of two biographical sketches of Smekal lamented the fact that, since most of his work was published only in the German language, Smekal may not have been given the full credit that he deserved for his work.[35,36] The journal Glastechnische Berichte had a single page obituary for Smekal in 1959.[37] A new biographical article on the 50th anniversary of Smekal's death lists his accomplishments.[38]

(a)   (b)

Figure 19. Adolph Smekal (1895 – 1959) (a) from the 1930s (American Institute of Physics, College Park, MD), (b) from his obituary in 1959, Reference 36. (Reprinted with permission of the German Society of Glass Technology.)

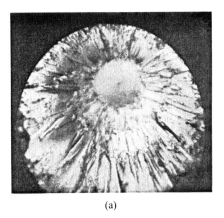

(a)   (b)

Figure 20. Fracture surface of fire-polished glass rods. (a) shows an internal origin and mirror, and (b) shows a surface origin and mirror. (Reference 30). (Reprinted with permission of the Society of Glass Technology.)

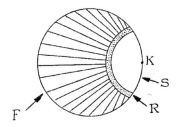

Figure 21. Schematic of a glass rod with a surface original break. As described by Smekal in 1936, K is the flaw at which fracture starts, S is the mirror, R the region of increasing roughness, and F is an inclined grooved surface. (Reference 30). (Reprinted with permission of the Society of Glass Technology.)

Despite all of Smekal's work and publications, it was a single short nine page paper by a visiting postdoctoral worker, also an Austrian, that was to have the greatest impact. Dr. Helmut Wallner, shown in Figure 22, had done his doctoral work on physics and mathematics at the University of Vienna, but then went to Smekal's institute in 1938. His now famous paper: "Lineinstrukturen an Bruchflächen" (Linear Structures on Fracture Surfaces)[39] was published in the fateful month of September 1939. This paper definitively and mathematically explained the cause and shapes of the curved ripple lines on glass fracture surfaces that had been commented on for decades by previous authors. The ripples were indeed caused by elastic waves interacting with the crack front as it propagated. Wallner even showed how the crack velocity could be analyzed from the arced lines. This was only one of two papers that Wallner ever wrote. He left the technical field shortly afterwards. Over the succeeding years, the terminology in the field changed, and what had been known as "rib lines" or "ripples" became "Wallner lines."

Figure 22. Helmut Wallner (1910-1984). (Courtesy of H. Richter.)

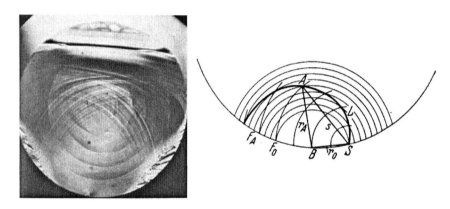

Figure 23. Wallner's figure 2 (left) of a fracture surface of a glass rod broken in four-point flexure.[39] His figure on the right illustrated how as a crack radiated outward (thin concentric semicircular lines) from an origin site B, elastic waves generated at a surface irregularity S interacted with the advancing crack front. (Reprinted with permission of Springer Science and Business Media.)

Smekal gave full credit[33,34] to Wallner's work and Smekal even differentiated between various classes of Wallner lines (e.g., secondary Wallner lines in Fig. 5, ref. 33, 1950). A new biography of Helmut Wallner was prepared by Dr. Herbert Richter in 2008.[40]

Beginning in Berlin in 1937 and continuing in Freiburg in the 1950s, Schardin (Figure 24), Struth, and colleagues[41,42,43,44] conducted experiments using very high speed spark-camera photography to measure the terminal velocities of cracks in glass that were struck by bullets. Time lapse photography with extremely small time intervals enabled them to monitor individual crack extensions

Figure 24. Professor Dr. Ing. Hubert Schardin (1902 – 1965). (Wikipedia)

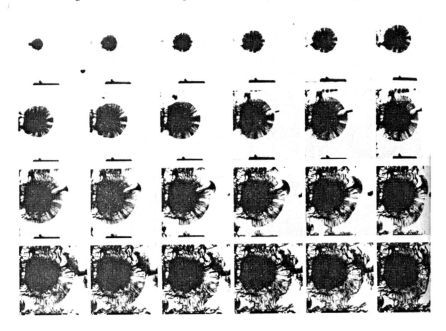

Figure 25. Schardin's time-lapse figure showing crack propagation from impact of a glass plate using a 24-spark camera at 300,000 frames per second. (From Reference 44 in 1955.)

as well as crack network expansions and damage-wave progressions with time (Figure 25). Terminal velocity varied from as low as 750 m/s for a flint glass, to 1500 m/s for soda lime silica, to 2200 m/s for fused silica. These velocities were later confirmed by Barstow and Edgerton with their own electric-spark camera system in 1939.[45] The terminal velocities were also confirmed by independent Wallner line analyses and some of Smekal's own results. Schardin came to the realization that the terminal velocity of glasses was about 0.5 to 0.6 times the Rayleigh surface elastic wave speed. Figure 26 is another example of his work, showing crack propagation in a four-point loaded bend bar. Schardin continued his work until 1962. His physics group at the University of Freiburg later evolved into the famous Institute for Material Mechanics. Reference 46 is a short obituary for Schardin.

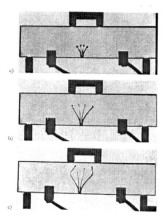

Figure 26. A high-speed photograph of cracks propagating in a glass beam loaded in four-point bending. Two ultrasonic transducers are attached to the bottom of the beam created pulsed waves that marked the fracture surface. This example of Schardin's work is from Kerkhof's 1974 book, Ref. 103. (Reprinted with permission of the German Society of Glass Technology).

Much later in the 1960s and 1970s, Field and colleagues at the Cavendish Laboratory of the University of Cambridge were able to confirm these terminal velocity speeds in glass using spark photography, ultrasonic fractography, and Wallner line analysis.[47] They extended the work to ceramics such as magnesium oxide, diamond, sapphire, and even lithium fluoride.

J. B. Murgatroyd, who worked for two glass companies in England, wrote about fracture surface markings in glass in 1942.[48] He correctly observed that overlapping parallel crack segments formed hackle lines where steps between the overlapping segments linked. He showed curved rib lines. Some of these were sharp "arrest lines," where a crack stopped and then resumed propagation on a different plane. On the other hand, many of the rib lines he showed were Wallner lines that were more gentle ripples on the fracture surface. Murgatroyd evidently was unaware of Wallner's 1939 paper, which may not be too surprising considering the events then taking place in Europe.

ADVANCES IN MICROSCOPY

At this point, it is appropriate to point out that nearly all examinations of fracture surfaces up to this time had been done with optical microscopes. The magnifications, the resolution limits, and the depth-of-field limits put constraints on what could be discerned and photographed. Interference microscopy enabled features smaller than the wavelength of light to be discerned, but only on very flat smooth surfaces. Hull[2] credits Tolansky and his group for leading work on this topic from 1943 for the next twenty years.[49] Most fracture surfaces are neither flat nor smooth. The advent of transmission electron microscopy (TEM) using replicas in the 1930s broke through the limitations of optical microscopy. In 1943, Gölz showed spectacular images of mist and hackle around fracture mirrors in glass rods as shown in Figure 27.[50] Poncelet[51] showed comparable images in 1958 and Peter in 1968.[52] The most systematic work was done by Beauchamp, who in 1971 studied various regions in the mirror.[53,54] His photos, some of which are shown in Figure 28, demonstrated that the formation of the mirror is a gradual progression of very localized crack-path deviations from the main plane. Hackle forms when small perturbations and micro branching of the crack plane create tiny "tongues" of overlapping segments on the fracture surface. Atomic force microscopy in the 1990s has not changed these conclusions.

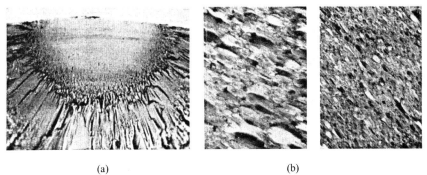

(a)   (b)

Figure 27. Gölz's 1939 photos of a glass fracture mirror. (a) is an optical image, originally at 26X (b) shows two TEM images originally at 2500X from the mist-hackle zone with the crack running from top left to lower right. (Reprinted with permission of Springer Science and Business Media.)

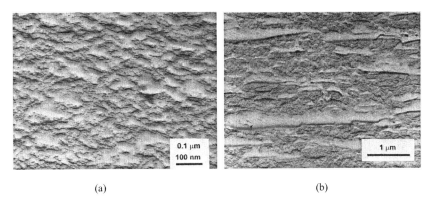

(a)   (b)

Figure 28. Beauchamp's TEM images from inside the fracture mirror in glass. (a) is one-third of the distance from the origin to the mist, and (b) is near the mist. The direction of crack propagation is from left to right. (Courtesy of E. Beauchamp.)

The advent of the scanning electron microscope in the mid 1960s revolutionized fracture surface examination due to the vastly improved depth of field, high magnification, and chemical analysis capability via energy dispersive spectroscopy. The heretofore elusive minute Griffith flaws now could be found and characterized, provided the examiner knew where to look! In addition, crack interactions with microstructure could now be studied more methodically.

WEIBULL THEORY AND FRACTURE MECHANICS

While the fractographers were documenting their findings, there were some important developments in materials science and mechanics. In 1939 Waloddi Weibull[55,56] presented his new distribution function to account for the variability in strength of ceramic bodies. Strength depended on the size of the body. He did not show a single flaw or even a schematic of a crack in his papers, but it is clear from his descriptions that he envisioned an isotropic body as having a distribution of cracks that controlled strength. One of his data examples was for the strength of porcelain rods. He used an arbitrary, but shrewd, function for the risk of rupture for volume elements in the body, without concerning himself about the size of the flaws or what stresses would be necessary to propagate the crack. While he did not cite Griffith's work, and modern fracture mechanics was still twenty years in the future, Weibull's work led to a dramatic improvement in our understanding of the strength of brittle materials. Fracture occurs from the origin location that has the worst combination of tension stress and flaw severity.

An important analytical paper was written in 1951 by Ellen Yoffe on the "Moving Griffith Crack."[57] The stress distribution around a moving crack was analyzed and compared to that around a static or slowly moving crack. It was shown that at high velocities (0.6 of the shear wave velocity), the peak stresses moved out of the plane of the crack. This explained why cracks began to wobble or even bifurcate locally once they moved at high velocity. These local small bifurcations account for the mist and hackle regions around a fracture mirror.

Starting in 1953 the phenomena of multiple breakages also was accounted for by analyses of elastic-wave propagation during fracture.[58,59,60] Once fracture occurs, there is an extremely rapid local change in the stresses that causes the propagation of elastic waves away from the fracture site. Miklowicz[58] and later Phillips[59] showed that unloading waves can reverberate and superimpose, and the phases of the waves (tension – compression) can switch after reflections. Kolsky[60] analyzed the matter further and said that the stress field in the section being broken is extremely uneven and that a complicated local wave pattern is set up near the fracture plane with longitudinal and flexural waves being generated. Kolsky showed that, for rods tested in flexure, the initial fracture from the origin penetrates about 70% of the cross section before it slows down as the crack approaches the opposite face. Fracture is completed only when the elastic waves reflected off the rod end faces and came back and reached the fracture plane.

In 1957, George Irwin[61] (Figure 29) revolutionized our understanding of the stresses and strains near the tip of a static crack. His work led to what is now known as the field of fracture mechanics. A superb 1997 book,[62] compiled on the anniversary of his 90th birthday, has a biography and a remarkable series of articles by experts in the field of fracture mechanics. They describe Irwin's breakthrough and the subsequent evolution of the field of fracture mechanics which was controversial in the beginning. There also are articles about Griffith and Weibull and their influence on Irwin. The effects of external stresses, crack dimensions, and specimen shape are contained in the stress intensity factor, K_I. The symbol K was named in honor of his colleague J. A. Kies; and the symbol, G, for strain energy release rate, was chosen in honor of A. A. Griffith. The stress intensity factor is directly related to the strain energy release rate and also the fracture surface energy. Fracture occurs when the stress intensity factor for a crack in a body reaches a critical value, K_{Ic}, also known as the fracture toughness. It was no longer necessary to be concerned about the exact peak stress (a singularity) at the tip of a sharp crack. Engineering compilations are available which show tables or equations for K_I for a

Figure 29. George Irwin (1907 – 1998), the founder of modern fracture mechanics. (Courtesy of the Clark School of Engineering, University of Maryland.)

variety of cracks and specimen shapes. Some materials have rising R-curves such that the fracture resistance varies with crack size and crack extension due to crack interactions with microstructure.

The advent of modern fracture mechanics in the 1960s had dramatic influences on fractographic analysis. More rigorous mathematical analysis could be applied to cracks of various sizes and shapes, in various loading conditions and stress states and it was no longer necessary to worry about estimates for the crack tip radius and stress concentration factors. One of the most important works on stress intensity shape factors was by Newman and Raju in 1979.[63],[64] They developed an accurate empirical formula for the K_I factors for semicircular and semielliptical surface flaws in beams in tension and flexure. This is a classic engineering problem. Many flaws such as machining grinding cracks may be modeled by semielliptical surface flaws, so their solution was very helpful. Prior to their work, there were a number of conflicting and incomplete stress intensity factor solutions. Fractographic analyses using stress intensity factor analyses have largely replaced analyses using fracture energies. An added impetus for the adoption of fracture mechanics was finding that slow crack growth velocities are strongly dependent upon the stress intensity factor.[65]

Poncelet, a Belgian-American metallurgical engineer, wrote papers in French in 1950[66] and in English in 1958[51] on the fracture of glass. These offered explanations for overall fracture patterns and some of the markings on fracture surfaces. His 1958 paper showed how crossing Wallner lines could be analytically modeled to give improved estimates of crack velocities. He also showed "ripple pairs" that are now known as "gull wings" from bubbles in glass, tempered glass markings, and electron micrographs of various regions inside a fracture mirror. He believed that an advancing crack moved forward as an uncoordinated series of jumps all along its front separating the breaking bonds from the lagging bonds. Superimposed elastic vibrations interact with the crack front and cause ripples or undulations. Major redirections of the stress field caused striations to form in the direction of crack propagation. It is worth noting here that many in the early literature used the term "striation" to describe such lines. The term "striations" may be confused with fatigue crack growth markings observed in metals, so it seems reasonable to use the alternative term "hackle" instead, as Preston suggested in 1926.

In addition to all the work being done in universities, some large glass companies were contributing to the science of fractographic analysis, but not all of their work was published. In 1936, Leighton Orr (Figure 30) began a thirty-six year career at the Pittsburgh Plate Glass (PPG) company as

Figure 30. Leighton Orr, who worked for many years at Pittsburgh Plate Glass. (Reprinted with permission ASME.)

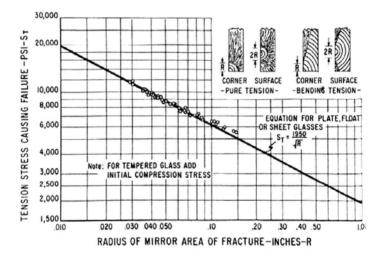

Figure 31. Orr's graphs of breaking stress versus mirror radius.[67] The inserts show how to measure the radius in plates tested in pure tension or bending. (Reprinted, with permission of ASTM Int.,100 Barr Harbor Drive, West Conshohocken, PA, 19428.)

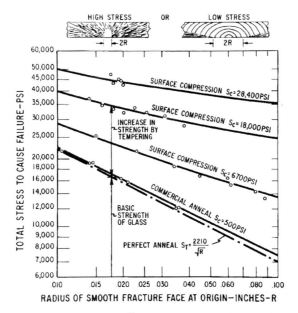

Figure 32. Orr's Figure from his 1972 paper[67] for a barium crown glass. Notice equation 2 is shown on the bottom right. (Reprinted with permission of ASTM, Int.)

head of the Physical Testing department of their research laboratory. He was unable or unwilling to publish much during his career, but upon the year of his retirement in 1972, he had one notable short publication about practical analysis of fracture in glass windows.[67] Figures 31 and 32 show two of his figures. This paper has a number of useful practical tips on testing and examining broken annealed or tempered plates. It also has some intriguing observations such as a minimum stress of 10 MPa is necessary for cracks to develop mist, hackle, or to branch. This is an interesting observation and was verified by the work of J. Quinn[68] twenty-seven years later. Orr's paper also has extensive practical information and data on the measurements of fracture mirror sizes and how to use the basic relationship relating the mirror radius to the stress at the origin at the instant of fracture. Orr's advice on how to interpret non circular mirrors and how to measure the mirror radii in tempered plates is especially valuable. On the basis of a review of the literature, plus conversations and correspondence[69] this author had with Orr, it became apparent that Orr was using the now standard relationship for stress and mirror size as early as 1942:

$$\sigma\, R^{\frac{1}{2}} = A \tag{2}$$

where R is the mirror radius, σ is the stress at the origin and across the mirror, and A is a constant known as the fracture mirror constant. This relationship has tremendous practical value, and there is nothing like it for metals fractography. It is not necessary to have prior information about how the

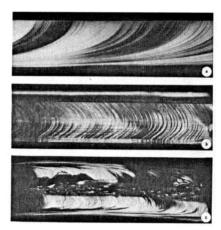

Figure 33. Orr showed the fracture surfaces of plates, (a) broken in bending; (b) broken with thermal stresses; and (c) broken tempered glass. (Reprinted with permission of ASTM, Int.)

specimen was loaded. A small mirror is proof that the specimen was strong and had a small strength-limiting flaw. Conversely, large mirrors mean the failure stress was low and there was a large flaw. In some instances, a specimen may be so weak that the mirror size is larger than the part cross-section, and hence the mirror markings are not visible. That, in and of itself, is valuable information. In other words, the existence of a mirror boundary means that the part was stressed to a moderate or high level. This knowledge was disseminated at various glass conferences such as the Bedford, Pennsylvania meetings of the American Ceramic Society in the 1950s. Relationships similar to eq. 2 were in use by many authors in the 1950s and 1960s, but with different exponents for the mirror radius term. Orr was the first to systematically use the equation with the one-half power to solve practical problems. He also recognized how residual stresses (as in tempered plates) altered the relationship, such as shown in Figure 32, and used that to practical advantage too. I did a careful review of the literature pertaining to fracture mirror size analysis in 2006,[70] and I concluded that eq. 2 should be deemed "Orr's equation." Although many associate eq. 2 with Johnson and Holloway[71] in 1966, the relationship was already in use for over 15 years. Their primary contribution was to use an energy analysis to give some theoretical underpinnings to the equation. After he retired, Orr did a great deal of consulting and prepared 950 failure analysis reports from 1972 until 2003 when he was 96 years old. These reports were donated by Orr to the University of Pittsburgh before his death in 2004. They were intended to be accessible to students and historians.[a] Orr's colleagues[b] arranged for these reports to be scanned and converted to digital pdf format. The reports are a treasure trove of practical information, and include notes on minor and major cases alike. For example, there are many interesting observations about the

---

[a] In January 2009, I went to the Brevier Engineering Library of the University of Pittsburgh and was disappointed to learn that the reports were not cataloged and they did not know where they were. On further investigation with the Director of Development for the Swanson School of Engineering, it was learned that the original paper reports had been removed to archives. They were not accessible to students or investigators such as me. Ms. Sonia Gill, Director of Marketing and Communications of the Swanson School of Engineering did kindly furnish me a CD with digital copies of the files.
[b] Robert Spindler, Cardinal Glass Industries, 2003.

failures of the hundreds of large windows in the John Hancock skyscraper building in Boston in the 1970s. An American Society for Mechanical Engineering web site has an obituary-summary of Orr's life[72] and has an award in his name.

Figure 34. Dr. Frederick Ernsberger (1920 - 2003) of PPG proved the existence of the elusive microscopic Griffith cracks. The image on the right shows his 1970s monograph on Polarized Light in Glass Research. The photo of Ernsberger is from inside the book. His proof that Griffith cracks existed actually came decades after Frank Preston had shown photos of large Griffith flaws on fracture surfaces in glass. (Courtesy of F. Ernsberger.)

Another PPG researcher, Dr. Fred Ernsberger, shown in Figure 34, joined the company in 1955 and by 1960 did classic scientific experiments on glass that finally detected the elusive submicroscopic Griffith flaws.[73,74] He used an ion-exchange treatment with lithium and potassium nitrates to create a differential thermal expansion strain that caused cracks to run stably from starter sources, as shown in Figure 35. A small uniaxial stress was superimposed on the plate so that the cracks grew in a preferred orientation, as shown in the figure. The starter source cracks were in fact minute Griffith flaws that manifested themselves as tiny kinks in the longer thermal expansion cracks. The ion exchange treatment effectively highlighted the 25 nm to 30 nm wide Griffith cracks. The interpretation was confirmed by electron microscope examination of replicas. The cracks were made even larger by the acid etching so that they could be photographed. Ernsberger said that the:

> "origins seem to be simple straight Griffith cracks; sometimes the crack is crescent shaped or a 'star' with three or more arms. All Griffith cracks so far identified appear to have resulted from accidental mechanical injury."

Ernsberger continued working on mechanical properties of glass and wrote several enjoyable reviews such as References 74 and 75. He reviewed all flaw types in glasses in 1974[74] and categorized them as internal (which rarely cause fracture), surface, or edge. Surface flaws can be from mechanical damage or from chemical flaws. Short biographical sketches of him are in Reference 75 in 1977 and in his monograph on polarized light in glass research.[76]

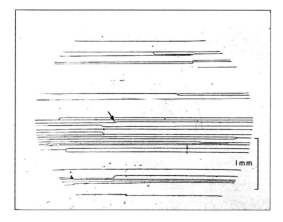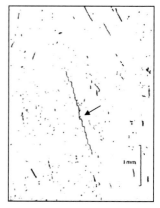

Figure 35.   Figures from Ernsberger's 1960 *Proc. Royal. Soc. London A* paper.[73] (a) shows long horizontal shrinkage cracks propagated from the tiny initial Griffith flaws which manifested themselves as small kinks, as marked by the arrow. The axis of the tensile stress was top to bottom in this view. (b) shows a pair of cracks propagated in opposite directions from a single origin (arrow) under the influence of uniaxial stress, the axis of which was rotated 90° at 1 minute intervals. (Reprinted with from the Royal Society of London.)

Figure 36.   Errol Shand (1893 – 1976).   (Courtesy of S. DeMartino, Corning, Inc.)

A third industrial scientist from this era was Errol Shand of Corning (Figure 36) who wrote a number of papers on strength and fractography of glass from 1954 to 1970.[77-81] In 1954, he expanded the Wallner lines analyses for crack velocity analysis to rectangular beams in bending.[77] He showed how cracks that might reach terminal velocities very quickly after propagating from a surface origin, but then slow down to only 5% -10% of the transverse wave velocity as the crack reaches the initially

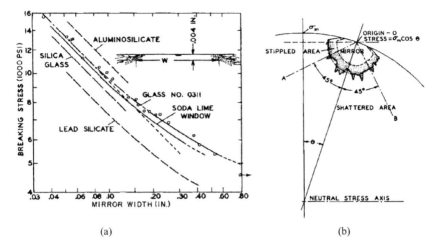

(a)   (b)

Figure 37. Shand's 1964 illustrations for mirror size (reference 80). (a) shows the correlation of the fracture stress to the mirror width with an insert showing the size should be measured just below the external surface, and (b), how to measure a mirror in a high-strength glass rod for an origin on the side, not at the peak stress location. The stress at the origin location should be used. Notice the mist and hackle regions are shown in the insert.

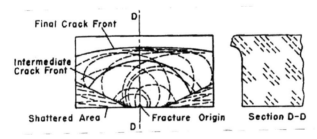

Figure 38. Figure by Shand (1954) showing a partial mirror in a rectangular beam broken in bending. The fracture surface view on the left illustrates hackle lines near the bottom and Wallner line ripples. The section view on the right emphasizes the compression curl.

compression-loaded half of the beam.[77] Shand was also concerned with the time-dependent changes in strength. Many of his mathematical analyses dealt with stress concentrations at the tip of sharp cracks. His equations are similar in form to the fracture mechanics expressions that were developed by Irwin, but Shand never used stress intensity factors in any of his papers. Since many of his final expressions included the crack tip radius, a very difficult parameter to measure or confirm, it was difficult to apply them in practice. He came very close to deriving eq. 2 for fracture mirror size analysis, but settled for a form with the crack tip radius and the mirror width, instead of the radius.[77,78] His 1959 paper[78] on measuring fracture mirrors had some excellent tips on how to make the measurements in bodies with stress gradients. He distinguished between the mirror–mist and mist-hackle boundaries. Shand also gave recommendations on lighting and on how many radii to measure in different directions. In 1961[79] he showed how a wedge-shaped tungsten carbide indenter tool could be used to make controlled cracks and scratches, a technique that would be widely used with Vickers and Knoop hardness indenters in later years. Papers in 1964[80] and 1967[81] recapitulated some of the earlier work, but also added some useful sketches of scratch-induced cracks, more guidance on how to measure fracture mirrors, and an interesting discussion of how to treat elliptically shaped flaws to compute an effective flaw size. An interesting testimonial to Errol Shand is on-line as Reference 82 which summarizes his career. It also notes that an award in his name has been set up by the local chapter of the United Way, a charitable organization. It states:

> "A personal glass testing laboratory in his home basement enabled him to make accurate studies of breakage phenomenon, and his judgment was much sought after in damage suits involving glass. His laboratory and library were bequeathed to Alfred University in New York State."

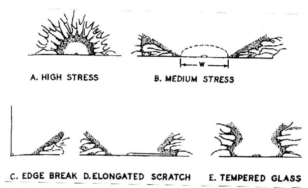

Figure 39. Shand's illustrations of various mirrors in beams in bending from Reference 81.

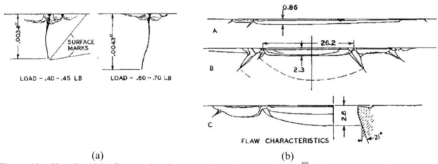

(a)   (b)

Figure 40. Shand's 1964 figures showing scratch-scoring type flaws.[80] (a) shows an end-on, cross-sectional view underneath a scored surface on the top. (b) shows side views underneath a scratch.

Figure 41. Prof. Van Derck Fréchette (author's private collection) and his famous 1990 book.

VAN FRECHETTE AND THE 1960s

One of the most influential brittle materials fractographers, Prof. V.D. Fréchette of Alfred University (Figure 41), began to write on the topic in 1965,[83] halfway through his career. His paper discussed the nature of fractography and the techniques and goals of observation of the fractured pieces. It reviewed classic fracture markings such as the mirror, mist, hackle, gull wings, Wallner lines, river patterns, cleavage, transverse and intergranular fracture in polycrystalline ceramics. Over the succeeding years, Fréchette expanded and refined his nomenclature system[84] (a step that was not without controversy, see Rice's commentary at the end of the paper). Others (e.g., Hull[2]) have rued the myriad terms that are used to describe fractographic features, and while there may not be universal agreement to Fréchette's system, it is logical and easy to use. Many of his terms have been adopted in documentary standards.[85,86,87,88] His crowning achievement was the first-ever textbook on fractography of brittle materials, *Failure Analysis of Brittle Materials*, that was published in 1990 at the end of his career.[89] Figure 41 - 43 show several illustrations from his book. This book popularized fractographic analysis and was used to train a generation of fractographers, this author included.

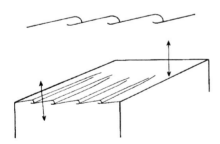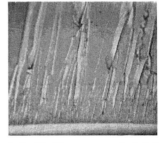

Figure 42. Illustration's from Fréchette's book showing twist hackle. The arrows on the left show how the change in the direction of maximum principal stress causes the crack front to split up.

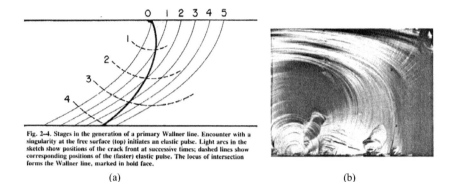

Fig. 2–4. Stages in the generation of a primary Wallner line. Encounter with a singularity at the free surface (top) initiates an elastic pulse. Light arcs in the sketch show positions of the crack front at successive times; dashed lines show corresponding positions of the (faster) elastic pulse. The locus of intersection forms the Wallner line, marked in bold face.

(a)   (b)

Figure 43. Figures from Fréchette's book. (a) shows the evolution of primary Wallner lines for a curved crack front advancing from left to right in a glass plate. (b) shows tertiary Wallner lines in a thick glass plate impacted on the lower left.

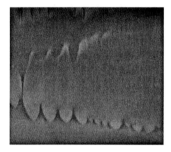

Figure 44. "Scarps" on a glass fracture surface from Fréchette's book. Scarps form when a crack outruns (or is overtaken by) a fluid or vapor. The direction of crack propagation is from bottom to top.

In 1977, Fréchette began a three-day summer hands-on fractography short course at Alfred University. He trained hundreds of engineers and scientists over the ensuing years. The course continues now as a four-day course, 34 years later. Fréchette and his students James Varner and Terry Michalske used fractographic analysis to study the effects of water on slow crack growth and interpreted "scarps," which are markings on a fracture surface associated with a crack outrunning or being overtaken by a fluid or a vapor as shown in Figure 44.[90,91,92,93] They also studied precracks and running cracks with ultrasonic fractography in fracture mechanics type specimens.[94,95] Varner described defects in glass from processing[96] and contact damage[97] with superb scanning electron microscopy photos of the latter. Fréchette was also well known for his participation in many court cases, some of which are documented in his book. He was a good oral story teller, and there is a chapter in his book with many fascinating case studies. Some of the cases he described were: "A Tale of Two Teapots," "A Case of Water Hammer," "An Ancient Art Explains a Modern Catastrophe," and "Panic in the Gym." One of his most famous cases involved glue chipping cracks in Boston skyscraper windows.[98] In addition to his book and the new fracture surface markings he and his students identified, Fréchette's primary contributions were to popularize fractographic analysis, bring consistency to the analyses, and show how scientific study of fracture solved many practical problems.

Johnson and Holloway (Figure 45) wrote an important paper in 1966 on fracture mirrors in glass.[71] This 1966 work was summarized later in an excellent review article on the fracture behavior of glass by Holloway.[99] They examined eq. 2 in detail and concluded that the hackle boundary is the

Figure 45. Professor Holloway from Reference 99. (Reprinted with permission of the Society of Glass Technology.)

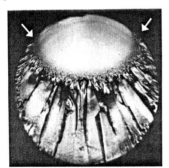
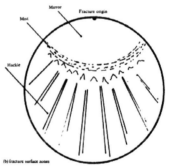

Figure 46. Fracture mirror figures from Ref. 99. Notice how their schematic drawing does not show the side cusps marked by this author with white arrows in the photo on the left. Kirchner's later analysis accounts for the surface cusps. (Reprinted with permission of the Society of Glass Technology.)

locus of points at which the rate of release of strain energy by the expanding fracture is sufficient to create continuously four new surfaces and to maintain the finite kinetic energy for the moving cracks. Two of their figures are in Figure 46. They also pointed out that the stress that should be used in eq. 2 is the local stress at the crack front periphery (the mirror boundary) which is not necessarily the maximum stress in the piece or the stress at the origin in the center of the mirror. So for example, these stresses are the same in a rod loaded in direct tension, but are quite different for large mirrors in rods loaded in flexure. This paper was important at the time since there was considerable doubt as to the correct criterion for the formation of the mirror features. Many had postulated that a critical velocity criterion was the key. This was eventually disproven by various investigators. For example, shortly after the Johnson and Holloway paper was published in 1966, Congleton and Petch[100] used Wallner line velocity analysis to show that crack branching occurred at various velocities in glass and sapphire. They argued that eq. (2) held in general and, more specifically that:

$$K_{lb} = \pi^{\frac{1}{2}} \sigma_f c^{\frac{1}{2}} \qquad (3)$$

where K_{lb} is the critical stress intensity at branching, σ_f is the fracture stress, and c is the semi crack length at branching. As will be described below, Kirchner's work in the 1970s and 80s verified that a stress intensity criterion gave much better fit to the actually observed mirror shapes.

Almost twenty years later in 1985, Holloway wrote a fascinating paper: "A Look at the History of Glass Strength."[14] He credited Griffith's work[12,13] as a starting point for a review of the strength of glass and commented that:

"Although Preston and Milligan evidently realized the significance of the work, …its implications were more often ignored and much that was important or useful in the original paper was soon dropped out of sight."

Holloway lamented about the poor development of the understanding of glass strength in the thirty years after Griffith's two papers in the 1920s.[14] He referred to the era as the "Dark Ages" and noted that many published papers were of mediocre quality. There were signs of intensive investigations at industrial laboratories that were never reported in the literature. A "Renaissance" occurred in the 1950s as long-range fundamental research was supported at company research and development laboratories and in universities. He cites Ernsberger's writings as part of the renaissance.

Schardin's work on dynamic crack propagation in Freiburg in the 1950s (described above) was continued by Kerkhof in the 1950s into the 1970s.[101,102,103,104,105,106,107] Kerkhof, shown in Figure 47, founded the Institute for Solid Mechanics in 1971 which in turn became the famous Institute for Materials Mechanics. This institute was a world leader in the new field of fracture mechanics. Much

Figure 47.   F. Kerkhof and his 1970 book.

of the work and many illustrations were included in Kerkhof's masterpiece book: Bruchvorgänge in Gläsern (Fracture Processes in Glasses) in 1970.[104] Among the many topics this institute studied was the dynamic propagation of cracks in glass. High-speed spark photography work continued, but a new technique "ultrasonic fractography" or "stress wave fractography" was invented by Kerkhof in 1953.[101] A transducer superimposed stress waves into a specimen while the crack was propagating. The slight perturbations to the direction of maximum principle tensile stress caused the propagating crack to form slight ripples (tertiary Wallner lines) on the fracture surface. Since the ultrasonic wave frequency was known, it was a simple matter to compute local crack velocities from the line spacings. Figure 48 shows examples. Kerkhof and colleagues were able to use this new approach to measure the terminal velocity of cracks.[103] They also were able to study crack shape evolution with crack extension and interactions of moving cracks with inclusions.[104] Other amazing experiments with high speed photography showed instances where cracks were driven well above the common terminal velocities to supersonic velocities.[107] The use of lower-frequency transducers allowed slower crack velocities to be measured, and Richter and Kerkhof[105,106] were able to study slow crack growth behavior in glasses. Sommer[108] used optical interference microscopy to study "lances" or hackle and proved that a 3.3 degree rotation in the maximum principal stress was necessary to trigger a lance. A summary of Kerkhof's career may be found on line.[109]

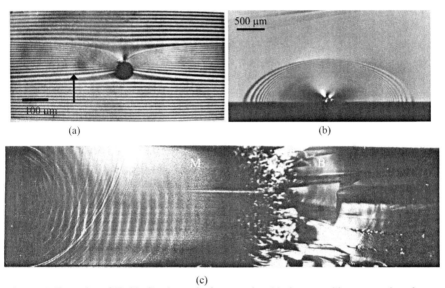

(a) (b)

(c)

Figure 48. Examples of Kerkhof's ultrasonic fractography. (a) shows markings on a glass fracture surface left by a crack passing by a void. The crack direction is shown by the arrow. The Wallner lines created by the external ultrasonic transducer show that the crack locally accelerated as it neared the void, but then slowed around it, but then snapped past the void. (b) shows an accelerating crack that grew from a surface flaw in a bend bar. The evolution of the shape into semielliptical curves of varying aspect ratio is completely accounted for by fracture mechanics analysis. (c) shows a crack accelerating from a notch on the left towards the right. M is for mist and B is for branching. (All courtesy of H. Richter.)

PROGRESS IN MATERIALS SCIENCE IN THE 1970s

Considerable time and energy has been spent over the years on whether cracks in brittle materials experience any plastic deformation at the very tip of the crack. The topic remained controversial for a long time until Hockey (Figure 49) and colleagues in 1975[110] and 1980[111] showed that, although dislocation-like features may be present in Si, Ge, SiC, and Al$_2$O$_3$, the concept of an atomically sharp crack provides a sound basis for the theory of fracture of brittle solids. Figures 50 and 51 show some of his transmission electron microscope images. Later work in the late 1980s with high-resolution transmission electron microscopy by H. Tanaka and Y. Bando confirmed that crack tips in SiC, Si, and sialon were indeed atomically sharp.[112,113]

In the meantime, intensive studies started in the early 1970s on the relationship between microstructure, strength, flaws, and fracture toughness of polycrystalline ceramics. Many of these studies began to exploit the capabilities of the scanning electron microscope which was becoming more readily available. Advances in ceramic processing led to finer-grained, fully-dense, strong ceramics that were more amenable to fracture analysis. Fractographic markings were clearer and fracture origins could easily be detected.

Figure 49. Bernard Hockey in 2008. (Author's photo)

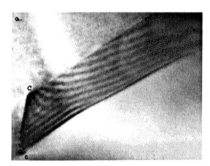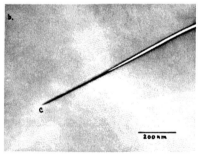

Figure 50. Hockey's TEM illustrations from Reference 110 showing radial cracks segments in single-crystal silicon. Moiré fringes are evident from mismatched portions of the diffracting crystal portions on opposite sides of the interface in (a). (Courtesy of B. Hockey.)

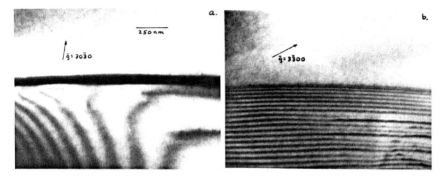

Figure 51. Hockey's TEM illustrations from Reference 110 showing the tip region of a lateral crack in single crystal alumina. There is no trace of microscopic slip. (Courtesy of B. Hockey.)

Henry Kirchner (Figure 52) wrote a number of papers starting in 1973 to 1987 while he was the owner of the Ceramic Finishing Company, a small private grinding shop and testing laboratory in State College, Pennsylvania.[114,115,116,117] He was particularly interested in the interpretation of fracture mirrors and the mirror size - strength relationship. He and his colleagues published many of the first fracture mirror images and mirror constant values for polycrystalline ceramics (e.g., high strength aluminas, silicon nitrides and silicon carbides) as shown in Figure 53. They showed that not all aluminas or silicon nitrides are alike. Variations in the microstructures led to differences in the fracture mirror constants. In addition to the published journal articles, in 1974 Gruver, Sotter and Kirchner published a thick comprehensive Summary Report[118] on a study done for the US Navy. The study covered the fracture surface markings, the mirrors, and even the flaws in a large number of alumina, silicon nitride and silicon carbide rods broken in flexure. This 100 page report was remarkable in that it included a huge supplemental appendix that had hundreds of photographs of the

Figure 52. Henry Kirchner (1923 – 2008) (Courtesy of James Kirchner).

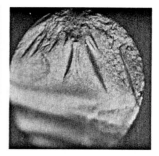

Figure 53. Fracture mirrors in 96% alumina, hot-pressed alumina, and hot-pressed silicon nitride from Kirchner, Reference 118.

entire fracture surface and flaw close-up of every single test specimen. Nothing like this had been published up to that time. Specimens were tested in air, in liquid nitrogen, and at elevated temperature in static load, delayed failure, or impact loadings. The authors even commented:[118]

> "Early attempts to identify the flaws at fracture origins were unsuccessful because, in weak bodies, the fracture surfaces are rather featureless and it was frequently impossible to determine which one of several observed flaws had acted as the fracture origin. The availability of stronger bodies and improved fractographic techniques have made it possible to identify the flaws at fracture origins in a large fraction of cases."

They commented that some recent publications had started to show isolated examples of fracture origins and flaws in the ceramic literature, but:

> "To some degree these publications are misleading because of the tendency to show outstanding flaws that illustrate particular points. In many cases, fractures originate in regions in which none of the features are outstanding and there is no definitive evidence that any particular feature represents the critical flaw."

These comments underscore three important milestones in the 1970s on the history of the fractography of ceramics:

1. Ceramics were being made stronger, often by hot-pressing. Therefore, fracture surfaces became easier to interpret. Discrete flaws could be found.
2. Scanning electron microscopy made it possible to photograph and characterize small flaws and even measure their sizes.
3. Flaw sizes could be compared to calculated flaw sizes based on fracture mechanics.

Most practitioners were by this time using Orr's equation (eq. 2) to analyze their data, but Kirchner was not satisfied. Distortions from the classic circular or semicircular mirror shapes were commonly attributed to stress gradients, but simple adjustments to account for the local stresses failed to account for the actual observed mirror shapes. In addition, stress adjustments could not account for the small cusps located at the surface of a fracture mirror that started at a surface flaw. Eventually Kirchner teamed with the fracture mechanics expert Prof. J. Conway at Pennsylvania State University and they published two definitive companion papers in 1987.[117] As shown in Figure 54, these showed that the actual shape of fracture mirrors in tensily or flexurally loaded rods was fully accounted for by a branching stress intensity factor criterion, rather than the more simple stress – mirror radius, Orr's equation. The stress intensity shape factors for semielliptical surface flaws in bending or tension were instrumental in Kirchner and Conway's analysis. Although a full accounting for actual mirror shapes and sizes was now possible, most practitioners continue to use Orr's simple equation since it is accurate for small mirrors in uniform tension or very gradual bending stress fields. It can be used as an effective approximation for many other configurations. I wrote a review of the history of fracture mirror analysis[70] and used it as a basis for a standard for fracture mirror size analysis.[119] Kirchner also published extensively on other polycrystalline ceramic topics such as slow crack growth markings,[120] flaw sizes and shapes,[121] the effects of grinding and single-point scratching on strength.[122,123] An obituary on Kirchner is accessible on line.[124]

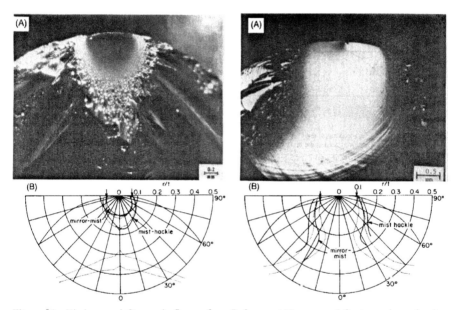

Figure 54. Kirchner and Conway's figures from Reference 117 compared fracture mirrors in glass rods tested in bending to predictions based on fracture mechanics that took the stress gradient and the expanding crack front size and shape into account.

One of the most productive teams ever was R. Rice, J. Mecholsky, Jr. (shown in Figure 55), S. Freiman, and their colleagues at the Naval Research Laboratory (NRL) in the 1970s. They produced an impressive body of publications and advanced the state of the art of fractographic analysis through the correlation of microstructure, flaw characterization, mechanical properties, and fracture mechanics. Roy Rice spent much of his forty plus years career investigating the relationship between

Figure 55.   Roy Rice (1934 - 2011) (Courtesy of Craig Rice.), left, and Jack Mecholsky, Jr. right (Courtesy of J. Mecholsky).

processing, microstructure, flaws, strength, and fracture toughness. The team did research on many topics, and their papers were well illustrated. For example, they studied fracture mirror constants for a wide range of glasses and ceramics. They noted that the mirror sizes were a multiple of the flaw size in many glasses and ceramics.[125,126,127] Figure 56 shows one of their famous schematics of a fracture mirror, where in this instance, the prospect of the initial flaw growing due to slow crack growth is noted.

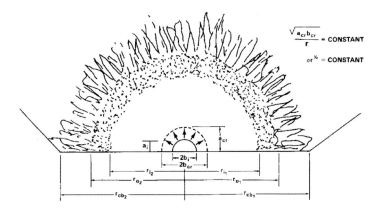

Figure 56. Schematic of a fracture mirror showing several mirror and branching boundaries, but also stable crack extension from the initial flaw. This figure was used in several papers to discuss the mirror to flaw size ratios. (Courtesy of J. Mecholsky.)

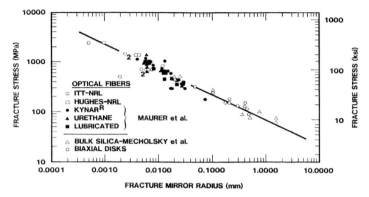

Figure 57. A typical graph of breaking strength versus fracture mirror size illustrating that Orr's equation, equation 2 of this paper, was applicable over a very broad range of strengths, mirror sizes, and specimen types including uniaxial and biaxial stressed test specimens. (Courtesy of J. Mecholsky.)

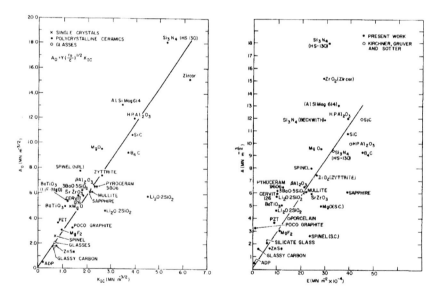

Figure 58. Mecholsky, Freiman and Rice[126] showed that a general trend exists for the fracture mirror constant to be a multiple of the fracture toughness (a), or the elastic modulus (b).

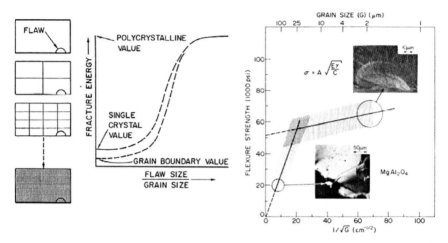

Figure 59. Two illustrations showing how fracture energy and strength can vary with grain size.[128] (Courtesy of R. Rice.)

Rice, Mecholsky, and Freiman identified many different flaw types in ceramics.[128,129,130,131] Of special interest was the effective fracture toughness at the scale of the flaw and the microstructure and the effective fracture energy or fracture toughness for flaws at the appropriate microstructural level.[132,133] Figure 59 shows that with coarse-grained materials, the critical flaw may be within a single large grain and the single-crystal fracture energy or fracture toughness is applicable. On the other hand, with smaller grain-sized materials, flaws are often a multiple of the grain size and a polycrystalline fracture energy applies. The strength - grain size trends will therefore vary depending upon the flaw to grain size ratio. A transition from single-crystal to polycrystalline fracture toughness controlled behavior occurs when flaw sizes are comparable in size to the grain size.

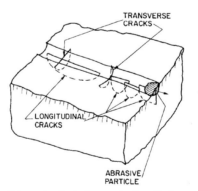

Figure 60. Schematic of grinding cracks from Reference 136.

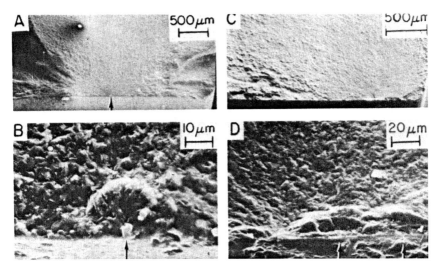

Figure 61. Grinding flaws in mullite from Reference 136. A and B show short transverse cracks that occur perpendicular to the grinding direction, and C and D show longitudinal cracks that are parallel to the grinding direction. These are usually more deleterious to strength. The difference in flaw size and shape account for the directionality of strength in test pieces: bend bars ground parallel to their length (i.e., longitudinally ground, are stronger than bars ground transversely to their length.

This team also wrote a fine series of papers on the nature of grinding cracks and their directionality.[134,135,136,137,138] They showed that the grinding cracks penetrated far deeper beneath the surface than the grinding striations (confirming Preston's conclusions), and that the size, shape and severity of grinding cracks were different for cracks parallel or perpendicular to the grinding axis. Figure 60 shows a schematic of cracks that can occur from grinding in a superbly illustrated paper by Rice and Mecholsky.[136] Figure 61 shows actual examples for mullite.[136] The paper also shows several dozen grinding flaws for a variety of ceramics.

Eventually this team dispersed. Freiman turned to management duties at the National Institute for Standards and Technology (NIST). Mecholsky went to Sandia National Laboratory, then Pennsylvania State University, and finally to the University of Florida. He investigated many fractographic topics including structural ceramics fracture origins, quantitative analysis, fractal analysis, single-crystal behavior, slow crack growth effects, manatee bones, and dental ceramics.

Rice retired from NRL and worked at W.R. Grace in the 1980s until he retired. He continued to investigate microstructure-property relationships. Figure 62 is from his keynote presentation on intergranular versus transgranular fracture at the third Alfred Fractography conference in 1995.[139] Rice was a prolific writer. His memorable "Ceramic Fracture Features, Observations, Mechanisms, and Uses" was an astonishing book length (99 pages) paper for a 1984 ASTM conference.[140] It had a wealth of information about ceramic fracture features and was very well-illustrated. An even longer 182 page article, "Microstructure Dependence of Mechanical Behavior of Ceramics" was published in 1977, but had far fewer illustrations.[141] In 1998, near the end of his career, Rice summarized much of this in his book on porosity in ceramics[142] and a second book in 2000 on grain- and particle-size effects on mechanical properties.[143] He discussed the value of fractographic analysis in his 1977 treatise paper on mechanical properties of ceramics:[141]

> "The most significant experimental procedure that can aid the understanding of mechanical properties is a study of fracture surfaces, especially to identify origins . . . It is indeed amazing the number of mechanical properties studies conducted that were extensively concerned directly or indirectly with the size and character of flaws and microstructure from which failure originated in which no attempt was made to experimentally observe and verify the predicted or implied flaw character."

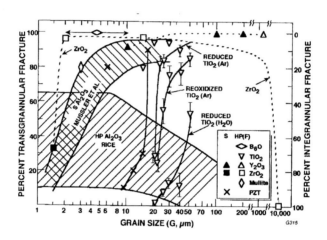

Figure 62.   Percent intergranular and transgranular fracture versus grain size from Rice's Ref. 139.

Rice's papers often distilled information from many sources in a quest to find general trends. A distinctive trait of his papers was complex summary figures such as Figure 62. He was meticulous and thorough, and he ferreted out much useful data from a myriad of sources. He expected no less from others, and woe to any oral presenter at a conference who was not prepared. Roy could be counted on to speak from the audience during the question-and-answer period and admonish the speaker for not being aware of an obscure reference.

Roy passed away on April 29, 2011 as this article was being written. I appreciate the many pictures that he gave me, particularly of single-crystal fracture mirrors that are included in my Guide book. After he retired in the 1990s, Roy spent many hours in the basement of the NIST library in Gaithersburg, digging through the archives for data on strength versus porosity, fracture toughness, or grain size. He used my office at NIST as a staging area for these forays, and I benefited from many interesting conversations. An obituary on Rice is on the American Ceramic Society web site.[144]

PROGRESS FROM 1970 - TODAY
Throughout the 1970s and 1980s, thousands of papers appeared with some degree of fractographic analysis. Some of the more systematic studies were done by laboratories and engineering firms that were refining ceramics for use in structural applications such as heat engines. Notable work during this era was done by Munz et al.[145] and Richerson.[146] There has been progress on many fronts

in the last decade and space limitations preclude mentioning more than just a few studies. A noteworthy recent paper about bioceramics was presented by Richter.[147] A similar article appeared in the last Alfred Conference series book.[148] After a review of ultrasonic fractography, it showed a number of ceramic hip joint ball fracture examples.

One of the current leading teams is Profs. K. Uematsu and S. Tanaka and their students at Nagaoka University in Japan. They have studied the formation of flaws in sintered ceramics by careful fabrication, innovative microscopy, microstructural analysis, and fractographic analysis.[149,150,151,152,153,154]

In recent years, I have collaborated with Drs. S. Scherrer and J. Quinn to solve challenging problems in the dental ceramics field.[155,156,157,158]

A team of Austrian researchers at the University of Leoben led by Prof. Robert Danzer has effectively applied fractographic analysis to solve many fascinating engineering problems. Prof. Danzer, Monika Hangl, Tanya Lube, Walter Harrer, and Peter Supanic, have written a number of articles[159,160,161,162,163,164] for the Alfred Conference and Slovakian Conference series described below.

New fractography tools have great promise to further expand our field and lead to new discoveries. These include but are not limited to atomic force microscopes, digital camera-microscopes, optical profilometers, computer programs that can create virtual three-dimensional images, programs that interpret roughness and fractal dimensions, and programs and microscopes that overcome the depth-of-field limitations of optical lenses. As an example, Atomic Force Microscopes have led to new insights about fracture mirrors in glasses and are an interesting complement to electron microscopy.[165,166,167] In my new area of interest, digital laser scanning of dental prosthetics is routine. Virtual three-dimensional models of structures can be tilted, rotated, manipulated and viewed from any perspective. At present, their resolution is not sufficient to record fine detail such as subtle Wallner lines on fracture surfaces, but it may not be very long before they do.

BOOKS ON FRACTOGRAPHY

Books specifically written on fractography of brittle material were rare until the 1980s. Kerkhof's book Fracture Processes in Glasses was published in 1970.[104] Volume 1 of the Fracture Mechanics of Ceramics conference series organized by Prof. Bradt in 1974 was subtitled "Concepts, Flaws and Fractography."[168] It had nine interesting papers on fractography including contributions from Ernsberger, Rice, Kirchner, and Richerson. A 1982 ASTM conference proceedings on ceramic and metal failures[169] had nine superb papers, including Rice's 99 page article. It also has a fascinating set of comments and recommendations by a panel of experts. Fréchette's 1990[89] book was a milestone and a masterpiece. Bradt and Tressler edited a book Fractography of Glass in 1994 that was a compilation of eight papers.[170] Hull's 1999 book Fractography, a masterpiece with a superb layout and outstanding illustrations, is about fractography in general; and ceramics and glasses share space with metals and polymers.[2] Morrell's Guide to Fractography of Brittle Materials[171] is a concise useful starting point for any aspiring fractographer, and it has some fascinating practical examples.

Conference proceedings books from the quadrennial Alfred University conference series on Fractography of Glasses and Ceramics were first published in 1986 by Professors Fréchette and Varner.[172] The present conference is the sixth in the series. The proceedings series are an impressive compilation of some of the best fractographic work in the last 25 years. A European conference series was organized by Prof. Jan Dusza of the Slovakian Academy of Sciences, Kosice, starting in 2001.[173] There have been three conferences with proceedings books so far, and a fourth is planned for 2013.

My Guide to Fractography of Ceramics and Glasses[1] was the first to be printed in color on glossy paper and the first available in digital form. There are 725 figures and schematics, such as

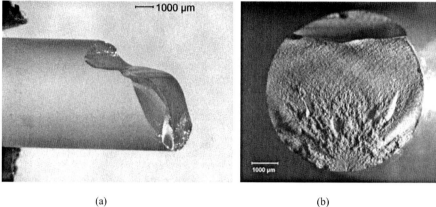

(a)   (b)

Figure 63. Two figures from the author's guide book. (a) shows a glass rod broken in flexure. The alignment and illumination were meticulously adjusted to show the compression curl, hackle lines, the fracture mirror, and even the flaw itself. (b) shows a silicon nitride rod also broken in flexure. Even at this low magnification, vicinal illumination revealed a telltale "V" shape in the mirror at the origin site. This is a telltale sign that the fracture origin was machining cracks from transverse grinding.

Figure 63. Most had not been published before. The Guide introduces some new terms, markings, and characterization procedures. The book has a strong practical slant. New terms such as "corner hackle," "T-crack intersections," and many dozens of carefully drawn schematics will help the next generation of fractographers learn our craft. The book has been widely disseminated and it is free. It is used in several courses including the annual summer 4-day short course Fracture Analysis of Glasses and Ceramics at Alfred University and the annual summer 3-day American Dental Association Foundation course on Dental Materials Fractography.

FRACTOGRAPHY STANDARDS

In my experience, documentary standards are developed once a discipline has matured to the point that some consistency in procedure is needed. In the late 1970s, I realized that there are two equally important pieces of information that may be gleaned from a strength test: the strength value and the fracture origin flaw. Shortly after we had developed the first standard test method for flexural strength testing of advanced ceramics, MIL STD 1942 (MR),[174] I decided that the next step should be to write a standardized procedure for finding and characterizing fracture origins. At that time, fractography was dismissed by some as subjective and interpretive. I spoke on this matter in my paper at the first Alfred conference on Fractography of Glasses and Ceramics in 1986.[175] Fréchette[84] had taken the first important steps for adopting a common nomenclature, but there was a need for further clarification and refinements.

My colleagues Dr. J. Swab and M. Slavin at the US Army Materials Research Laboratory in Watertown, MA, and I started with a Military Handbook (MIL HDBK) 790 in 1992.[86,176,177] It recommended fractographic analysis procedures and nomenclature for finding and characterizing fracture origins in advanced ceramics. It was a bold step at the time. A fracture origin was characterized by what the flaw was, where it was located, and its approximate size. A simple fracture mechanics analysis was suggested to verify that the appropriate flaw had been found. An atlas of flaw types was included. We evaluated the MIL HDBK's effectiveness with a full scale, seventeen laboratory Versailles Advanced Materials and Standards (VAMAS) international round robin.[177,178] Lessons learned from the exercise led to the creation and adoption in 1996 of the much more comprehensive ASTM C 1322, Standard Practice for Fractography and Characterization of Fracture Origins in Advanced Ceramics.[87] A comparable European Committee for Standard document for ceramics was written by R. Morrell of the National Physical Laboratory in the United Kingdom in 2004.[88]

In the meantime, in 1993, ASTM C 1256, Standard Practice for Interpreting Glass Fracture Surface Features, written by S. DeMartino of Corning, Inc., was adopted by ASTM committee C-14 on Glass and Glass Products.[85] It is a short, well-illustrated document that describes some classic fracture markings in glass, and uses Van Fréchette's nomenclature.

Over the years, considerable effort has been expended on refining and standardizing test methods for evaluating fracture toughness, K_{Ic}. Fractographic analysis is a critical aspect of several of the test methods. In 1992 – 1994, Robert Gettings of NIST, Jakob Kübler of the Swiss Federal Research Laboratory (EMPA), and I organized a major VAMAS international round robin on fractographic analysis of precracks in the surface crack in flexure (SCF) test method.[179,180] Knoop indentations were used to create tiny semielliptical-surface flaws in flexure test specimens. After the indentation damage zones were removed by polishing, the specimens were broken in flexure. The semielliptical-precracks such as shown in Figure 64 were measured on the fracture surface by fractographic methods. Fracture toughness was computed from the fracture load, the precrack size, and an appropriate stress intensity shape factor. Three materials, two silicon nitrides and one zirconia, were evaluated by twenty laboratories. We discovered that the fracture toughness outcomes were surprisingly consistent. Errors or uncertainties in the precrack size measurements had only a small effect on the calculated fracture toughness. This was due in part on the K_{Ic} dependence on the square root of the crack size. Furthermore, there was also a curious compensating effect that of the stress intensity shape factor, Y. Errors in crack size were mitigated by a compensating shift in Y. The results of this successful round robin led to the adoption of the SCF method as one of three in the fracture toughness standard ASTM C 1421 in 1999[181,182] and ISO standard 18756 in 2003.[183] The SCF round robin results also were instrumental in the creation of the world's first standard reference

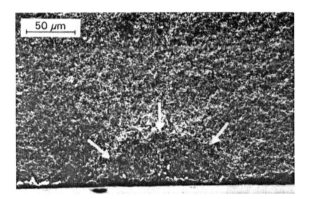

Figure 64. Fracture surface of a Knoop indentation induced semielliptical surface crack for the SCF method. The test specimen, a hot pressed silicon nitride, that was used for NIST SRM 2100 for K_{Ic}.

material for fracture toughness, K_{Ic}, NIST Standard Reference Material (SRM) 2100. It is a set of five test specimens with a certified fracture toughness of 4.57 MPa$\sqrt{m}$ ± 0.11 MPa$\sqrt{m}$ at the 95% confidence level.[184,185]

My final fractographic standard was ASTM C 1678[119] for the analysis of fracture mirror sizes in ceramics and glasses, which was adopted in 2008. It was an outgrowth of the Guide to Practice. When I compiled all published results, I realized there was a wide divergence of fracture mirror constants. A summary of my research into this matter and the rationales for the chosen procedures in C 1678 are in my paper at the 2006 Alfred Fractography conference.[70] Shand, Kirchner, Smekal, Preston, and Orr might have been surprised that such a procedure could be standardized, but I suspect they would be pleased. Rice wrote in 1984:[140]

"Standardization of the measurement criteria (by analysis of topography for example, to more consistently define the onset of mist and other features) deserves serious attention."

Controversies will probably continue about fracture mirrors and the basic physics for their formation, but we now have well-defined procedures that will help us solve practical problems and will bring consistency to the field.

CONCLUSIONS

This history is a story of the evolution of materials science, theoretical developments on the strength of brittle materials, the fabrication of stronger materials, advancements in microscopy, standardization, and the key people who developed our craft. Fractographic analysis is an objective scientific discipline that may take time to master, but unlocks the secrets of where, how, and why fracture occurred.

Figure 65. Fracture surface of a leucite porcelain bend bar processed to have a minimal amount of leucite crystals. The origin is a bubble-pore that almost touched the tensile surface. One can discern the mirror, mist, and hackle, Wallner lines, gull wings from microscopic bubbles and leucite sites. Note how these tiny features trigger early mist well inside the mirror, prior to its general formation for the mirror boundary. (Unpublished, J. Quinn).

ACKNOWLEDGEMENTS
I learned much from the late Professor Van Derck Fréchette of Alfred University. I have had many fruitful collaborations and discussions with Professor James Varner, Alfred University; Mr. Roy Rice, NRL; Dr. Roger Morrell, National Physical Laboratory, UK; Prof. Richard Bradt, University of Alabama; and Dr. Susanne Scherrer of the University of Geneva. Finally, I dearly miss Dr. Janet B. Quinn of the American Dental Association Foundation (ADAF).[186,187] We collaborated on engineering problems for 39 years. She brought a fresh perspective and set of eyes to our field. She had a growing body of publications on fractographic analysis, and had she not died so prematurely in 2008, I am sure she would have made great contributions in our field. Figure 65 is a spectacular photo (especially when seen in color) that I found in her files. I continue her work in her memory and acknowledge the support of NIST, ADAF, and the National Institute of Health with NIH Grant R01-DE17983.

REFERENCES

[1] G. D. Quinn, *NIST Recommended Practice Guide for the Fractography of Ceramics and Glasses*, Special Publication 960-16, National Institute of Standards and Technology, May, 2007.

[2] D. Hull, *Fractography, Observing, Measuring and Interpreting Fracture Surface Topography*, Cambridge University Press, Cambridge, 1999.

[3] *ASM Handbook, Vol. 11, Fractography*, ASM Int., Materials Park, OH, 1987.

[4] *ASM Handbook, Vol. 12, Failure Analysis and Prevention*, ASM Int., Materials Park, OH, 2002.

[5] *Metals Handbook, Vol. 9, Fractography and Atlas of Fractographs*, 8th edition, American Society of Metals, Metals Park, OH, 1974.

[6] G. D. Quinn, "A History of the Fractography of Brittle Materials," in *Fractography of Advanced Ceramics, III*, ed. J. Dusza, R. Danzer, R. Morrell and G. Quinn, TransTech Publ., Zurich, 2009. *Key Engineering Materials* Vol. 409 pp. 1-16 (2009).

[7] R. Hooke, *Micrographia: Some Physiological Descriptions of Minute Bodies Made by Magnifying Lenses*, Royal Society, London, 1665.

[8] C. Brodmann, *Nachrichten von der Gessellschaft der Wissenschaften zu Göttingen, Mathematisch-Physikalische Klasse*, Vol. 1, 44 – 58 (1894).

[9] Ch. De Freminville, *Revue de Metallurgie*, Vol. 4, 833 – 884 (1907).

[10] Ch. De Freminville, *ibid*, Vol. 11, 971 – 1056 (1914).

[11] Ch. De Freminville, *Trans. ASME*, Vol. 41, 907 – 923 (1919).

[12] A. A. Griffith, *Proc. Trans Roy Soc.*, Vol. A221, 163 – 198 (1920).

[13] A. A. Griffith, in *Proc. 1st Int. Congress on Appl. Mech.*, eds. C. B. Biezeno and J. M. Burgers, 55 – 63 (1924).

[14] D. G. Holloway, pp. 1 – 19 in *Strength of Inorganic Glass*, ed. C. Kurkjian, Plenum, NY. 1985.

[15] B. Cotterell, pp. 105- 122 in *Fracture Research in Retrospect*, ed. H.P. Rossmanth, Balkema, Rotterdam, 1997.

[16] F. W. Preston, *Trans. Optical Society*, Vol. 23, No. 3, 141-164 (1921-1922).

[17] L. G. Ghering, J. C. Turnbull, and F. W. Preston, *Bull. Am. Ceram. Soc.*, Vol. 19, No. 8, 290 – 294 (1940).

[18] F. W. Preston, *J. Soc. Glass Techn.*, Vol. 10, 234 – 269 (1926).

[19] F. W. Preston, *ibid.*, Vol. 11, 3 – 10 (1927).

[20] F. W. Preston, *J. Amer. Ceram. Soc.*, Vol. 18, No. 6, 175-176 (1935).

[21] F. W. Preston, *Bull. Amer. Ceram. Soc.*, Vol. 18, No. 2, 35 – 60 (1939).

[22] F. W. Preston, *J. Amer. Ceram. Soc.*, Vol. 25, No. 11, 427 – 434 (1942).

[23] F. W. Preston, *J. Appl. Phys.*, Vol. 13, 623 – 634 (1942).

[24] Obituary, Frank W. Preston, 1896 – 1989, *Bull. Amer. Ceram. Soc.*, Vol. 68, No. 11, 1900 (1989).

[25] S. M. Wiederhorn, *J. Amer. Ceram. Soc.*, Vol. 75, No. 10, 2643 (1992).

[26] www.americanglassresearch.com.

[27] A. Smekal, *Glastechn. Ber.*, Vol. 13, No. 7, 222-232 (1935).

[28] A. Smekal, *ibid*, Vol. 13, No. 5, 141 – 151 (1935).

[29] A. Smekal, *Zeitschrift Phys.*, Vol. 103, 495 – 525 (1936) .

[30] A. Smekal, *J. Soc. Glass Techn.*, Vol. 20, 432 – 448 (1936).

[31] A. Smekal, *Phys. Zeitschr.*, Vol. 41, 475 – 480 (1940).

[32] A. Smekal, *Ergibnisse der Exakten Naturwissenschaften,* Vol. 15, 106 – 188 (1936).

[33] A. Smekal, *Glastechn. Ber.*, Vol. 23, No. 3, 57 – 67 (1950).

[34] A. G. Smekal, *Ostereichische Ingenieur Archiv*, Vol. 7, 49 – 70 (1953).

[35] A. Momber, *Forsch. Ingenieur.*, Vol. 70, 114 – 119 (2006).

[36] H. Rumpf, *Chem. Ingenieur Techn.*, Vol. 31, No. 11, 697 – 705 (1959).

[37] Adolf Smekal, *Glastechnische Berichte*, Vol. 32, No 4, page 180 (1959).

[38] A.W. Momber, *J. Mat Sci.*, Vol. 45, No. 3, 750-758 (2010).

[39] H. Wallner, *Z. Phys.* Vol. 114, 368 – 378 (1939).

[40] H. G. Richter, *DGG Jour.*, Vol. 7, No. 3, 1 – 3 (2008).

[41] H. Schardin and W. Struth, *Glastechn. Ber.*, Vol. 16, No. 7, 29 – 231 (1938).

[42] H. Schardin, *Glastech. Ber.*, Vol. 23, No. 12, 325 – 336 (1950).

[43] H. Schardin, L. Mücke, and W. Struth, *Glastechn. Ber.,* Vol. 27, No. 5, 141 – 147 (1954).

[44] H. Schardin, L. Mücke, and W. Struth, *The Glass Industry*, Vol. 36, No. 3, 133 – 139 (1955).

[45] F. E. Barstow and H. E. Edgerton, *J. Amer. Ceram. Soc.*, Vol. 22, No. 9, 302 – 307 (1939).

[46] In Memoriam, Hubert Schardin, *Int. J. Fract.*, Vol. 1, No. 3, ii (1965).

[47] J. E. Field, *Contemp. Phys.*, Vol. 12, No. 1-2, 1 – 31 (1971).

[48] J. B. Murgatroyd, *J. Soc. Glass Techn.,* Vol. 26, 155 – 171 (1942).

[49] S. Tolansky, *Surface Microphotography*, Longmans, Green and Co. Ltd, London, 1960.

[50] E. Gölz, *Zeitschrift Physik*, Vol. 120, 773 – 777 (1943).

[51] E. F. Poncelet, *J. Soc. Glass Techn.*, Vol. 42, 279T – 288T (1958).

[52] K. Peter, *Glastechn. Ber.*, Vol. 41, 442 – 445 (1968).

[53] E. K. Beauchamp, Sandia Laboratories Research Report, SC-RR-70-766, Jan. 1971.

[54] E. K. Beauchamp, pp. 409 – 446 in *Fractography of Glasses and Ceramics III,* eds. J. Varner, et. al., *Ceramic Transactions,* Vol. 64, American Ceramic Society, Westerville, OH, 1996.

[55] W. Weibull, *Ing. Vetensk Akad. Proc.,* Vol. 151, No. 153, 293 - 297 (1939).

[56] W. Weibull, *J. Appl. Mech.*, Vol. 18, 293 - 297(1951).

[57] E. H. Yoffe, *Phil. Mag.*, Vol. 330, 739 -750 (1951).

[58] J. Miklowicz, *J. Appl. Mech.*, Vol. 20, March, 122 - 130 (1953).

[59] J. W. Phillips, *Int. J. Solids Struct.*, Vol. 6, 1403 - 1412 (1970).

[60] H. Kolsky, *Trans. Soc. Rheol.,* Vol. 20, No., 3 441 - 454 (1976).

[61] G. R. Irwin, *J. Appl. Mech.,* Vol. 24, No. 3, 361 - 364 (1957).

[62] H. P. Rossmanith, ed., *Fracture Research in Retrospect, An Anniversary Volume in Honour of George R. Irwin's 90th Birthday,* Balkema Press, Rotterdam, 1997.

[63] J. C. Newman, Jr., and I. S. Raju, *NASA Tech. Paper 1578,* NASA-Langley Res. Ctr., VA, 1979.

[64] J. C. Newman, Jr., and I. S. Raju, *Eng. Fract. Mech.*, Vol. 15, No. 1-2, 185 – 192 (1981).

[65] S. M. Wiederhorn and L. H. Bolz, *J. Amer. Ceram. Soc.*, Vol. 53, No. 8, 543 – 548 (1970).

[66] E. Poncelet, *Verres et Refractaires*, Vol. 4, 158 – 171 (1950).

[67] L. Orr, *Mater. Res. Standards*, Vol. 12, No. 1, 21 – 23 (1972).

[68] J. B. Quinn, *J. Am. Ceram. Soc.*, Vol. 82, No. 8, 2126 – 2132 (1999).

[69] Private Letter, L. Orr to G. D. Quinn, December 15, 2001.

[70] G. D. Quinn, pp. 163 – 190 in *Fractography of Glasses and Ceramics,* Vol. 5, eds. J. R. Varner, G. D. Quinn, M. Wightman, American Ceramic Society, Westerville, OH, 2007.

[71] J. W. Johnson and D. G. Holloway, *Phil. Mag.* Vol. 14, 731 – 743 (1966).
[72] http://www.asmenews.org/archives/backissues/jun04/features/604orr.html.
[73] F. M. Ernsberger, *Proc. Roy. Soc. London A,* Vol. 257, 213 – 223 (1960).
[74] F. M. Ernsberger, pp. 161 – 173 in *Fracture Mechanics of Ceramics, Vol. 1*, eds. R. C. Bradt, D. P. H. Hasselman, and F. F. Lange, et al., Plenum, NY, 1975.
[75] F. M. Ernsberger, *J. Noncryst. Sol.*, Vol. 25, 295 – 321 (1977).
[76] F. M. Ernsberger, *Polarized Light in Glass Research*, PPG Industries, Pittsburg, PA, 1972.
[77] E. B. Shand, *J. Amer. Ceram. Soc.*, Vol. 37, No. 12, 559 – 572 (1954).
[78] E. B. Shand, *ibid.*, Vol. 42, No. 10, 474 – 477 (1959).
[79] E. B. Shand, *ibid.*, Vol. 44, No. 9, 451 – 455 (1961).
[80] E. B. Shand, *ibid.*, Vol. 48, No. 1, 43 – 48 (1964).
[81] E. B. Shand, *The Glass Industry*, April, 190 – 194 (1967).
[82] "History of the Errol Shand Award," www.uwst.org/ShandAwardHistory.asp, 2011.
[83] V. D. Fréchette, *Proc. Brit. Ceram. Soc.,* Vol. 5, 97 – 106 (1965).
[84] V. D. Fréchette, pp. 104 – 109 in *Fractography of Ceramic and Metal Failures,* ASTM STP 827, J. J. Mecholsky, Jr., and S. R. Powell, Jr. eds., ASTM, West Conshohocken, PA, 1984.
[85] C 1256-93, Standard Practice for Interpreting Glass Fracture Surface Features, ASTM, West Conshohocken, PA, 1993.
[86] MIL HDBK 790, Fractography and Characterization of Fracture Origins in Advanced Structural Ceramics, U. S. Army Research Laboratory, Watertown MA, June 1992.
[87] C 1322-92, Standard Practice for Fractography and Characterization of Fracture Origins in Advanced Ceramics, ASTM, West Conshohocken, PA, 1992.
[88] EN 843-6, Advanced Technical Ceramics, Monolithic Ceramics, Part 6: Guidelines for Fractographic Examination, European Committee for Standardization, Brussels, 2004.
[89] V. D. Fréchette, *Failure Analysis of Brittle Materials*, American Ceramic Society, Westerville, OH, 1990.
[90] V. D. Fréchette, pp. 71 – 76 in *Fractography of Glasses and Ceramics, Advances in Ceramics*, Vol. 22, eds. J. Varner, and V. Fréchette, American Ceramic Society, Westerville, OH, 1988.
[91] T. A. Michalske, pp. 121 – 136 in *Fractography of Ceramic and Metal Failures*, ASTM Special Technical Publication 827, eds., J. J. Mecholsky, Jr. and S. R. Powell, Jr., ASTM, West Conshohocken, PA, 1984.
[92] V. D. Fréchette and J. R. Varner, *J. Appl. Phys.*, Vol. 42, No. 5, 1983 – 1984 (1971).
[93] C. L. Quackenbush and V. D. Fréchette, *J. Amer. Ceram. Soc.*, Vol. 61, No. 9, 402 – 406 (1978).
[94] V. D. Fréchette, pp. 245 – 245, in *Advances in Ceramics*, Vol. 22, *Fractography in Glasses and Ceramics*, American Ceramic Society, Westerville, OH, 1988.
[95] T. A. Michalske, M. Singh, and V. D. Fréchette, pp. 3 – 12 in *Fracture Mechanics for Ceramics, Rocks, and Concrete*, ASTM STP 745, eds. S. W. Freiman and E. R. Fuller, 1980.
[96] J. R. Varner, pp. 389 – 406 in *Strength of Inorganic Glass*, ed. C. Kurkjian, Plenum, NY, 1985.
[97] J. R. Varner and H. J. Oel, *J. Non-Crystal. Solids*, Vol. 19, 321 – 333 (1975).
[98] V. D. Fréchette and M. Donovan, pp. 407 – 411 in *Fractography of Glasses and Ceramics, II, Ceramic Transactions,* eds. V. Fréchette, and J. Varner, eds., American Ceramic Society, Westerville, OH, 1988.
[99] D. G. Holloway, *Glass Technology*, Vol. 27, No. 4, 120 – 133 (1986).
[100] J. Congleton and N. J. Petch, *Phil. Mag.*, Vol. 16, No. 142, 749 - 760 (1967).
[101] F. Kerkhof, *Naturwiss*, Vol. 40, 478 (1953).
[102] F. Kerkhof, *Glastechn. Ber.,* Vol. 28, No. 2, 57 – 58 (1955).
[103] F. Kerkhof, *Glastechn. Ber.*, Vol. 35, No. 6, 267 – 272 (1962).

[104] F. Kerkhof, *Bruchvorgänge in Gläsern,* Verlag derDeutschen Glastechnischen Gessellshcaft, Frankfurt am Main, 1970.

[105] F. Kerkhof and H. Richter, pp. 463 – 473 in *Fracture 1969,* 1969.

[106] H. Richter, pp. 219 – 229 in *Strength of Inorganic Glass,* ed. C. Kurkjian, Plenum, 1986.

[107] S. Winkler, D. A. Shockey, and D. R. Curran, *Int. J. Fract.,* Vol. 6, No. 2, 151 – 158 (1970).

[108] E. Sommer, *Eng. Fract. Mech.,* Vol. 1, 539 – 546 (1969).

[109] www.iwm.fhg.de/pdf/presse/FraunhoferIWM_TodKerkhof.pdf.

[110] B. Hockey and B. Lawn, *J. Mat. Sci.,* Vol. 10, 1275-1284 (1975).

[111] B. R. Lawn, B. J. Hockey, and S. M. Wiederhorn, *J. Mat. Sci.,* Vol. 15, 1207 – 1223 (1980).

[112] H. Tanaka and Y. Bando, *J. Am. Ceram. Soc.,* Vol. 73, No. 3, 761 – 763 (1990).

[113] H. Tanaka and Y. Bando, Y. Inomata, and M. Mitomo, *J. Am. Ceram. Soc.,* Vol. 71, No. 1, C32-C33 (1988).

[114] H. P. Kirchner, and R. M. Gruver, *Phil. Mag.,* Vol. 27, No. 6, 1433 – 1446 (1973).

[115] H. P. Kirchner, and R. M. Gruver, pp. 309 – 321 in *Fracture Mechanics of Ceramics I,* eds., R. C. Bradt, et al., Plenum, NY, 1974.

[116] H. P. Kirchner, R. M. Gruver, and W. A. Sotter, *Phil. Mag.,* Vol. 33, No. 5, 775 – 780 (1976).

[117] H. P. Kirchner and J. C. Conway, Jr., *J. Amer. Ceram. Soc.,* Vol. 70, No. 6, 413 – 418 and 419 – 425 (1987).

[118] R. M. Gruver, W. A. Sotter and H. P. Kirchner, "Fractography of Ceramics," Ceramic Finishing Company Summary Report to the Naval Systems Command of the Department of the U. S. Navy, 22 November 1974.

[119] C 1678-08, Practice for Fractographic Analysis of Fracture Mirror Sizes in Ceramics and Glasses, ASTM, West Conshohocken, PA, 2008.

[120] H. P. Kirchner, R. M. Gruver, and D. M. Richard, *J. Mat. Sci.,* Vol. 14, 2713 – 2720 (1979).

[121] H. P. Kirchner, R. M. Gruver, and W. A. Sotter, *Mater. Sci. Eng.,* Vol. 22, 147 – 156 (1976).

[122] H. P. Kirchner, R. M. Gruver, and R. E. Walker, pp. 353 – 363 in *The Science of Ceramic Machining and Surface Finishing,* eds. S. Schneider and R. W. Rice, National Bureau of Standards, Gaithersburg, MD, 1972.

[123] H. P. Kirchner, *J. Amer. Ceram. Soc.,* Vol. 67, No. 2, 127 – 132 (1984).

[124] H. P. Kirchner, www.ceramics.org/in-memoriam.

[125] J. J. Mecholsky, Jr., R. W. Rice, and S. W. Freiman, *J. Amer. Ceram. Soc.,* Vol. 57, No. 10, 440 – 443, (1974).

[126] J. J. Mecholsky, Jr., S. W. Freiman, and R. W. Rice, *J. Mat. Sci.,* Vol. 11, 1310 – 1319 (1976).

[127] R. W. Rice, *J. Amer. Ceram. Soc.,* Vol. 62, No. 9-10, 533 – 535 (1979).

[128] R. W. Rice, pp. 323 343 in *Fracture Mechanics of Ceramics, Vol. 1,* eds. R.C. Bradt et al., Plenum, NY, 1975.

[129] R. W. Rice, S. W. Freiman, and J. J. Mecholsky, Jr., pp. 669 – 687 in *Ceramics for High Performance Applications,* II, eds. J. J. Burke et al., Brook Hill Publ., Chestnut Hill, MA, 1977.

[130] R. W. Rice, pp. 303 – 319 in *Processing of Crystalline Ceramics,* eds., H. Palmour et al., Plenum, NY, 1978.

[131] R. W. Rice, *J. Mat. Sci.,* Vol. 19, 895 – 914 (1984).

[132] J. J. Mecholsky and S. W. Freiman, pp. 136 – 150 in *Fracture Mechanics Applied to Brittle Materials, ASTM STP 678,* ed., S. W. Freiman, American Society for Testing and Materials, Philadelphia, PA, 1979.

[133] R. W. Rice, S. W. Freiman, and J. J. Mecholsky, Jr., *J. Amer. Ceram. Soc.,* Vol. 63, No. 3-4, 129 – 136 (1980).

[134] J. J. Mecholsky, Jr., S. W. Freiman, and R. W. Rice, *ibid,* Vol. 60, No. 3-4, 114 – 117 (1977).

[135] R. W. Rice, J. J. Mecholsky, Jr., and P. F. Becher, *J. Mat Sci.,* Vol. 16, 853 – 862 (1981).

[136] R. W. Rice and J. J. Mecholsky, Jr., pp. 351- 378 in *The Science of Ceramic Machining and Surface Finishing*, SP 562, II, eds. B. J. Hockey and R. W. Rice, National Bureau of Standards, Gaithersburg, MD.

[137] R. W. Rice, *J. Am. Ceram. Soc.*, Vol. 76, No. 4, 1068 – 1070 (1993).

[138] R. W. Rice, *J. Eur. Ceram. Soc.*, Vol. 22, 1411 – 1424 (2002).

[139] R. W. Rice, pp. 1 – 52 in *Fractography of Glasses and Ceramics*, III, eds. J. R. Varner, V. D. Fréchette, and G. D. Quinn, *Ceramic Transactions*, Vol. 64 (1996).

[140] R. W. Rice, pp. 5 – 103 in *Fractography of Ceramic and Metal Failures, ASTM STP 827*, eds. J. J. Mecholsky, Jr. and S. R. Powell, Jr., ASTM, Philadelphia, PA, 1984.

[141] R. W. Rice, pp. 199 – 381, in *Treatise on Materials Science and Technology*, Vol. 11, Academic Press, NY, 1977.

[142] R. W. Rice, *Porosity of Ceramics*, Marcel Dekker, New York, 1998.

[143] R. W. Rice, *Mechanical Properties of Ceramics and Composites, Grain and Particle Effects*, Marcel Dekker, New York, 2000.

[144] Roy Rice, www.ceramics.org/in-memoriam.

[145] D. Munz, O. Rosenfelder, K. Goebells, and H. Reiter, pp. 265–283 in *Fracture Mechanics of Ceramics, Vol. 1*, eds. R. Bradt, D. Hasselman, and F. Lange, Plenum Press, NY, 1986.

[146] D. W. Richerson, *Modern Ceramic Engineering*, Marcel Dekker Inc., NY, 1982.

[147] H. G. Richter, pp. 157 – 180 in *Key Engineering Materials*, Vol. 223 (2002).

[148] S. Hecht-Mijic and H. G. Richter, pp. 313-337 in *Fractography of Glasses and Ceramics V*, Ceramic Transactions Vol. 199, eds, J. R. Varner, G. D. Quinn, M. Wightman,American Ceramic Society, Westerville, OH, 2007.

[149] K. Sato, S. Tanaka, N. Uchida, and K. Uematsu, pp. 225 – 228 in *Key Engineering Materials*, Vols. 264 – 268, TransTech. Publ., Switzerland, 2004.

[150] H. Abe, M. Naito, T. Hotta, N. Shinohara, and K. Uematsu, *J. Amer. Ceram. Soc.*, Vol. 86, No. 6, 1019 – 1021 (2003).

[151] Y. Zhang, M. Inoue, N. Uchida, and K. Uematsu, *J.. Mater. Res.*, Vol. 14, No. 8, 3370 – 3374 (1999).

[152] H. Takahashi, N. Shinohara, K. Uematsu, and T. Junichiro, *J. Am. Ceram. Soc.*, Vol 79, No. 4, 843 – 848 (1996).

[153] N. Shinohara, M. Okumiya, T. Hotta, K. Nakahira, M. Naito, and K. Uematsu, *J. Am. Ceram. Soc.*, Vol. 82, No. 12, 3441-46 (1999).

[154] S. Nakamura, S./ Tanaka, R. Furushima, K. Sato, and K. Uematsu, *J. Ceram. Soc. Japan*, Vol. 117, No. 6, 742 – 747 (2009).

[155] J. B. Quinn, S. S. Scherrer, and G. D. Quinn, pp. 253 – 270 in *Fractography of Glasses and Ceramics V*, eds, J. R. Varner, G. D. Quinn, M. Wightman, American Ceramic Society, Westerville, OH, 2007.

[156] S. S. Scherrer, J. B. Quinn, G. D. Quinn, and J. R. Kelly, *Int. J. Prosthodont.*, Vol. 19, No. 2, 151 – 158 (2006).

[157] S. S. Scherrer, J. B. Quinn, G. D. Quinn, and H. W. A. Wiscott, *Dental Materials*, Vol. 23, No. 11, 1397-1404 (2006).

[158] J. B. Quinn and G. D. Quinn, *Dental Materials*, Vol. 26, No. 6, 589 – 599 (2010).

[159] P. Supancic, pp. 69-78 in *Fractography of Advanced Ceramics*, ed. J. Dusza, Trans Tech Publ. Zurich, CH, (2002).

[160] P. Supancic, Z. Wang, W. Harrer, K. Reichmann, and R. Danzer, pp. 46-53 in *Fractography of Advanced Ceramics, II*, ed. J. Dusza, R. Danzer and R. Morrell, Trans Tech Publ. Zurich, CH (2005).

[161] W. Harrer, R. Danzer, P. Supancic, and T. Lube, pp. 176-184 in *Fractography of Advanced Ceramics, III*, ed. J. Dusza, R. Danzer, R. Morrell, and G. Quinn, Trans Tech Publ. Zurich, CH (2009).

[162] R. Morrell, W. Harrer, R. Danzer, and K. Berroth, pp. 304-307 in *Fractography of Advanced Ceramics, III*, ed. J. Dusza, R. Danzer, R. Morrell, and G. Quinn, Trans Tech Publ. Zurich, CH (2009).

[163] R. Danzer, A. Platzer, P. Supancic, and Z. Wang, pp. 231 – 241 in *Fractography of Glasses and Ceramics V*, eds., J. Varner, G. Quinn, and M. Wightman, *Ceramic Trans.* Vol. 199, American Ceramic Society, Westerville, OH, 2007.

[164] M. Hangl, pp. 133-156 *Fractography of Glasses and Ceramics IV*, eds., J. Varner and G. Quinn, *Ceramic Trans.* Vol. 122, American Ceramic Society, Westerville, OH, 2001.

[165] C. Wunsche, E. Radlein, and G. H. Frischat, *Glastech. Ber. Glass Sci. Technol.*, Vol. 72, No. 2, 49 – 54 (1999).

[166] D. M. Kulawansa, L. C. Jensen, S. C. Langford, J. T. Dickinson, and Y. Watanabe, *J. Mater. Res.*, Vol. 9, No. 2, 476 – 485 (1994).

[167] D. Hull, *J. Mat. Sci.*, Vol. 31, 1829 – 1841 (1996).

[168] *Fracture Mechanics of Ceramics, Vol., 1, Concepts, Flaws, and Fractography*, eds R. C. Bradt, D. P. H. Hasselman and F. F. Lange, Plenum, NY, 1975.

[169] *Fractography of Ceramic and Metal Failures, ASTM, STP 827*, eds. J. J. Mecholsky, Jr. and S. R. Powell, Jr., ASTM, Philadelphia, PA, 1984.

[170] *Fractography of Glass*, eds. R. C. Bradt and R. E. Tressler, Plenum, NY, 1994.

[171] R. Morrell, *Fractography of Brittle Materials, Measurement Good Practice Guide 15,* National Physical Laboratory, Teddington, UK, 1999.

[172] *Fractography of Glasses and Ceramics*, Eds., J. R. Varner and V. D. Fréchette, American Ceramic Society, Westerville, OH, 1988.

[173] *Fractography of Advanced Ceramics*, ed. J. Dusza, Trans Tech Publ., Zurich, 2002.

[174] MIL STD 1942 (MR), "Flexural Strength of High Performance Ceramics at Ambient Temperature," U.S. Army Materials and Mechanics Research Center, Watertown, MA, 21 Nov. 1983.

[175] G. D. Quinn, pp. 319 – 334 in *Fractography of Glasses and Ceramics*, Eds. J. Varner and V. Fréchette, American Ceramic Society, Westerville, Ohio, (1988).

[176] G. D. Quinn, J. J. Swab, and M. J. Slavin, pp. 309 – 362 in *Fractography of Glasses and Ceramics II*, eds. V. D. Fréchette and J. R. Varner, American Ceramic Society, Westerville, Ohio, 1991.

[177] J. J. Swab, M. J. Slavin, and G. D. Quinn, pp. 936 – 943 in *4th International Symposium on Ceramic Materials and Components for Engines*, eds. R. Carlsson, et al., Elsevier, England, 1992.

[178] J. J. Swab and G. D. Quinn, pp. 55 -70 in *Fractography of Glasses and Ceramics III*, eds., J. R. Varner, V. D. Fréchette, and G. D. Quinn, ACS, Westerville, OH, 1996.

[179] G. D. Quinn, J. Gettings, and J. J. Kübler, pp. 107 – 144, in *Fractography of Glasses and Ceramics III*, eds., J. R. Varner, V. D. Fréchette, and G. D. Quinn, ACS, Westerville, OH, 1996.

[180] G. D. Quinn, R. J. Gettings, and J. J. Kübler, *Ceram. Eng. and Sci. Proc.*, Vol. 15, No. 5, 846-855, (1994).

[181] C 1421-99, Standard Test Method for the Determination of Fracture Toughness of Advanced Ceramics, ASTM, West Conshohocken, PA, 1999.

[182] J. A. Salem, G. D. Quinn, and M. G. Jenkins, pp. 531- 554 in *Fracture Mechanics of Glasses and Ceramics, Vol. 14*, eds. R. C. Bradt, D. Munz, M. Sakai, and K. W. White, 2005, Klewer/Plenum, NY, (2005).

[183] ISO 18756, Fine Ceramics (Advanced Ceramics, Advanced Technical Ceramics) – Determination of Fracture Toughness of Monolithic Ceramics at Room Temperature by the Surface Crack in Flexure (SCF) Method, International Organization for Standards, Geneva, SW, (2003).

[184] SRM 2100, Fracture Toughness of Ceramics, NIST, Gaithersburg, MD, SRM Office, (1999).

[185] G. D. Quinn, K. Xu, J. A. Salem, and J. J. Swab, pp. 499 – 530 in *Fracture Mechanics of Glasses and Ceramics, Vol. 14*, eds. R. C. Bradt, D. Munz, M. Sakai, and K. W. White, Klewer/Plenum, NY, (2005).

[186] Obituary, Janet B. Quinn, *Dental Materials*, Vol. 25, No. 1, 2-3 (2009).
[187] Janet B. Quinn, www.ceramics.org/in-memoriam.

THE EFFECT OF DEFECTS AND MATERIALS TEXTURE ON THE FRACTURE OF LOW-PRESSURE INJECTION MOULDED ALUMINA COMPONENTS

Tanja Lube,[1] Roger Morrell,[2] Irina Kraleva[3]
[1]Institut für Struktur- und Funktionskeramik, Montanuniversitaet Leoben, Franz-Josef-Straße 18, A-8700 Leoben, Austria
[2]National Physical Laboratory, Hampton Road, TW11 0LW Teddington, UK
[3]Materials Center Leoben Forschung GmbH, Roseggerstraße 12, A-8700 Leoben, Austria

ABSTRACT

The strength of tube shaped components made by low-pressure injection moulding of a 96% alumina ceramic was measured using 3-point- and 4-point bending tests. Weibull statistics were applied to describe the strength distributions. Fracture surfaces were investigated using stereo-microscopy and SEM. Thin sections of the components were made in different orientations. A mineralogical method was used to investigate the microstructure with respect to possible textures due to the manufacturing process.

The Weibull moduli of the strength distribution of both specimen types were statistically identical. Fractography revealed that three different types of defects were responsible for failure of the 3-point as well as the 4-point bending specimens: small surface defects led to high strength values, internal pores resulted in strength values spread over the whole measured strength range and huge surface cracks resulted in low strengths. The specimens with the same defect type from both test configurations were pooled to retrieve strength distributions for each defect population. This statistical analysis shows that superior components can be produced, if one or two defect types can be avoided. The investigation of the thin sections revealed a strongly inhomogeneous grain orientation distribution throughout the cross section of the components. This texture results in residual stresses on the macro scale.

INTRODUCTION

Ceramic components fail by brittle fracture that usually start at defects (fracture origins) within the components,[1] Strength σ_f and the defect size a_c are related via the Griffith relation.[2] Since the defect sizes vary in different components, the strength also varies. The Weibull distribution[3, 4] is most commonly used to describe the probability of failure of a ceramic component or specimen. In many cases this distribution describes the material behaviour adequately, but exceptions do exist.[5]

The Weibull distribution is often represented on specific scales (in a "Weibull diagram") so that the data lie on a straight line. Deviations from Weibull behaviour manifest themselves as deviations from this linear behaviour: if, for example, two different defect types are responsible for failure, two line segments with different slopes can be expected in the diagram.[6] But due to the small number of specimens that are generally used to determine the distribution, such deviations can also be caused by the sampling procedure.[5] Only an investigation of all fracture origins makes it possible to decide, if multiple defect populations are present or if the non-linear behaviour is caused by statistical artifacts.[7]

During material and component development it is crucial to identify the defects that cause failure. Since many defects are typical for specific processing or machining procedures their identification provides key knowledge for improving the product quality. An appropriate statistical treatment can help to evaluate the effects of the elimination of one defect population on the overall reliability of the product.

The present investigation demonstrates how fractography together with thin-section microscopy can explain the fracture behaviour of low-pressure injection moulded alumina components. The knowledge on the specific defects and material texture can assist in the interpretation of strength statistics.

SPECIMENS AND EXPERIMENTAL PROCEDURE

Investigated Specimens
The material for this investigation was provided in the form of cylinder shaped alumina components. They have an outer diameter of ~12 mm, an inner diameter of ~3.5 mm and a length of ~84 mm. They were produced by low-pressure injection moulding of a 96% alumina ceramic. An alumina powder originating from the Bayer process with platelet morphology was used. The platelet base plane is normal to the c-axis of the alumina crystal.[8] The sintered parts were machined to final outer diameter by a diamond grinding process to achieve the desired degree of geometrical precision and to improve surface quality. This was done by centreless grinding on a through-feed machine with diamond wheels. Several passes through the machine were required, since each pass could only take off a relative small amount of material. Coarse diamond grit (90 μm) and a fine grit (50 μm) were used for wet surface grinding. The regulating circumferential velocity was 20 m/s and through feed speed was 1.2 m/min. The components were tested without further modification.

Strength Testing
Strength testing was performed on the machined components in 3-point- and 4-point bending using a Zwick Z010 universal testing machine. The support span for the 3-point bend tests was 73 mm. The specimens were positioned on V-blocks to minimize contact stresses at the support points. The 4-point bending tests were performed using a support span of 72 mm and a loading span of 24 mm. This geometry was chosen to reduce the fracture loads and therefore the contact loads as compared to the tests with loading at quarter points. A semi-articulating fixture with rollers with a diameter of 10 mm was used. For the fracture tests, care was taken to test all components in a similar orientation with respect to the position of the injection of the ceramic suspension. The crosshead speed of 5 mm/min was chosen in order to achieve fracture within 5s to 15s.[9, 10]
The 3- and 4-point bending strength was evaluated assuming elastic beam bending[11] using

$$\sigma = g \cdot \frac{8 F_{\max} L d_a}{\pi \left(d_a^4 - d_i^4 \right)} \tag{1}$$

$F_{\max}$ is the fracture load, L is the support distance. The inner diameter d_i and the outer diameter d_a were calculated as the mean of the values in loading direction and normal to loading direction. The constant g depends on the loading geometry: $g = 1$ for 3-point bending and $g = 2/3$ for 4-point bending.
After fracture, the position of the primary fracture with respect to the support point was determined. The fracture surfaces of all broken specimens were investigated using a stereo microscope (Olympus SHZ10 with ColorView IIIu digital image acquisition). For detailed inspection of the fracture origins a scanning electron microscope (SEM Zeiss EVO MA 25) was used.

Microstructural and Texture Evaluation
In order to investigate the general texture in the components a geological thin section method[12, 13] was used. A thin slice of material taken in appropriate orientation from the component was polished down to a thickness of 30 μm. This thin section was then observed at low magnification in a microscope equipped with crossed polars and a λ-tint plate set at 45° to the polarization axes of the polars.

The individual corundum grains of hexagonal crystallographic symmetry are optically uniaxially negative with refractive indices of 1.760 and 1.768 for the extraordinary and ordinary rays, respectively. Thus when a corundum crystal c-axis is orientated parallel to the λ-tint plate slow (γ) direction, its colour is shifted from first-order red towards yellow, and when orientated with its c-axis perpendicular to the plate slow direction, it appears blue. In intermediate orientations, or when viewed down the c-axis, the crystal appears red. Even though the grain size of individual crystals is rather less than the thickness of the thin slice, when a strong degree of alignment of the individual grains occurs, there is a net colour change visible because of optical translucence.

Statistical Evaluation and Data Pooling

Weibull distributions were fitted to the measured strength values according to ENV 843-5[14] using the maximum likelihood procedure. Biased values are reported for the Weibull modulus.

If different defect types are responsible for failure, a separate evaluation of all specimens with one defect type is useful to investigate the influence of each defect population on the overall strength distribution. This can be done by pooling the data for each defect type from both the 3- and 4-point bend specimens as suggested in textbooks.[15] The 4-point bend geometry is taken as a reference.

i.) In a first step, the individual 3-point-bending strength values, $\sigma_{i,3PB}$, were transformed to equivalent 4-point-bending strength values, $\sigma_{i,4PB}$, by using the size extrapolation,[6]

$$\sigma_{i,4PB} = \sigma_{i,3PB} \cdot \left(\frac{V_{eff,3PB}}{V_{eff,4PB}}\right)^{1/m_0} \tag{2}$$

where $V_{eff,3PB}$ and $V_{eff,4PB}$ are the effective volumes of 3 and 4-point specimens respectively. If surface defects are responsible for failure, the effective volumes have to be exchanged by effective surfaces in eq. (2). Using the relations for beams[15] or rods[16] in literature, the effective volume ratio for bending specimens with identical cross section but different loading geometries (i.e. 3-point bending on a support span L_{3PB} and 4-point bending on a support span L_{4PB} with the load applied at the third points) can be evaluated to be:

$$\frac{V_{eff,3PB}}{V_{eff,4PB}} = \frac{3}{3+m} \frac{L_{3PB}}{L_{4PB}} \tag{3}$$

The same expression results for the calculation of the effective surface ratio $S_{eff,3PB}/S_{eff,4PB}$.

As a guess value for the Weibull modulus m_0 the value for the relevant defect type in the 3-point bend specimens was used.

ii.) Then, in a second step, a new Weibull modulus m_{i-1} and characteristic strength $\sigma_{0,i=1}$ are determined by the usual maximum likelihood procedure.

iii.) With the new m steps i.) and ii.) are repeated until m_i remains constant.

This procedure can be performed for each defect type and leads to several different Weibull distributions, one for each defect type.

FRACTURE ORIGINS, STRENGTH STATISTICS AND MICROSTRUCTURE
Fracture Origins

Three different types of fracture surfaces were identified by fractography. Fractures of type 1 (Figure 1a) show the typical features of ceramic fracture surfaces. The detailed SEM picture, Figure 1b, reveals the defect that caused fracture: in this case there are *grooves*, probably caused by machining, and below them a shallow median grinding crack. These fracture origins are not always at the point on the surface with the highest tensile stress. Some are offset towards the neutral axis where stresses are lower. This indicates that such defects are frequent and at all positions of the periphery of the shafts.

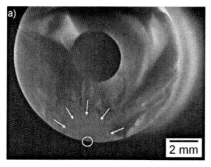 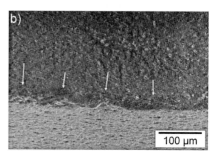

Figure 1. (a) Fracture surface of a specimen that failed due to a type 1 defect, regular fracture mirror. The border of the fracture mirror is indicated by arrows, the position of the fracture origin is marked. (b) Type 1 fracture origin: machining groove and shallow surface crack.

Type 2 fracture surfaces, Figure 2a, have a much larger fracture mirror. The defect that caused failure is located in the bulk of the cross section, close to the middle of the wall of the shaft. A detailed picture taken in the SEM shows that it is a *pore*, Figure 2b. Fracture surfaces of this type also show a seam at a distance of approx. 2 mm from the inner surface of the shafts. The pores are always located at this seam. Most probably they are generated during the injection moulding process.

Type 3 fracture surfaces, Figure 3a, also show a large fracture mirror that is related to a surface defect. This type of fracture is caused by a large *semi-elliptical surface crack* that is inclined to the fracture surface, Figure 3b. In the weakest 3-point bend specimen the failure causing crack was about 700 µm wide and more that 400 µm deep.

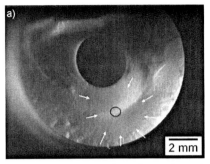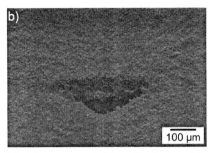

Figure 2. (a) Fracture surface of a specimen that failed due to a type 2 defect. The border of the fracture mirror is marked by arrows; the position of the fracture origin is marked. A discontinuity is visible in the middle of the wall of the tube. (b) Type 2 fracture origin: pore at the seam at mid-wall thickness.

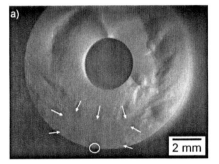

Figure 3. (a) Fracture surface of a specimen that failed due to a type 3 defect. The border of the fracture mirror is marked by arrows; the position of the fracture origin is marked. (b) Type 3 fracture origin: large pre-crack at the surface.

Influence of Texture on the Fracture

When a transverse thin section of the rod material from near the rod centre is examined (Figure 4a), it is immediately clear that there is a primary alignment of the alumina grains with their c-axes radial, but this occurs in two concentric annular zones separated by a narrow central annular band in which the c-axes are arranged tangentially. It is noteworthy also that a number of cracks have developed traversing this central zone, Figure 4b. It is most probable that residual stresses exist within the cylinder walls resulting from thermal expansion anisotropy of corundum. The expansion coefficient in the c-axis direction is about $1 \times 10^{-6}\ °C^{-1}$ greater than that in the a-axis direction.[8] In the current case it can be readily envisaged that the central annular zone will become stressed in hoop tension compared with zones on either side.

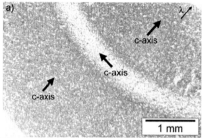
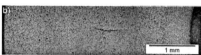

Figure 4. (a) Blue component of the colour image of a thin transverse cross-section of a test piece. White regions correspond to high blue intensities, grey areas to high red intensities. The slow direction of the λ-tint plate is indicated by an arrow. The bright band in the centre shows that the alumina platelets are orientated with their c-axes in tangential direction in this region, while they are predominantly radial orientated in the dark regions. (b) Polished cross section of a test piece with a crack in the region corresponding to the bright band in (a).

In terms of the influence of the role of the residual stress on the path of cracks during bending fracture, it can be seen that for fractures from the external tube surface (type 1, grooves), that occur at high stress, the residual tensile stress plays little role, until the crack has passed around the central hole and slows as the far side from the origin is approached.

For fractures that occur from defects in the central tensile zone (type 2, pores), strong hackle develops as the crack front moves radially outwards from the defect and especially tangentially along the high-tensile zone. The parts of the crack front moving towards the central hole are much slower moving with less strong hackle developing.

For low energy fractures from surface cracks (type 3, cracks), there is very little radiating hackle observed as the crack accelerates inwards, but once the tensile zone is approached, those parts of the crack front approaching this zone obliquely are accelerated sharply producing hackle 'wing' markings that move tangentially, while those parts that are moving radially inwards pass quickly through the tensile zone and emerge into the inner zone without producing hackle. It is also clear that the 'wing' zones move symmetrically around the central hole into the compressive side of the test-piece, resulting in very inhomogeneous crack front geometry, and leading to complex topography.

Strength Distributions

The strength distributions of the 4-point and the 3-point bending specimens are plotted in a Weibull diagram, Figure 5a, the parameters of the distributions are summarized in Table I. The strength values of the 4-point bend specimens (open symbols) are lower than those of the 3-point bending specimens (full symbols). Different symbols are used to indentify specimens that broke due to the three defects types described above. In both Weibull distributions it can be recognized, that type 1 fractures (grooves) are predominantly found at high strength values, type 3 defects (cracks) govern the strength at low values and type 2 defects (pores) spread over the whole strength range. The strength values of one specific defect type from both test geometries are thus used to determine the strength distributions for this defect type according to the data pooling procedure described above. The resulting parameters are also found in Table I.

The data pooling also provides a means to estimate the effect that can be achieved if one or two defect types can be avoided or if components containing such defects can be sorted out. In Figure 5b strength distributions for 4-point bending geometry for different sub-sets of specimens are plotted. From the complete set of specimens containing all types of defects, more than 4% of all specimens fail at a stress of 150 MPa. If the large type 3 machining defects can be avoided and only type 2 pores and type 1 defects were present in the components, less than 0.75% would fail at this stress. If the pores can be avoided too, only one out of 4288 specimens breaks at 150 MPa. This effect is mainly caused by the increase of the Weibull modulus from 5.5, for the combined defect population, to 11.4 for type 1 groove defects only.

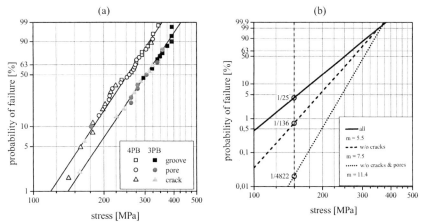

Figure 5. (a) Strength distributions measured with 3-point and 4-point bending tests. Failures due to the three different defect types are indicated by different symbols. (b) The effect of avoiding type 3 defects (cracks), dashed line, or type 3 and type 2 defects (cracks and pores), dotted line, on the strength distribution of the components. The strength distributions are plotted for 4-point bending geometry. The number of specimens that fail for each distribution at 150 MPa is indicated in the graph.

Table I. Results of the Weibull evaluation. Numbers in square brackets indicate 90% confidence intervals. The pooled evaluations refer to the geometry used in the 4-point bending tests.

	3-point bending tests	4-point bending tests	without type 3 (4PB geometry)	without type 3 and type 2 (4PB geometry)
number of specimens	25	31	39	17
char. strength σ_0 [MPa]	324.6 [303.3 - 347.7]	270.7 [255.2 - 287.4]	288.9 [278.0 – 300.3]	316.3 [303.6 – 329.7]
Weibull modulus m	5.4 [4.0 – 6.8]	5.5 [4.2 - 6.8]	7.5 [5.9 – 8.9]	11.4 [7.5 – 14.6]

SUMMARY AND CONCLUSIONS

The strength of injection moulded alumina components was tested by three and four point bending of the as-machined components. The strength distributions of both specimen sets have a low Weibull modulus (approx. $m = 5.5$) and show linear trends if plotted in a Weibull diagram. Fractography revealed that three different types of defects were responsible for failure of both the three-point and the four-point bending specimens. Small scale machining damage (type 1, grooves) causes failure at high stresses, pores near the mid-wall thickness (type 2) give strength values that span the measured strength range and severe machining or handling damage at the outer surface (type 3) leads to low strength failures. A statistical analysis of the fracture strength results indicates that the low Weibull modulus is mainly caused by the type 3 defects. Since these defects can easily be avoided, an increase of the components strength and a decrease of the scatter of strength can readily be achieved. An even superior behaviour could result if the pores can additionally be removed from the components.

The fracture surface of specimens containing pores and large machining cracks have untypical fracture mirrors. An investigation of a transverse thin section using a geological method showed that the material has a distinct texture. The alumina crystals have a preferred orientation with their c-axis in tangential direction in an annular zone at mid wall-thickness of the rods. This leads to macroscopic hoop tensile residual stress in this region. The existence of this stress explains the extraordinary shape of the fracture mirrors in specimens that contain either pores in the tensile stress zone or large defects which cause large fracture mirrors.

REFERENCES
[1]R. Danzer, et al., Fracture of Ceramics, *Advanced Engineering Materials*, **10**, 275-98 (2008).
[2]D. Gross and T. Seelig, Fracture Mechanics, pp. 319, Springer, Berlin, 2006.
[3]W. Weibull, A Statistical Theory of the Strength of Materials, pp. 45, Vol. 151, Generalstabens Litografiska Anstalts Förlag, Stockholm, 1939.
[4]W. Weibull, A Statistical Distribution Function of Wide Applicability, *J. Appl. Mech.*, **18**, 293-98 (1951).
[5]R. Danzer, et al., Fracture Statistics of Ceramics - Weibull Statistics and Deviations from Weibull Statistics, *Engineering Fracture Mechanics*, **74**, 2919-32 (2007).
[6]D. Munz and T. Fett, Ceramics, pp. 298, Vol. 36, Springer, Berlin, Heidelberg, 1999.
[7]R. Danzer, Mechanical Failure of Advanced Ceramics: The Value of Fractography, *Key Engineering Materials*, **223**, 1-18 (2002).
[8]W. H. Gitzen, Alumina as a Ceramic Material, pp. 253, The American Ceramic Society, Westerville, Ohio, 1970.
[9]EN 843-1, Advanced Technical Ceramics - Monolithic Ceramics - Mechanical Properties at Room Temperature: Part 1 - Determination of Flexural Strength, 2006.
[10]ASTM C 1684-08, Standard Test Methods for Flexural Strength of Advanced Ceramics at Room Temperature - Cylindrical Rod Strength, 2008.
[11]W. C. Young, Roark's Formulas for Stress and Strain, McGraw-Hill, New York, 1989.
[12]D. J. Clinton, R. Morrell, and M. McNamee, Textures in High-Alumina Engineering Ceramics, *Brit. Ceram. Trans.*, **85**, 175-79 (1986).
[13]M. M. Raith, P. Raase, and J. Reinhardt, Guide to Thin Section Microscopy, http://www.dmg-home.de/lehrmaterialien.html, 2011.
[14]EN 843-5, Advanced Technical Ceramics - Monolithic Ceramics - Mechanical Properties at Room Temperature: Part 5 - Statistical Evaluation, 2006.
[15]J. B. Wachtman, Mechanical Properties of Ceramics, pp. 448, Wiley-Interscience, New York, Chichester, 1996.
[16]G. D. Quinn, Weibull Effective Volumes and Surfaces for Cylindrical Rods Loaded in Flexure, *J. Am. Ceram. Soc.*, **86**, 475-79 (2003).

SINGLE EDGE V-NOTCH FRACTOGRAPHY OF ENGINEERING CERAMICS

Jasbir Singh Aujla & Kevin Kibble
University of Wolverhampton
Department of Engineering
Telford, England TF2 9NT, UK

ABSTRACT
There are numerous methods to measure the fracture toughness of engineering ceramics. One of the more recent developments has been the Single Edge V-Notch Beam technique. (SEVNB). This method is effectively a development of the Single Edge Notch Beam (SENB) technique but with the addition of a second notch which is polished-in using diamond paste and a razor blade. Apparatus was developed which was used to polish in the second notch under controlled conditions. Tests were conducted on seven different engineering ceramics; i.e. Coors AD999, GEC lithium alumino-silicate (LAS), Pyrex borosilicate, gas pressure sintered silicon nitride, magnesium partially stabilized zirconia (PSZ) and reaction bonded silicon carbide (RBSiC). Results showed that a final notch root radius down to 3 μm could be achieved. For the RBSiC there were additional semi-elliptical cracks, around 20-30 μm deep, found at the base of the secondary polished notch.

Keywords: Fracture toughness, Single Edge V-Notch Beam, Indentation, Notching, Diamond, Ceramics.

INTRODUCTION
There is considerable interest in the Single Edge V-Notch Beam technique, as part of the CEN TS 14425-5 method for the determination of fracture toughness evaluation of monolithic ceramics. The method itself may be perceived as a modification to the standard Single Edge Notch Beam (SENB). In SENB a flexure bar of rectangular cross-section is tested usually in 4 point bend loading.[1]

In conventional SENB, fatigue pre-cracking is used to produce a sharp crack at the end of the notch. However, when this is applied to ceramic materials failure often occurs in the pre-cracking stage and thus the margin of control in pre-cracking is too small.

To overcome the difficulties of fatigue pre-cracking other pre-cracking methods such as thermal shock, wedge loading and Single Edge Pre-crack Beam (SEPB) have been applied.[2] However, with all of these techniques there are difficulties in achieving consistent and repeatable pre-cracks, resulting in sample wastage. The Single Edge V-Notch Beam overcomes this problem by introducing a finer second notch at the base of the primary notch. The second notch is polished-in using a razor blade and diamond paste, leaving a notch with a small notch root radius.

NOTCH WIDTH
SENB specimens are usually prepared by creating the notch using a diamond edge saw. The notch tip radius ρ, is approximately equal to the half width of the notch slit (saw blade thickness). The fracture toughness for SENB, for four point flexure, is evaluated using Equation 1.[3]

$$K_{IC} = \frac{P(S_1 - S_2)a^{\frac{1}{2}}}{2BW^2}Y$$

(1)

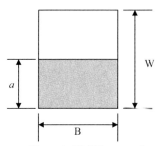

Figure 1. F(a/W) nomenclature.

Where P is the load, S_1 and S_2 are the outer and inner spans or the bending fixture, B and W are the width and thickness of the sample bar, a is the depth of the SENB notch and Y is the geometric factor, Figure 1. Tests carried out on ceramics without a pre-cracking stage show that K_{IC} calculated from the standard formula result in higher values with increasing notch root radius.[4] Considerable variation occurs depending on the particular ceramic material. Apparent fracture toughness values that are calculated from maximum load decrease with decreasing notch radii, eventually becoming constant, i.e. yielding the true K_{IC} value, below some critical notch root radius ρ^*. This critical radius, ρ^*, is strongly dependant on the microstructure of the material.

NOTCH ROOT RADIUS AND FRACTURE TOUGHNESS ESTIMATION
For certain ceramics saw cutting during the machining of notches induces cracking at the root radius, it is then assumed that cracks of this type constitute very sharp 'zero width' cracks, the assumption is that the machining damage has pre-cracked the bar, but the extent of the machining damage at the notch root is variable due to machining practice and material used.

Davidge & Tappin[5] showed that machining of notches can induce cracking at the base of the notch. It was shown that using a large grained alumina, under certain machining conditions the SENB fracture toughness was comparable to values obtained using wedge pre-cracking. However, a drawback is that sufficient root cracking does not occur during machining in all ceramics. In addition this work was conducted with coarse grained material and therefore this material will be less sensitive to notch root radius.

Similarly, Seshadri & Srinivasan[6] found that the root cracking associated with diamond wheel grinding (DWM) actually acted as pre-cracks for alpha silicon carbide. However, work by Chia et al[7] proved that silicon carbide test bars, machined with different wheel conditions and a lower rate of material removal showed minimal notch root damage and higher calculated K_{IC} values. Therefore the assumption of the existence of sufficient root cracking with DWM cannot be relied on.

THE DEVELOPMENT OF V-NOTCHING
The development of the SEVNB technique, by Nishida et al[4], was prompted as a result of the problems associated with the Single Edge Pre-crack Beam technique. The problems that Nishida highlighted were:
a. The pop-in propagation is difficult to control, leading to a pre-crack geometry frequently outside the tolerance dictated by the Japanese Industrial Standard rule.[8]
b. The determination of the pre-crack length on the fracture surface after the fracture test is often uncertain and can be interpreted as a source of errors in K_{IC} calculations.

c. Fracture toughness values by SEPB are sometimes influenced by the residual compressive stress beneath the indentation impression from which the pop-in pre-crack propagates.

The fundamental feature of SEVNB is the introduction of a secondary notch, beginning at the base of the SENB notch. By using a razor blade and diamond paste, a sharper notch can be 'polished-in.' No allowance is required for the residual stress. This aspect is particularly appealing when compared with indentation fracture methods where the residual stress induced from the indentation, has to be included in the analysis and where there is uncertainty regarding calibration factors.

NOTCH INTRODUCTION IN SEVNB

The main objective of the work reported in this paper was to develop a simple, reliable, cost effective yet accurate technique for determining K_{IC} for the characterization of engineering ceramics. The methods used by Kübler[9] involved manual polishing of the notch into the bar. The method used by Nishida et al[4] was semi-manual and involved the fixture shown in Figure 2. Kübler proposed, and built, a notching machine in order to automate notching. With any test sample it is important that each specimen is the same (within a tolerance), but regulation of the force and frequency using these manual methods is virtually impossible. In introducing the second v-notch, parameter control is vital. Therefore, what is required is a fixture that provides control and constraint which significantly minimizes the possibility of parameter error, and hence sample replication is consistent.

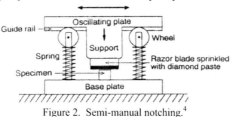

Figure 2. Semi-manual notching.[4]

DEVELOPED FIXTURE

The developed fixture is shown in Figure 3. A number of key features are:
1. Blade load control using two interchangeable springs of 2 N deadload.
2. Frequency control via a motor which is monitored using a tachometer. This gives a frequency tolerance of less than 1% at 1 and 2 Hz, when maintaining a motor speed tolerance of ± 0.4 rpm.
3. Digital counter: establishes the exact number of cycles completed.
4. Traverse length control: ensures that wear of the razor blade is constant across the set length.
5. Lightweight aluminum alloy and glass reinforced nylon crack linkage system, and a polymer tray in order to minimize inertia effects during oscillations, hence a motor of low torque is used.

Figure 3. Developed apparatus.

EXPERIMENTAL PROCEDURE

The materials used were GEC lithium alumino-silicate, borosilicate glass, gas pressure sintered silicon nitride, magnesium partially stabilized zirconia and reaction bonded silicon carbide. The magnesium partially stabilized zirconia was supplied by Morgan Crucible (Stourport-on-Severn, West Midlands, UK). The RBSiC was supplied by Prof. Kevin Kibble (University of Wolverhampton), the gas pressure sintered silicon nitride (GPS Si_3N_4) was supplied by EMPA (Swiss National Laboratory), and the remaining ceramics supplied by Dr. Roger Morrell, at NPL (Middlesex, UK) The test bars were longitudinally ground using a Jones & Shipman 1400 surface grinder, using a BX 400 diamond wheel, to 3 mm x 4 mm x 58 mm with a depth of cut that did not exceed 0.03 mm per pass. Finish grinding at 0.002 mm per pass, removing the last 0.1 mm of material, with a final pass of zero cut. The edges were chamfered in order to eliminate edge effects.

For most of the bars the primary notch was introduced using a 0.5 mm thick BX 400 diamond wheel removing 0.01 mm per cycle and then 0.001 mm per cycle for the final 0.002 mm, whilst maintaining $F(a/w)$ at 0.5. Test specimens, where only the primary notch was introduced, i.e. SENB, were tested to give a comparison with SEVNB test specimen results.

The SEVNB bars were prepared so that $F(a/w)$ was maintained at 0.5. The razor notch was introduced by filling the primary notch with 1 μm diamond paste, and polishing-in at 1 Hz using a static load of 2 N. Table 1 shows the notching parameters. The assumption of constant blade wear was made. Fracture toughness of the SEVNB bars was evaluated using the following equation:[9]

$$K_{IC} = \frac{P}{B\sqrt{W}} \times \frac{S_1 - S_2}{W} \times \frac{3\sqrt{\alpha}}{2(1-\alpha)^{3/2}} \times Y \tag{2}$$

$$Y = 1.9887 - 1.326\alpha - (3.49 - 068\alpha + 1.35\ \alpha^2)\ \alpha\ (1 - \alpha)(1 + \alpha)^{-2} \tag{3}$$

Where α is the ratio of the notch depth to the bar thickness, S_1 and S_2 are the outer and inner spans of the bending fixture.

Table I. SEVNB Notching Parameters

Material	Number of roughing cycles	Number of finishing cycles
GEC lithium alumino-silicate	1200	300
Pyrex borosilicate	1200	300
RBSiC	1200	300
PSZ	1200	300
GPS Si_3N_4	1800	300
Coors AD999	1200	300
96% Alumina	1200	300

RESULTS & DISCUSSION

The modified fixture clearly demonstrates that notch tip radii of the order of 5 μm can be consistently achieved. In Figs. 4 to 9, it can be seen that the polished-in notch is introduced with no associated damage to the surrounding material. This is clear evidence that the SEVNB technique can consistently yield sharp notches that are comparable to the microstructure of the material. This is best exemplified in Figure 7, for the PSZ.

Figure 4 shows machining damage at the base of the primary notch, however, the polished-in notch extends beyond any influence of machining damage. A fine tip radius is also clear in the borosilicate glass, in Figure 6. Similarly, there is associated machining damage on the sides of the primary machined-in notch but none from the secondary polished in notch, thus the effect of machining damage, that is unavoidable from primary notching machining, is circumvented.

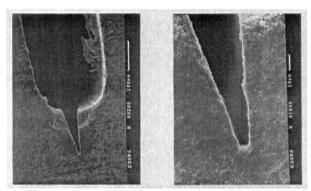

Figure 4. Coors AD999. V-notch exceeds machining damage
and has a notch root radius of approximately 4 microns.

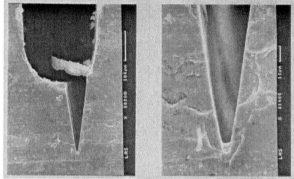

Figure 5. GEC lithium alumino-silicate. V-notch root radius of approximately 3 microns.

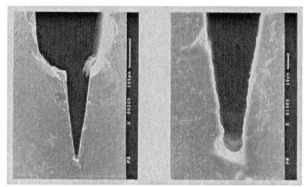

Figure 6. Borosilicate glass. V-notch root radius of approximately 5 microns.

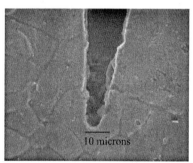

Figure 7. V-notched PSZ. Note v-notch ends within the grain.

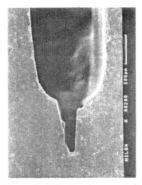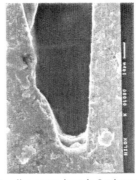

Figure 8. 96% alumina, v-notch root radius approximately 8 microns.

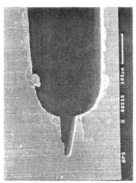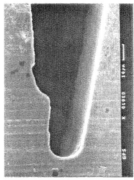

Figure 9. GPS Si_3N_4 EMPA samples, v-notch root radius approximately 10 microns.

Table II shows the results from the developed technique, in comparison with SENB (for selected materials). The greater accuracy in fracture toughness for SEVNB in comparison to SENB can be seen. For the selected materials, there is a reduction in variation in SEVNB when compared to SENB.

Table II. Fracture Toughness Results

	96% Alumina	RBSiC	PSZ	LAS	Coors AD999	Borosilicate	GPS Si₃N₄
Mean							
SENB	4.45	3.61	4.27	N/A	N/A	N/A	N/A
SEVNB	3.88	2.98	4.24	1.49	5.55	0.72	5.55
Standard Deviation							
SENB	0.53	0.34	1.06	N/A	N/A	N/A	N/A
SEVNB	0.35	0.24	0.25	0.09	0.20	0.10	0.20
Coefficent of Variation							
SENB	11.91	9.42	24.82	N/A	N/A	N/A	N/A
SEVNB	9.02	8.05	5.89	6.04	3.60	13.88	3.60
Sample Sizes							
SENB	22	14	14	Zero	Zero	Zero	Zero
SEVNB	12	14	14	7	5	6	5

From Table II, the consistency of some materials is better than others, demonstrated by low coefficient of variation values. For example, the results for GPS Si_3N_4, the mean is 5.55 MPa m$^{1/2}$ with a standard deviation of 0.2 and this is comparable with 'Round Robin' values from Kübler[9] where the average fracture toughness was 5.35 MPa m$^{1/2}$, with a standard deviation of 0.33, for the same material.

It is paramount for any testing technique to be able to demonstrate consistency. Unfortunately, due to the brittle nature of engineering ceramics, not all testing techniques are sensitive enough to highlight variance due to processing.

The results for the LAS also show a very low coefficient of variation, despite the very similar mean fracture toughness values.

FRACTOGRAPHY

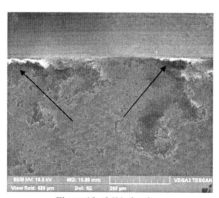

Figure 10. 96% alumina.

Figure 10 shows a scanning electron microscope image of the base of the polished notch for the 96% alumina sample. Below the line of the v-notch, there are microstructural defects which may be attributed to processing, as indicated by the arrows.

Figure 11. RBSiC showing semi-elliptical cracks induced
from the secondary polished-in notch step.

From Figure 11, it can be seen that below the v-notching line, other flaws can exist. These may
be attributed to the v-notching process, for example in RBSiC, there were small semi-elliptical flaws
20 to 30 microns deep.

Figure 12. Additional crack at the base of the polished notch of LAS as indicated by the arrow.

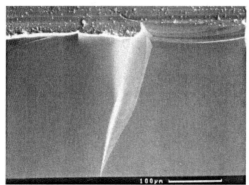

Figure 13. Additional crack in LAS sample.

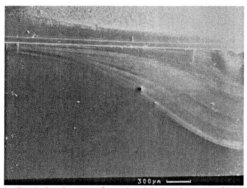

Figure 14. Crack development from the right hand corner of LAS sample.

The glass-ceramic LAS sample shows an additional crack, approximately 260 microns long, at the base of the polishing notch (Figure 13). However, it is believed that this crack developed in fracture rather than at the time of notching, as discussed with G. Quinn.[10] This additional crack had a length of approximately 260 microns. The other area of interest is the crack development from the right hand edge of the test bar, shown in Figure 14. There may be an initiation site at the right edge of the test bar which accounts for the crack development pattern in fracture as seen in figure 14.

CONCLUSIONS
A SEVNB fixture has been designed and built that provides a high degree of parameter control. The importance of this control is that the sample preparation is more reproducible and hence reliable when compared to other v-notching fixtures. Using this SEVNB notching fixture resulted in notch root radii of 3 μm for lithium alumino-silicate and the magnesium partially stabilized zirconia, 4 μm for the Coors AD999, 5 μm for the Pyrex borosilicate, 8 μm for a 96% alumina, and 10 μm for gas pressure sintered silicon nitride. This shows that the polished notch root radius is comparable to the microstructural features of a given material, indeed in partially stabilized zirconia, the root radius was significantly less than the grain size of the material. An interesting point is the comparison between the

96% alumina to the 99% alumina with 99% alumina v-notch root radius half that of the 96% alumina. This may be attributed to the superior microstructure of the 99% alumina. When compared to SENB, the SEVNB method shows reduced scatter, notches that are comparable to microstructural features, and more accurate values of fracture toughness.

It is also important to note that below the v-notching line, other flaws can exist, and these may be attributed to the v-notching process, for example in RBSiC, there were small semi-elliptical flaws 20 to 30 microns deep.

REFERENCES

[1]Damani, R (et al). Critical notch-root radius effect in SENB-S fracture toughness testing. *Journal of the European Ceramic Society*. **16** 695-702 1996.
[2]Sakai, M & Bradt, R.C., Fracture toughness testing of brittle materials. *International Materials Review*. **38** [2] 53-78 1993.
[3]ASM Engineered Materials Handbook. pp603, volume 4 Toughness, Hardness and Wear. 1991. ASM International.
[4]Nishida, T (et al)., Effect of notch root radius on the fracture toughness of fine grained alumina. *Journal of the American Ceramic Society*. **77** [2] 606-608 1994.
[5]Davidge, R.W & Tappin, G., Effects of temperature and environment on the strength of two €polycrystalline aluminas. *Proceedings of the British Ceramic Society*. [15] 47-60 1970.
[6]Srinivasan, M & Seshadri, S.G. Presentation at the 37th Pac. Coast Regl Mtg of the American Ceramic Society. 1984
[7]Chia, K.Y (et al)., Notching techniques used in SENB fracture toughness testing . *Ceramic Engineering and Science Proceedings*. **7** 795-801 1986
[8]JIS., Testing methods for fracture toughness of high performance ceramics. Japanese Industrial Standard. JIS R 1607 – 1990.
[9]Kübler, J., SEVNB method. VAMAS TWP#3/ ESIS TC6 round robin. 1998
[10]Quinn, G. Private Communication. 6th June 2011.

ROUGHNESS OF SILICA GLASS SUB-CRITICAL FRACTURE SURFACES

Gaël Pallares
Laboratoire de Physique de la Matière Condensée et Nanostructures, Université Claude Bernard Lyon I and CNRS
Villeurbanne, France

Frédéric Lechenault, Matthieu George
Laboratoire des Colloïdes, Verres et Nanomatériaux, CNRS, Université Montpellier 2
Montpellier, France

Elisabeth Bouchaud
CEA, IRAMIS, SPEC, 91191
Gif sur Yvette, France

Cindy L. Rountree
CEA, IRAMIS, SPCSI, Groupe Systèmes Complexes et Fracture
91191 Gif sur Yvette, France

Matteo Ciccotti
Laboratoire de Physico-Chimie des Polymères et des Milieux Dispersés, CNRS, ESPCI
Paris, France

ABSTRACT
We study the roughness properties of fracture surfaces obtained by sub-critical fracture of silica glass in a double-cleavage-drilled-compression (DCDC) setup under controlled velocity, stress intensity factor and relative humidity. We extract the topography of these surfaces using an atomic force microscope and study their roughness properties using two different indicators. On the one hand, we find that the root-mean square roughness R_q decreases linearly with the stress intensity factor K_I, or equivalently with the logarithm of the crack velocity at fixed relative humidity, as previously reported, while no significant dependence on relative humidity at fixed K_I is observed. On the other hand, we study the 1D height-height correlation functions. Recent work on fracture surfaces has shown that the lower cut-off in the asymptotic self-affine regime of this function is related to the size of the process zone. Here we show that the cut-off we find is caused by the finite size of the probe.

INTRODUCTION
Thanks to the development of local-probe microscopy, a lot of work has been devoted to the study of topographical properties of fracture surfaces of brittle materials, and in particular to a characterization of their roughness at the nano-scale. Wünsche et al[1] have shown that the usual classification into mirror (although the name itself loses its meaning at scales smaller than optical wavelengths!), mist and hackle zones is still relevant at such length scales, and the signature of these regimes can be found in the off-plane root-mean-square (RMS) roughness. In this paper, we focus on the mirror zone in glass. For such a smooth surface, the RMS roughness can be as small as a fraction of a nanometer for micron-size images. Since glass, and especially pure SiO_2, is a poor electric conductor, techniques like Scanning Tunnelling Microscopy are unavailable; the fracture surfaces are imaged using Atomic Force Micropscopy (AFM). The question that naturally arises in this context is that of the existence of a process zone at the tip of the crack, and its relationship with the properties of the post-mortem topography. Earlier AFM measurements on glass fracture surfaces have uncovered the

self-affine geometry of the nanoscale roughness, as revealed by linear regimes in the 1D height-height correlation function evaluated in the direction of the crack front. The slope ζ of this function, called the roughness exponent, has been investigated in various materials under different experimental conditions,[2] In the case of silica glass in stress-corrosion, Bonamy, et. al.[3] have shown that at scales smaller than a characteristic cut-off ξ, the roughness exponent ζ is close to the "universal" value of 0.8. Moreover, they report evidence that the cut-off length scale provides a good estimate for the size of the damage zone in glass, defined as the region where the behaviour is no longer linearly elastic in the vicinity of the crack tip. The authors report the crack-velocity dependence of this length scale, which decreases from 80 to 20nm as the velocity is increased from 10^{-10} up to 10^{-5}m/s.

In this paper, we study fracture surfaces of silica glass broken in a double-cleavage-drilled-compression (DCDC) setup.[4,5,6,7] The crack propagates in the stress-corrosion regime at velocities ranging from 10^{-10} up to 10^{-6}m/s. In these experiments, the relative humidity RH, stress intensity factor K_I and temperature T are accurately controlled. Topographic maps of the post-mortem surfaces are obtained using AFM in tapping mode. We study two statistical roughness estimators of these maps: the RMS roughness R_q and the 1D height-height correlation function. On the one hand, we show that, as previously reported, R_q decreases linearly with the stress intensity factor K_I, or equivalently with the logarithm of the velocity, when the relative humidity is kept constant. Moreover, we found no significant dependence of this quantity on the relative humidity at fixed K_I. This establishes the fact that roughness depends on relative humidity only through variations of K_I. On the other hand, the computation of the 1D height-height correlation function yields a small-scale roughness exponent ζ and a cut-off length-scale ξ similar to that observed in.[8,9,10] Spatial length scale ξ is found to be an increasing function of K_I, in apparent contradiction with earlier results.[3,11] This could be due to significant differences in the humidity (78% versus 40%, as evidenced in ref.[12] However, following the procedure detailed in [13], while using prior measurements of the characteristic features of the AFM tip, we show that the measured spatial length ξ just results from the typical smoothing of the AFM tips.

EXPERIMENTAL PROCEDURE

The DCDC sample is a prism of dimensions $2w \times 2t \times 2L$ (5x5x25mm^3) with a cylindrical cross hole of radius R ($R = 500$ μm) drilled through the specimen (thickness $2t$), in order to trigger the propagation of two symmetric cracks of length a. As shown in Figure 1, the x and z axes correspond respectively to the direction of crack propagation and the direction parallel to the crack front. Cracks are initiated and propagated in pure fused-silica glass (Corning 7940 and Corning 7980, Corning, USA) at constant temperature $T = 31.0 \pm 2.0$ °C in a waterproof chamber under an atmosphere composed of air and water vapor at a given relative humidity level RH. For an applied stress σ, the stress intensity factor K_I can be evaluated from the value of the load and the crack length using Eq.   (1)[6]:

$$\frac{\sigma \sqrt{\pi R}}{K_I} = \left[c_0 + c_1 \frac{w}{R} + c_2 \left(\frac{w}{R} \right)^2 \right] + \left[c_3 + c_4 \frac{w}{R} + c_5 \left(\frac{w}{R} \right)^2 \right] \frac{a}{R} \tag{1}$$

with the set of parameters : $c_0 = 0.3156$, $c_1 = 0.7350$, $c_2 = 0.0346$, $c_3 = -0.4093$, $c_4 = 0.3794$, and $c_5 = -0.0257$ for $2.5 \leq w/R \leq 5$ and $w < a < L-2w$. In moist atmosphere, the stress corrosion by water molecules at the crack tip induces a subcritical crack propagation velocity.[14,22] By coupling optical and atomic force microscopy, the crack propagation velocity can be measured in a range from 10^{-5} m/s down to 10^{-12} m/s.

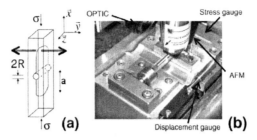

Figure 1. Experimental setup: (a) sketch of the
DCDC geometry (b) photograph of the experiment

Eventually, the specimens are fractured into two pieces, and the stress intensity factor K_I is determined as a function of the propagation distance according to Eq.   (1). On the post-mortem fracture surfaces, the areas associated to a given K_I are delimited by two arrest lines; optical micrographs were taken to identify these zones. We scan the topography of the fracture surface with the AFM (Veeco Nanoscope Dimension V) in tapping mode, parallel to the crack front, for a range of stress intensity factors and relative humidity, with tips having a nominal radius of $R_{tip} \sim 10$nm (DNP, VeecoMetrologyInc, Camarillo, CA, MPP-11100-10). The scanned areas are 2, 3 and 5µm-wide squares, and the images contain 512×512 pixels. The mean 1st-order plane is subtracted from each topographic image, yielding the topographic $y(x,z)$ data that we used to evaluate the 1D height-height correlation functions along the fast-scan direction. Prior to the estimation of the RMS roughness, a 1st-order line fit was also subtracted from each individual scan line in order to reduce typical vertical offset bias. An example of experimental map used to evaluate the RMS roughness is displayed in Figure 2.

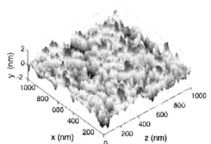

Figure 2. Experimental fracture surface (1×1 µm²) of fused-silica glass obtained by AFM.

RESULTS
Quadratic roughness
First we study the RMS roughness R_q of the resulting fracture surfaces, defined as:

$$R_q = \sqrt{\langle y(x,z)^2 \rangle_{x,z}}$$ (2)

Before analyzing these results, it should be emphasized that these surfaces are expected to be self-affine to some extent.[15,21] Consistent evaluation of R_q requires careful analysis. In order to illustrate the potential problem, we can consider a simple self-affine object, namely a random walk. Since Einstein's early work on this topic,[16] we know that the variance of the position of a random walker increases linearly in time and hence the RMS goes like the square root of time: the longer the walk, the larger the explored region. This scaling involves the exponent 1/2, which characterizes the roughness of the trajectory. In the situation at hand, it is then intuitive that the larger the image, the larger the RMS roughness, scaling with a finite exponent characteristic of the large-scale properties of the surface.

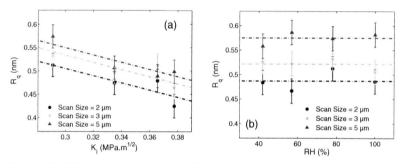

Figure 3. (a) RMS roughness as a function of stress intensity factor K_I at $RH = 78\pm2\%$ - (b) RMS roughness as a function of relative humidity RH for a given $K_I = 0.291\pm0.008$MPa.$\sqrt{m}$

We have represented the measured values of R_q for the three scan sizes and for four different values of K_I at fixed RH in Fig. 3(a), and at fixed K_I for four values of RH in Fig. 3(b). It is clear that, keeping everything else constant, an increase in scan size significantly increases the RMS, which is indicative of non-trivial, possibly self-affine, scaling properties at large scales. A quantitative understanding of this feature will be provided in the next section. For any given scan size, we can draw two conclusions from this set of data. First, the RMS roughness of the broken samples at fixed relative humidity decreases linearly with the stress intensity factor K_I. This result is consistent with previous measurement in silica glass.[17] Second, at fixed K_I, the RMS roughness does not appear to display any significant trend as the relative humidity is varied, and thus seems to be rather insensitive to this parameter.

Height-Height Correlation Function
The roughness features of the profile $y(z)$, parallel to the crack front (and along the fast scan lines), are extracted from the 1D height-height correlation function defined as:

$$\Delta y(\Delta z) \equiv \sqrt{\langle (y(z+\Delta z)-y(z))^2 \rangle}_{x,z} \qquad (3)$$

We have represented this function computed for a value of stress intensity factor $K_I = 0.335\pm0.008$ MPa.m$^{1/2}$ at fixed relative humidity $RH = 78\pm2$ % in Figure 4. For all values of K_I, these functions have a similar overall shape as those reported in,[18] exhibiting two distinct regimes at small and large scales. We can extract a small-scale exponent ranging from 0.62 to 0.78, thus close to the universal value of 0.8.[2] The large-scale behavior can also be described by a power law with a small exponent, consistently in the range 0.14 to 0.25, even though slow convergence towards the classical logarithmic behavior cannot be excluded. However, the scatter in these sets of exponents does not yield any systematic dependence on K_I and can be interpreted as experimental noise. The relevant features of these curves hence reduce to their vertical position and the value of the cross-over length scale ξ between the small and large scale regimes. At this point, it should be emphasized that the vertical position of these curves, assuming self affinity, is related to the associated large-scale topothesy l. Physically, this parameter provides information about the amplitude of the roughness. As a matter of fact, simple self-affinity translates into the following shape for the correlation function:

$$\Delta y(\Delta z) = \Delta z^\zeta l^{1-\zeta} \qquad (4)$$

where ζ is the roughness, or Hurst, exponent.

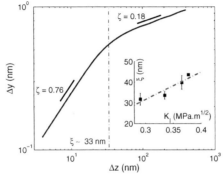

Figure 4. Experimental height-height correlation function computed on an image (2x2 μm²) obtained at RH = 78±2% and for K_I = 0.335±0.008MPa.m$^{1/2}$. Inset: Cut-off length scale ξ as a function of stress intensity factor K_I at RH=78±2%

DISCUSSION OF RESULTS

It has been shown recently[19] that scanning a self-affine profile with a finite-sized probe, like an AFM, results in spurious small-scale features in the height-height correlation function, in particular the emergence of a cut-off length scale and a large apparent exponent at small scales, much like what we observe here. The spurious cut-off Δz_c was shown to scale in the following way:

$$\Delta z_c \approx \left(R_{tip}^{\gamma-1} l^{1-\zeta}\right)^{\frac{1}{\gamma-\zeta}} \qquad (5)$$

where the shape of the tip is modelled by a smooth function $f_{R_{tip}}^{[\gamma]}(x) \equiv \dfrac{|x|^\gamma}{2R_{tip}^{\gamma-1}}$, R_{tip} is its typical size, γ is a shape parameter allowing for partial tip flattening, l is the topothesy of the original profile and ζ its

Hurst exponent. In the case at hand, we found that at ⊠xed RH, the topothesy decreases with K_I, which should lead to a decrease in the cut-off, assuming a constant shape and size of the tip, provided it originates from the aforementioned artifact. This is however not what we observe (cf. the inset of Figure 4). It is worth mentioning here that the data reported by Bonamy, et al.[3] are compatible with this scenario. Nevertheless, we have performed tip-characterization measurements in order to extract the size and shape of the probe used in these experiments prior and after scanning. Very surprisingly, the radii extracted from these measurements appear to blunt dramatically during the scanning process. A typical value R_{tip} = 30nm is measured before scanning, compatible with the nominal radius indicated by the manufacturer, with a shape parameter γ = 2.5. When measured after acquiring several (more than 15) images though, our measurements provide values as high as R_{tip} = 70nm and shape parameters γ = 4. Following the procedure detailed in [19] with the measured values for these parameters, we were able to reproduce our set of data together with the behavior of the cut-off, assuming a constant self-affine exponent ζ and keeping the original topothesy as the only ⊠tting parameter. The procedure is the following. A numerical self-affine pro⊠le is generated using the spectral method detailed in [20] for a given exponent. It is then numerically scanned by a tip with the measured shape. The exponent is chosen so as to reproduce the asymptotic behavior of the measured correlation function. The topothesy is then adjusted so that the correlation function of the scanned pro⊠les reproduces the original one as tightly as possible. In Figure 5, the correlation functions computed on 3x3 µm² images obtained at RH = 78±2% for K_I = 0.291±0.008MPa.√m (blue circles and red squares) and K_I = 0.378±0.008MPa.√m (green stars) are shown. They are compared with smoothed height-height correlation functions obtained for a numerical pro⊠le of constant self-affine exponent ζ = 0.13 with topothesy values respectively of l = 0.50, 0.47 and 0.39nm (solid lines).

It has to be emphasized that we measured the tip's characteristics prior to this adjustment, and they are critical in this procedure: a small change in R ~ 10nm or γ ~ 0.5 would make it impossible to reproduce the data. The fact that the presence and behavior of the cut-off ξ is compatible with a tip-smoothing artefact prevents further interpretation of the small-scale region in the correlation function, which consequently falls below the resolution of the AFM for this kind of measurement. Moreover, we suspect that the interpretation given by Bonamy, et al.[3] only translates into the regular evolution of the cut-off explicated through Eq. **(5)** due to the decrease in the topothesy with K_I.

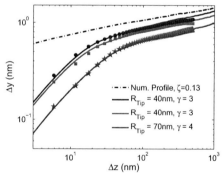

Figure 5. Experimental height-height correlation functions obtained for a silica-glass fracture surface. Original and smoothed height-height correlation functions obtained for a numerical pro⊠le with ζ = 0.13 and for different values of tip radius R_{tip}, shape parameters γ and topothesy l (cf. the text for more details)

CONCLUSION

We have performed roughness characterization of experimental fracture surfaces obtained under controlled conditions. We found that the RMS roughness of these surfaces is a reliable and robust estimator since its dependence on stress intensity factor reproduces previously reported results, and because it is substantially not dependent on the details of the AFM tip. We then extracted the characteristic features of the 1D height-height correlation function. We showed that its large-scale behavior is relevant to the physics of the problem, since it properly quantifies the way in which the RMS roughness scales with the image size. However, we showed that its small-scale decay and the cross-over length scale at which this decay settles are spurious consequences of the finite size of the probe used in the measurements, which prevents further interpretation of these quantities as characteristics of the process zone, as was previously suggested. These conclusions are in agreement with two recent results showing respectively the validity of linear-elastic fracture mechanics down to 10 nm distance from the crack tip[23] and the penetration of water around the crack tip during propagation in a region smaller than 10 nm.[24]

ACKNOWLEDGEMENT

We thank C. Ottina for providing us a part of the samples, D. Bonamy and A. Grimaldi for fruitful discussions. All co-authors of this work wish to acknowledge the financial support of ANR Grant "Corcosil" No. ANR-07-BLAN-0261-02.

REFERENCES

[1] C. Wünsche, E. Rädlein and GH. Frischat, Glass fracture surfaces seen with an atomic force microscope, *Fresenius J. Anal. Chem.*, **358**, 349–351 (1997).

[2] E. Bouchaud, Scaling properties of cracks, *J. Phys. Condens. Matter*, **9**, 4319–4344 (1997).

[3] D. Bonamy, L. Ponson, S. Prades, E. Bouchaud and C. Guillot, Scaling exponents for fracture surfaces in homogeneous glass and glassy ceramics, *Phys. Rev. Lett.*, **97**, 135504 (2006).

[4] T. Fett, G. Rizzi and D. Munz, T-stress solution for DCDC specimens, *Eng. Frac. Mech.*, **72**, 145-149 (2005).

[5] MY. He, MR. Turner and AG. Evans, Analysis of the double cleavage drilled compression specimen for interface fracture energy measurements over a range of mode mixities, Acta Meta. et Mater., **43**, 3453-3458 (1995).

[6] G. Pallares, L. Ponson, A. Grimaldi, M. George, G. Prevot and M. Ciccotti, Crack opening profile in DCDC specimen, *Int. J. Fracture*, **156**, 11–20 (2009).

[7] C.L. Rountree; D. Bonamy, D. Dalmas. S. Prades, R.K. Kalia, C. Guillot, E. Bouchaud. "Fracture in glass via molecular dynamics simulations and atomic force microscopy experiments" *Phys Chem Glasses-B*, **51**, 127-132 (2010).

[8] G. Mourot, S. Morel, E. Bouchaud and G. Valentin, Scaling properties of mortar fracture surfaces, *Int. J. Fracture*, **140**, 39-54, (2006).

[9] M. Hinojosa, E. Reyes-Melo, C. Guerra, V. Gonzalez, and U. Ortiz, Scaling properties of slow fracture in glass: from deterministic to irregular topography, *Int. J. Fracture*, **151**, 81-93 (2008).

[10] L. Ponson, D. Bonamy and L. Barbier, Cleaved surface of i-AlPdMn quasicrystals: Influence of the local temperature elevation at the crack tip on the fracture surface roughness, *Phys. Rev. B*, **74**, 184205 (2006).

[11] P. Daguier, B. Nghiem, E. Bouchaud, and F. Creuzet, Pinning and depinning of crack fronts in heterogeneous material, *Phys. Rev. Lett.*, **78**, 1062-1065, (1998).

[12] C. Ottina "Aspect de la rupture en corrosion sous contrainte" Undergraduate thesis at Institut Supérieur de Mécanique de Paris and CEA Saclay (2007).

[13]P. Mazeran, L. Odoni and J. Loubet, Curvature radius analysis for scanning probe microscopy, *Surf. Science*, **585**, 25-37 (2005).

[14]SM. Wiederhorn, Influence of water vapor on crack propagation in soda-lime glass, *J. Am. Cer. Soc.*, **50**, 407 (1967).

[15]E. Bouchaud, The morphology of fracture surfaces: A tool for understanding crack propagation in complex materials, *Surf. Rev. and Lett.*, **10**, 797-814 (2003).

[16]Einstein, A. The theory of the Brownian Motion, *Ann. Der Phys.*, **19**, 371-381 (1906).

[17]SM. Wiederhorn, J. Lopez-Cepero, J. Wallace, JP. Guin and T. Fett, Roughness of glass surfaces formed by sub-critical crack growth, *J. Non-Cryst. Solids*, **353**, 1582-159 (2007).

[18]D. Dalmas, A. Lelarge and D. Vandembroucq, Crack Propagation through Phase-Separated Glasses: Effect of the Characteristic Size of Disorder, *Phys. Rev. Lett.*, **101**, 255501 (2008).

[19]F. Lechenault, G. Pallares, M. George, CL. Rountree, E. Bouchaud, and M. Ciccotti, Effects of finite probe size on self-affine roughness measurements, *Phys. Rev. Lett.*, **104**, 025502 (2010).

[20]R. Voss, Fractals in nature: from characterization to simulation, *Springer-Verlag*, (1988).

[21]D. Bonamy, Intermittency and roughening in the failure of brittle heterogeneous materials, *J. Phys. D: Appl. Phys.*, **42** 214014 (2009).

[22]M. Ciccotti, Stress-corrosion mechanisms in silicate glasses, *J. Phys. D: Appl. Phys.*, **42** 214006 (2009).

[23]K. Han, M. Ciccotti and S. Roux S, Measuring nanoscale stress intensity factors with an atomic force microscope, *EPL*, **89**, 66003 (2010).

[24]F. Lechenault, C.L. Rountree, F. Cousin, J.P. Bouchaud, L. Ponson, E. Bouchaud, Evidence of Deep Water Penetration in Silica during Stress Corrosion Fracture, *Phys. Rev. Lett.*, **106**, 165504 (2011).

FRACTOGRAPHY IN THE DEVELOPMENT OF ION-EXCHANGED COVER GLASS

G.S. Glaesemann, T.M. Gross, J.F. Bayne, and J.J. Price
Corning Incorporated

ABSTRACT

Many mobile devices are transitioning to glass as a protective cover because plastic covers scratch easily. Cover glass must be thin and able to withstand the rigors of everyday use. Early devices employed soda-lime glass strengthened by ion exchange. Soda-lime glass is inexpensive and strong after ion-exchange, but it is not that much more damage resistant than its non-ion-exchanged condition due to the fact that the exchange layer is thin, no more than 15 microns deep. An ion-exchangeable glass has been formulated and manufactured for cover glass applications where the depth of compression is 45 to 50 microns with a compressive stress of about 750 MPa. Key to this development was the role of fractography in identifying the key mechanical attributes necessary for use as a protective cover glass. In particular, fractographic evidence suggested that the predominant failure mode for cover glass is mechanical damage caused by sharp contact events. Therefore, the primary mechanical attribute of the glass is damage resistance as opposed to initial strength. Consequently, the ideal cover glass has both a deep depth of compression as well as a high compressive stress.

INTRODUCTION

Cover glass for mobile devices is almost exclusively chemically strengthened glass. Until a few years ago soda-lime silicate glass was the primary glass composition for chemically strengthened cover glass. Soda-lime silicate glass exchanges in such a way that the compressive layer is no more than 15 microns deep and the maximum compressive stress is in the range of 600 MPa. Indeed one can achieve quite high strengths with soda-lime silicate glass. However, in the case of cover glass for mobile devices, it is clear that a high measured strength is not the only mechanical attribute of interest. The mechanical attributes for cover glass should be based on observed failure modes, and fractography is a key tool for such a determination. Armed with this knowledge one can design the optimal glass.

SURFACE DAMAGE AFTER IN-SERVICE USE

Several cell phones with soda-lime silicate cover glass were examined after experiencing in-service use. The glass surface has both scratches and individual impact sites. The impact sites shown in Figure 1 are typical and exhibit fractographic features indicative of sharp contact damage.[1]

There is a zone of plastic deformation from which lateral cracks emanate. These lateral cracks can form during the application of contact pressure or in response to the local residual stress field that accompanies plastic deformation.[1] The lateral cracks in Figure 1 are severe enough to grow and intersect the surface such that chips are formed. If severe enough, the resulting surface features can be seen visually.

Of particular importance are the subsurface cracks that form beneath the damage zone in that these cracks are strength controlling. In Figure 2 the fracture surface from a field failure is shown where a median crack extends 300 microns beneath the surface impact site. Note that this crack system can sometimes be difficult to observe from the surface. It is also important to note that the subsurface flaw systems control the strength of the glass, not the plastic zone; therefore, the measurement of surface roughness is not predictive of strength in this case. The image in Figure 2 was chosen because it clearly shows subsurface cracking. However, not all damage sites are this clear, as subsurface cracking is often quite complex.

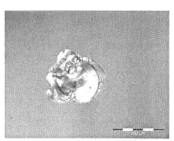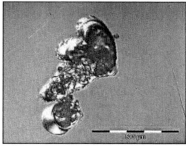

Figure 1. Typical in-service impact damage to the surface
of ion-exchanged soda-lime silicate cover glass.

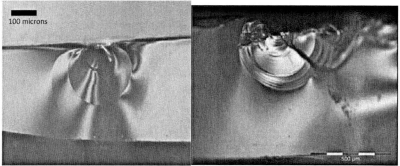

Figure 2. Subsurface median cracks beneath impact sites of cover glass field failures.

A scratch from a mobile phone cover glass is shown in Figure 3. Contact was initiated on the left end of the scratch as evidenced by the presence of just a trough of plastic deformation. As the object slid from left to right across the surface the contact pressure increased until lateral cracks were initiated. This type of contact has every potential for forming subsurface median/radial crack systems at higher applied loads.

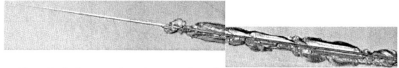

Figure 3. Surface scratch on the cover glass of a mobile electronic device. This figure is
a composite of two images and the width of the scratch at its widest is about 50 micrometers.

It is important to note that identifying the contact event as "sharp contact" has more to do with how the glass responds to the contact event than the actual contacting object. In most cases one will never retrieve the object that contacted the cover glass. The key evidence for sharp contact is the presence of permanent deformation. In addition, one looks for lateral crack formation, associated spalling, and subsurface crack systems.

REPLICATING FIELD DAMAGE USING CONTROLLED-FLAW-INTRODUCTION METHODS

Pressing a Vickers diamond indenter into the surface of an ion-exchanged soda-lime silicate glass replicates the observable surface damage in Figure 1 quite well. Figure 4 shows the same extensive lateral cracking about the damage zone after applying a 500 gram indentation load. The size of the overall damage site is about the same as well. Thus, one is justified in using controlled flaws produced by well-known diamond indentation methods in studying the damage mechanics of ion-exchanged glass.

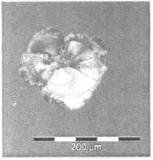

Figure 4. Replicating field damage by Vickers diamond indentation in ion-exchanged soda-lime silicate glass. The indentation load was 500 grams.

Similarly, one can replicate scratch-induced damage using diamond indenters. Figure 5 shows a scratch produced by pressing a Knoop indenter into ion-exchanged soda-lime silicate glass and increasing the load to about 1 N during movement.[2] The surface features mimic closely those found on the naturally occurring scratch in Figure 3. An examination of subsurface features reveals classic median/radial type crack systems that provide a primary means for weakening.

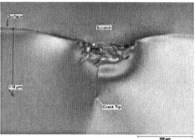

Figure 5. Scratch damage in ion-exchanged soda-lime silicate glass produced by a Knoop diamond and the ramped-scratch method. The long axis of the Knoop indenter was aligned with the scratch direction and the peak ramp load was 1N. The total width of the composite image is about 1.3 millimeters and the width of the scratch at its widest is about 230 micrometers. The cross-sectional view shows a median crack from an 8N scratch that is 118 microns deep.

DAMAGE RESISTANT GLASS FOR MOBILE DEVICES

Fractographic analyses of glass surfaces of hand-held electronic devices inspected after field use indicate that the predominant failure mode for glass cover panels is damage from sharp contact. In 2008 an alkali-aluminosilicate glass (Corning® Gorilla® Glass Code 2317) was developed as a protective cover glass for mobile devices. The goal was to provide damage resistance and high strength by way of a compression layer depth of 45 to 50 microns while maintaining maximum compressive stress of around 750 MPa. Note that ion-exchange, for this thin glass application, is limited by frangibility.[3] The center tension created by ion exchange drives the fragmentation behavior at failure and, consequently, is intentionally limited.

The ability of this glass to resist in-use damage was characterized using diamond indentation techniques. One such indentation test strategy is to measure the threshold load for the formation of strength-limiting flaws.[4] This test method consists of pressing and releasing a pyramid indenter from the glass surface until a threshold for radial crack formation is reached. The indentation load at which these flaws form can be considered a measure of resistance to crack initiation and is dependent on glass composition, compressive stress and depth of compression, as well as the indenter type.

A load-stepped indentation study was performed on both glasses and the results are shown in Figure 6. The soda-lime silicate glass has a 10 micron compression layer with a maximum compression of 500 MPa. At a low load of 250 grams classic indentation impressions were produced with no observable crack development in specimens of ion-exchanged soda-lime silicate glass and Corning Code 2317 alkali-aluminosilicate glass. At 400 grams, radial cracks emanate from the corners of the impressions in soda-lime silicate glass; and at 500 grams, significant lateral cracking occurs. The impression in the alkali-aluminosilicate glass is still well contained in the compressive layer, and no cracks develop at these loads. At higher loads of 3000 grams, the soda-lime silicate glass is showing radial cracking many times the impression size accompanied by lateral cracks. The alkali-aluminosilicate glass does not form crack systems until one reaches 4000 to 5000 grams. Thus, for purposes of damage resistance, one can see the importance of a deep compression layer. Note that both glasses in the non-ion-exchanged state will generate such flaws during unloading from a peak load of approximately 300 grams.

Figure 6 a. Vickers indentation to 250 grams generates a region of plastically deformed glass without crack formation in both glasses. (left- alkali-aluminosilicate glass, right- soda-lime silicate glass)

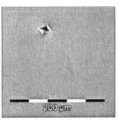 

Figure 6 b. Vickers indentation to 400 grams. The ion-exchanged soda-lime silicate glass has small radial cracks emanating from the damage site.

Figure 6 c. Lateral cracking in ion-exchanged soda-lime silicate glass at 500 grams indentation load leading to spalling and chipping.

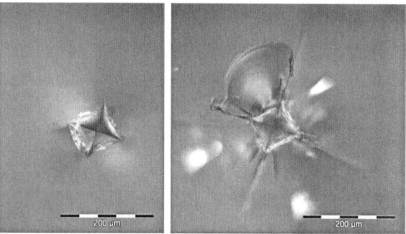

Figure 6 d. At 5000 grams radial cracks are just forming on the ion-exchanged alkali-aluminosilicate glass (Corning® 2317).

Strength testing glass after damage introduction is an effective means for assessing the extent of the formation of strength-limiting flaws and is a useful tool in understanding damage mechanics of ion-exchanged glasses. In the development of the alkali-aluminosilicate glass, several such damage introduction techniques were used to characterize retained strength. First, 10-mm-long scratches were placed on the surface of each glass at a constant load with a Vickers diamond indenter with loads ranging from 0.25 N to 3 N. Strength testing was performed using the ring-on-ring method, where the scratch was placed in the center of the load ring, and the results are summarized in Figure 7. At 0.25 N, the ion-exchanged SLS glass has weakened significantly, and the glass surface shows well-defined lateral cracks emanating from the edges of the scratch groove. Lateral cracking on the ion-exchanged alkali-aluminosilicate glass is delayed until 1 N, but at this load the more extensive compressive layer is minimizing the impact of crack formation. At a 3 N scratch load, the damage is extensive enough that the alkali-aluminosilicate glass is significantly weakened.

Recent research by Gross[5] explores the nature of weakening in scratches placed in ion-exchanged glass. One key observation is that lateral cracks form first when scratching these glasses with a Vickers diamond indenter. These lateral cracks effectively shield the formation of more severe median/radial cracks. Also, when lateral cracks initiate in soda-lime silicate glass, they do so beneath the shallow ion-exchange layer and cause the severe weakening at 25 grams shown in Figure 7. Note that lateral cracks are not generally considered to be strength limiting, but in this case they are. With a deeper exchange layer, lateral cracks form within the compression layer and the weakening effect is minimized until higher loads are reached. Note that, at higher scratch loads, the median/radial crack system does appear, as shown previously in Figure 5. Thus, the deep compression layer affects the flaw initiation process as well as the overall stress field on the flaw during loading to failure.

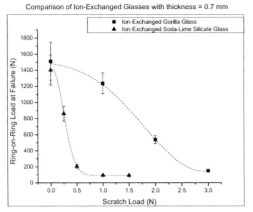

Figure 7. Strength after Vickers diamond scratching of 0.7 mm ion-exchanged glass.

Another damage introduction method used was particle abrasion. This study used SiC particles, because, when performed correctly, one can produce individual impact sites that look very much like Vickers indents, except that the features are much smaller, see Figure 8. SiC particles are sharp in comparison to the rather blunt Vickers and Knoop indenters, and one would expect flaw systems to initiate more readily. The severity of the abrasion was controlled by using several grit sizes and abrasive pressure levels. A simple metric for severity, "abrasion severity," was set to the square root of the abrasive particle size times the abrasion blast pressure. The strength comparison in Figure 9 shows results similar to the scratch results in Figure 7. The damage mechanics for particle impact on ion-exchanged glass are not entirely understood, but preliminary results show evidence of median/radial cracks extending beneath the impact sites, which indicates the deeper exchange layer affects both the initial depth of these flaws and the applied stresses required for failure. For example, at an "Abrasion Severity" of 27, a SiC grit size of 90 and a blast pressure of 35 kPa produce median/radial cracks about 20 micrometers deep. This is deep enough to penetrate the soda-lime silicate glass compression layer, but is still well within that for Corning® 2317.

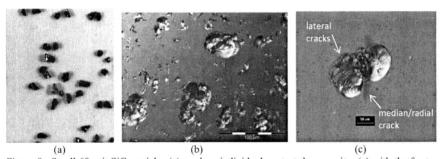

Figure 8. Small 60-grit SiC particles (a) produce individual contact damage sites (c) with the fracture features of Vickers indents (c), but at a much smaller size and, conceivably, lower load. The bar in (c) represents 50 microns.

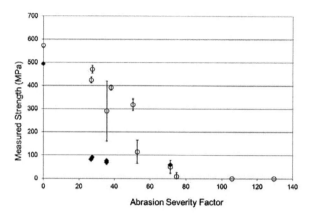

Figure 9. Strength ion-exchanged soda-lime silicate glass (closed diamond) and Corning® 2317 glass (open circle) after particle abrasion. An "Abrasion Severity" of 27 was created with 90 grit SiC particles and a blast pressure of 35 kPa.

SUMMARY

Failure analyses of mobile-device cover-glass panels have shown that failures can be attributed to characteristic "sharp contact." Thus, one is justified in using classic diamond indentation techniques in representing such damage in the laboratory for the purposes of cover-glass development. However, since real-life damage can be quite complex, one should use a wide range of damage introduction methods to understand the response of ion-exchanged glass. For example, the underlying damage mechanics of scratches in ion-exchanged glass are quite different than that of non-ion-exchanged glass. With these tools, the damage resistance of an alkali-aluminosilicate glass, ion-exchanged to the frangibility limit, was characterized and benchmarked against conventional soda-lime silicate glass.

ACKNOWLEDGEMENTS

The authors gratefully acknowledge the assistance of Don Clard and Kevin Reiman.

REFERENCES

[1]R.F. Cook and G.M. Pharr, "Direct Observation and Analysis of Indentation Cracking in Glasses and Ceramics", *J. Amer. Ceram. Soc.* **73** 787 (1990).

[2]V. Le Houérou, J.-C. Sangleboeuf, S. Dériano, T. Rouxel and G. Duisit, "Surface damage of soda–lime–silica glasses," *J. Non-Cryst. Solids* **316** 54-63 (2003).

[3]R. Tandon and S.J. Glass, "Controlling the fragmentation behavior of stressed glass," in *Fracture Mechanics of Ceramics, Vol. 14: Active Materials, Nanoscale Materials, Composites, Glass, and Fundamentals*, edited by R.C. Brandt, D. Munz, M. Sakai, and K.W. White (Springer, New York, 2005), pp. 77-91.

[4]D. J. Morris, S. B. Myers, and R. F. Cook, "Indentation Crack Initiation In Ion-Exchanged Glass", *J. Mater. Sci.* **39** 2399 – 2410 (2004).

[5]T.M. Gross, "Scratch Damage in Ion-Exchanged Alkali Aluminosilicate Glass: Crack Evolution and the Dependence of Lateral Cracking Threshold on Contact Geometry," in Fractography of Glasses and Ceramics VI, this publication.

CRACK BRANCHING IN GLASS INTERFACE DRIVEN BY DYNAMIC LOADING

Hwun Park and Weinong Chen
School of Aeronautics and Astronautics, Purdue University
West Lafayette, Indiana, USA

ABSTRACT

Crack branching is one of the main characteristics of failure in glass and has been discussed in many studies. However, the exact criteria of its initiation and factors determining its shape have not been explained clearly. We developed new methods to generate crack branching at specific positions with consistent shapes in glass samples having epoxy adhesive layers. Single crack propagation driven by dynamic loading is delayed at the layer in glass, and then the crack branches into multiple cracks. In our previous study, we showed that the number of branched cracks depends on the loading conditions and the delay duration of the crack propagation caused by the adhesive layer. We varied the dimensions of glass samples and conducted impact experiments to see the effects of stress distributions onto moving cracks. Reflected stress waves that encounter and interact with the cracks highly affects the direction and spreading angle of branched cracks. Transverse tensile stress waves intersect the moving cracks and broaden the branching angle. To obtain full-field information we simulated crack propagation and branching with a peridynamics code. In terms of crack speeds and branching positions, the simulation results show good agreements with the experimental results.

INTRODUCTION

A crack initiated by dynamic loading in brittle materials branches into multiple cracks in specific conditions. This crack branching is a common phenomenon observed in dynamic crack propagation and still remains as an open topic. The physics and criteria of crack branching have been investigated in many studies.[1-3]

The speed of a moving crack has a limitation. The theoretical maximum crack speed known as Rayleigh wave speed, is not achieved due to the inertia effect and material characteristics. It is known that a moving crack can be kinked and branched into multiple cracks when it reaches the maximum speed. However, the suggested values of critical velocities and kink angles from analytical studies are different from the observed values from experiments.[2,3] It is obvious that a crack begins to branch in order to dissipate excessive energy.[0] Each branched crack should run slower after branching because more energy is dissipated through the new fracture surfaces while the supplying strain energy remains the same. However, crack speeds hardly decrease after branching even though the rate of energy dissipation linearly increases with the number of cracks.[2] The review by Ramulu and Kobayashi described that the observation results of crack branching were inconsistent, because the branching depended on the loading conditions, geometry of specimens and material properties.[2]

In a brittle amorphous material such as glass, the estimation of paths of branched cracks is challenging because the crack paths are not decided by the material structure and the crack branching depends on pre-existing micro-cracks ahead of cracks.[4] Also, the direction of each branched crack is affected by stress distributions surrounding the crack tip. The interactions between stress waves and crack tips are one of primary parameters to decide the directions. Ravi-Chandar and Knauss showed that the longitudinal stress waves parallel to the crack growth affect the shape of crack branching. The branched cracks spread widely if the cracks encounter stress wave propagating in the opposite direction.[4]

Generating consistent crack branching is a prerequisite for the experimental investigation of crack branching. We suggested a method to create consistent crack branching in glass with dynamic loading in our previous study.[5] A notched glass having an adhesive layer is impacted with a plastic projectile. A single crack initiates at the notch and propagates into the adhesive layer. If the adhesive layer has a certain thickness, crack branching happens while the crack propagates through the layer. The number of cracks and the shape of branching depend on the loading condition and the thickness of the adhesive layer.[6] We present the dependency of the shape of branched cracks on stress propagation in this study.

As analytical methods have not solved the crack branching problems successfully, experimental analyses have been primary methods to investigate the problems. Numerical simulations based on continuum mechanics are not proper tools to simulate the growth of cracks because the spatial derivatives to represent displacements require continuity of materials. Recently, peridynamics based on the computation of pairwise force between particles was suggested to simulate discontinuities representing material failure.[7] Also the crack branching was successfully simulated with peridynamics.[8] However, a closed correlation between the results from peridynamic simulation and the results from experiments has not been presented. In this study, we conduct peridynamic simulation to correlate our experimental results on crack branching with the peridynamic simulation.

EXPERIMENT

Detailed description of the test set-ups and procedures have been previously disclosed.[5,6] To create different stress distributions, we changed the dimensions of samples as shown in Fig. 1. The length in the y direction of Type 2 sample is half that of the Type 1 glass sample. The shorter distance between the boundary and the centerline (where cracks are expected to propagate) for Type 2 induces earlier wave reflection than for the broader Type 1 geometry. These reflected tensile waves interact with growing cracks during the observation. To measure the stress propagation, strain gages were attached on the glass samples as described in Fig. 1. The gages were rosette gages to record stress propagation in the three directions. Although the detailed stress distributions in the whole body were not obtained, the wave propagation at specific positions were measured with the strain gages.

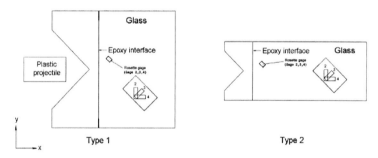

Figure 1. Dimensions of two types of glass samples

Images of crack branching recorded by a high-speed camera are presented in Fig. 2. The kinetic energy of projectiles in both experiments, considering as a loading condition, were similar as 2780 J and 2530 J for Type 1 and Type 2 respectively. The cracks took 50 μs to cross the adhesive layers. The numbers of main cracks were eight or nine, which were similar for each case. However, the spreading angles of branched cracks were different from each other as 60° and 90°. Multiple tests were conducted and the consistency in the number of branched cracks and the difference in the spreading angles was confirmed.[9]

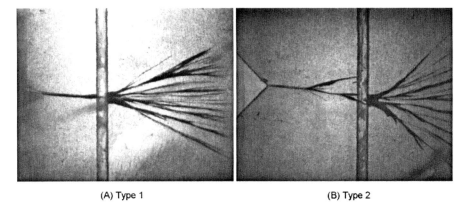

(A) Type 1   (B) Type 2

Figure 2. Crack branching patterns of two types of glass samples

The stress histories measured in both samples are presented in Fig. 3. The negative value represents compressive stress. The stress histories recorded by the Gage 4 in the x direction do not show much difference in both samples. However, the stress in the y direction recorded with the Gage 2 and 3 in the Type 2 samples shows tensile waves after 150 µs. This tensile wave interacts with the growing crack in the transverse direction and pulls the crack in the y direction. These interactions eventually broaden the spread of branched cracks.

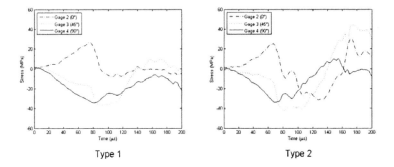

Type 1   Type 2

Figure 3. Stress variation of two types of glass samples

The stress propagation in the glass sample was generated by oblique impact with a projectile. The resultant stress profile in the sample was expected to be quite complex. Therefore, the measured stress history at a specific position cannot represent the full stress distribution. However, the stress histories from strain gages show distinct tensile waves and these waves intersect the running crack in the transverse direction. These interactions are believed to have affected the direction of crack growth and to have caused the observed wide spread of crack branching in the Type 2 case. To observe the interaction of the stress fields with the crack fronts in real-time, full-field techniques such as photoelastic observation would be preferred. However, optical investigation is not a proper method on glass because of the low optical constant of glass[4]. Therefore, we chose numerical simulation as a means to obtain full-field information on the stress profile and crack paths.

SIMULATION

Numerical analysis is another way to obtain full-field information on stress profiles and crack patterns. Peridynamics is a mathematical reformulation of a continuum model which enables inclusion of discontinuities in materials. Peridynamics does not require any spatial derivatives of medium or prerequisites for continuity. Instead, peridynamics employs integration of particles in a medium to compute the pairwise force between particles, which validate the equation of motion over discontinuities. The discontinuity in the medium can be simply formulated to set the force between particles as a zero. With computation of motions of all particles, the peridynamics simulates successfully dynamic crack propagation.[7] Ha and Bobaru simulated crack branching with the peridynamics and their results showed reasonable agreements with experimental results for crack speeds.[0] However, in terms of loading and boundary conditions, they did not present correlation between their simulation results and others' experiment results.

We used PDLAMMPS as a peridynamic code and Prototype Micro-elastic Brittle model suggested by Silling and Askari. Young's modulus and surface fracture energy of glass are the only properties assigned to the model.[7] The simulation results show good agreements in terms of crack speeds and the positions of crack branching as shown in Fig. 4 and 5. However, the thickness of the adhesive layer of the model is different from the sample in the experiment. The size of grid spacing of the peridynamic model restricts the size of adhesive layer. The grid spacing in the simulation is 0.8 mm and the corresponding thickness of the adhesive layer in the model is 3.2 mm.

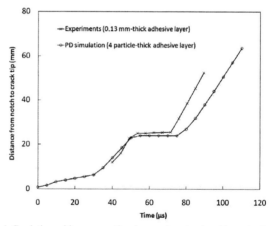

Figure 4. Crack tip positions versus time in experiment and peridynamic simulation

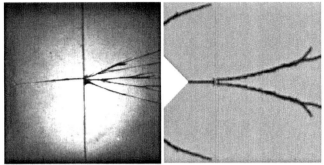

Figure 5. Crack branching shapes in experiment and peridynamic simulation

Although this peridynamic simulation generates crack branching at the same position, it only produces two branched cracks and the spreading angle is smaller than observed experimentally. The relatively large grid spacing (0.8 mm) may be one of the reason to limit the number of cracks. Considering the sharp tip of cracks in glass, this grid spacing does not provide sufficient resolutions. Also, the PMB model is considered to oversimplify the physics of dynamic fracture. The material properties are not constant under dynamic loading and rather depend on strain rate.[0,0] Future work will include refinements of the peridynamic model to improve the simulations.

CONCLUSIONS

The tensile stress reflected from boundaries interacts with moving cracks in the transverse direction, which causes the branched cracks to spread in the large angle. Numerical simulation based on peridynamic shows close results with experimental results in terms of crack speeds and crack branching positions. In future works, we will improve the simulation to create more than two branched cracks.

REFERENCES

[1]E.H. Yoffe, The Moving Griffith Crack, *Philosophical Magazine*, **42**, (1952).
[2]M. Ramulu, A.S. Kobayashi, Mechanics of Crack Curving and Branching – a Dynamic Fracture Analysis, *Int. J. Frac.*, **27**,187-201 (1985).
[3]K. Ravi-Chandar, W.G. Knauss, An Experimental Investigation into Dynamic Fracture: III. On Steady-state Crack Propagation and Crack Branching, *Int. J. Frac.*, **26**, 141-154 (1984).
[4]K. Ravi-Chandar, Dynamic Fracture, *Elsevier*, (2004).
[5]H. Park, W. Chen, Effect of an Interface on Dynamic Crack Propagation, *Am. Cer. Soc., 34th Int. Conf. Adv. Cer. Comp.*, (2010).
[6]H. Park, W. Chen, Experimental Investigation on Dynamic Crack Propagation Perpendicularly through Interface in Glass, *J. App. Mech*, **78**, 051013-1 (2011).
[7]S.A. Silling, E. Askari, A Meshfree Method Based on the Peridynamic Model of Solid Mechanics, *Comp. Struc.*, **83**, 1526-35 (2005).
[8]Y.D Ha, F. Bobaru, Studies of Dynamic Crack Propagation and Crack Branching with Peridynamics, *Int. J. Frac.*, **162**, 229-44 (2010).
[9]H. Park, Ph. D. Thesis, *School of Aeronautics and Astronautics, Purdue University* (2011).

A STRAIN ENERGY RELEASE RATE CRITERION FOR FORMATION OF THE MIRROR/MIST BOUNDARY, DEPICTION ON THE (G,R) vs. ΔC UNIVERSAL ENERGY DIAGRAM

Richard C. Bradt
Dept. of Metallurgical and Materials Engineering
The University of Alabama
Tuscaloosa, AL 35487-0202 USA

ABSTRACT
Fracture surface of glass lacks a specific criterion for formation. This paper considers that boundary formation criterion through the empirical "mirror constant" law which when rearranged suggests that a strain energy release rate criterion is appropriate. Experimental observations by Richter indicate that the mirror/mist boundary forms within the terminal crack velocity regime and at a stress intensity, K_I, that is greater than K_{IC}, the fracture toughness. Compilations of mirror constants and the fracture toughness values for silicate glasses by Mecholsky and by Quinn indicate that the mirror constant, $A_{m/m}$ is consistently ~2.6x the K_{IC} value and has the same dimensions, $MPa \cdot m^{1/2}$. The mirror/mist boundary forms at a $K_I \sim 2\ MPa \cdot m^{1/2}$ for a soda lime silicate glass, nearly the same as the $A_{m/m}$. Mist formation is amenable to presentation on a universal (G,R) vs. ΔC energy diagram, where G is the strain energy release rate, R is the crack growth resistance and ΔC is the crack growth. It explains the energy requirements and the kinetic energy of the crack at the formation of the mirror/mist boundary.

INTRODUCTION

The topographical features on a glass fracture surface near the origin are known as the mirror, the mist and the hackle. These have been documented for at least half a century, early by Shand[1] and then by Johnson and Holloway.[2] An example of their appearance from Quinn[3] is shown in Figure 1.

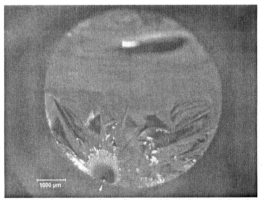

Figure 1. The fracture surface of a cylindrical glass rod showing the fracture origin at the arrow, the surrounding dark fracture mirror, the mist region, the hackle focusing back to the origin and the region of bifurcation.

Three distinct fracture surface regions may be physically described. The mirror is a very smooth, flat planar region of radius $R_{m/m}$. The crack front accelerates through this region in a planar manner. The mirror is surrounded by a narrow concentric band known as the mist region. The mist surface region has been described as appearing like the condensation of droplets to form a sort of dimpled appearance. In a radial direction away from the origin, the width of the mist is much less than the mirror radius. The width of the mist region is $< R_{m/m}$. The mist is followed by hackle lines which radiate away from the origin. The hackle fracture surface is actually at an angle to the mirror plane. The radial hackle lines may be traced back through the mist and the mirror to focus at the fracture origin and identify the location of the flaw which initiated the fracture. Along the mist/hackle boundary the mist penetrates the hackle for different extents along its circumference. That boundary is not as distinct and well defined as the mirror/mist boundary. For that reason, the mist/hackle boundary is not addressed in this paper, but is discussed in an excellent review by Mecholsky.[4]

In the fracture mirror region the crack front extends in a smooth, almost perfectly planar fashion. However, the mist region has been the subject of speculation as to just what is the physical process that leads to its formation and appearance. Various proposed mechanisms have been discussed by Lawn and Wilshaw[5] and also by Hull.[6] There are two favorites, both of which are suggested to occur in front of the advancing crack. One mechanism is the formation of microcracks ahead of the advancing crack front and the other is the development of microvoids ahead of the advancing crack front. Neither appears very plausible as the droplet-like mist surface does not have a very crack-like appearance. Its two opposing fracture surfaces are complementary, that is a depression in one corresponds to a protrusion in the other, as opposed to depression/depression pairs that one might expect from a crack intersecting microvoids. This author rejects both of those two speculative physical mechanisms for mist formation because of the lack of any direct experimental evidence.

THE MIRROR CONSTANT EQUATION

As a result of many authors presenting data for the mirror/mist boundary formation, it is well accepted that $R_{m/m}$, the radius of the fracture mirror when it breaks away from the planar stable crack growth to initiate formation of the mist band, follows the empirical equation:

$$\sigma_f \, R^{1/2}_{m/m} = A_{m/m} \tag{1}$$

where the σ_f is the fracture stress or the strength, the $R_{m/m}$ is the radius of the fracture mirror region and the quantity $A_{m/m}$ is known as the fracture mirror constant. Both Quinn[3] in his fractography text and Mecholsky[4] in his review article tabulate experimental mirror constants for many glasses. It is appropriate to note that the dimensions of the mirror constant are MPa·m$^{1/2}$. It is the same dimensions as those of the stress intensity. It is also necessary to note that several researchers Abdel Latif, et al[7] and Cantwell and Roulin-Moloney[8] have suggested that the mirror constant $A_{m/m}$ may not really be a constant, but rather only applies over a limited range. However, Michalske[9] has compiled an extensive amount of strength data and mirror radius measurements for fused SiO_2 which are presented in Figure (2). There is little doubt that Equation (1) is capable of describing fracture mirror sizes on glass fracture surfaces over a wide range of strengths. It must be added that Mecholsky[4] has extended the application of the form of Equation (1) to describe the mist/hackle boundary and also to include the phenomenon of crack branching, although perhaps not quite so convincingly as for the mirror/mist boundary. Notably, none of these present a criterion for the formation of the mirror/mist boundary.

Since Equation (1), although empirical, describes the mirror/mist boundary radius, $R_{m/m}$, it is informative to rearrange that equation to express $R_{m/m}$ to gain a further understanding of the criterion for the formation of the fracture mirror/mist boundary at $R_{m/m}$. Since the fracture mirror radius appears in Equation (1), it is only natural to rearrange that equation to obtain an expression for it, $R_{m/m}$. Once all of the terms in Equation (1) are moved to the right of the equal sign, then squaring the equation gives an expression for the mirror radius, $R_{m/m}$ that is:

$$R_{m/m} = A^2_{m/m} / \sigma_f^2 \tag{2}$$

Further modification of Equation (2) is even more informative. If both the numerator and the denominator on the right side of Equation (2) are multiplied by (1/2E) it yields the following form of Equation (2):

$$R_{m/m} = (1/2E) A^2_{m/m} / (\sigma_f^2/2E) = (1/2E) A^2_{m/m} / (½ \sigma_f^2/E) \tag{3}$$

where the denominator of the right side term in Equation (3) is more familiar as the strain energy density in the specimen:

$$½ \sigma_f^2/E \tag{4}$$

which is the stored elastic strain energy per unit volume in the specimen at fracture. This suggests that the mirror/mist boundary radius is determined by the elastic strain energy in the specimen. The numerator is ½ the mirror constant squared divided by the elastic modulus, E of the glass of interest.

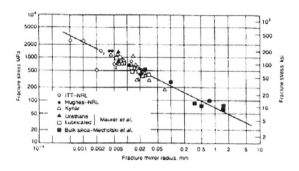

Figure 2. The fracture mirror constant equation, strength vs. fracture mirror radius, for fused SiO_2 after Michalske[9] in his review paper.

This result, namely that formation of the mirror/mist boundary is inversely related to the stored elastic strain energy in the specimen at fracture should not be surprising. Recently there was a conference whose proceedings were edited by Sih and Gdoutos[10] that was devoted to elastic strain energy density damage in materials. Equation (4) is the strain energy density per unit volume. The mirror radius, $R_{m/m}$, is inversely related to the strain energy density in a glass specimen at fracture.

As the strain energy density is proportional to the strength squared, σ_f^2, it is desirable to consider if there are other instances where either the strain energy density, or σ_f^2 form relates to the mirror or to the mist region. One such study is that of Ball, et al,[11] where the width of the mist region and also the roughness of the mist region were studied and shown to be proportional to the σ_f^2. Although Ball et al,[11] did not specifically address the criterion for the formation of the mirror/mist boundary, the paper nonetheless prompted Roy Rice[12] in 1984 to make the following statement and pose the following question after the paper presentation, "You have nicely demonstrated in more detail the widely observed general inverse trend between failure and density of fracture features, mist in this case. You have also noted a correlation with strain energy density, which I feel makes much sense. Have you quantitatively tested such a relation?" The authors replied that σ_f^2 is proportional to the strain energy density, so that for those mist features related to σ_f^2 the strain energy density would be expected to relate as well. It is necessary to pursue this association and examine the effect of the strain energy density on the formation of the mirror/mist boundary for evidently the strain energy density, or the release of that strain energy during fracture, has a profound effect on the fracture surface topography, formation of the mirror/mist/hackle features and their boundaries as well.

STRAIN ENERGY DENSITY AND THE STRAIN ENERGY RELEASE RATE

The strain energy release directly influences fracture surface topography as a crack extends and forms new fracture surfaces. Some of that elastic strain energy is converted to fracture surface energy and forms the fracture surface features and some supplies kinetic energy to the crack. Therefore it is important to examine the way that strain energy is released as a crack extends through a glass object or specimen. It evolves as a natural result of the Griffith Equation derivation. To graphically present this energy release, it is desirable to examine its evolution through the Griffith energy balance derivation. It can be found in more detail in any advanced text addressing fracture mechanics.

Many variations of the Griffith Equation derivation appear, however, the simplest for the objective at hand is for a constant load applied to an infinite plate in tension, where the load is free to descend when a center crack is introduced into the specimen, as derived by Hertzberg.[13] The energy balance for that fixed load specimen configuration can then be written as:

$$U_T = U_O + W_L + E_S + S_F \qquad (5)$$

where the U_T is the total energy of the plate, crack, load system, U_O is the energy of a crack free plate, W_L is the work done by the system by the introduction of a crack, E_S is the change in the elastic strain energy in the plate from the introduction of a crack and S_F is the surface energy required to form the crack. Incorporating $W_L = -2\,E_S$, and the appropriate expressions for both E_S and S_F, yields:

$$U_T = U_O - (\pi\,\sigma^2\,C^2\,t\,/\,E) + 4\,C\,t\,\gamma \qquad (6)$$

where t is the plate thickness, E is the elastic modulus of the specimen and γ is the surface energy of the material. Differentiating Equation (6) with respect to C and setting the derivative equal to zero, then rearranging the terms and canceling out the plate thickness, t, yields:

$$(\pi\,\sigma_f^2\,C\,/\,E) = 2\,\gamma \qquad (7)$$

This specifies the fracture criterion, so the σ in Equation (6) is replaced by the strength, σ_f, as it is in the Griffith Equation. This enables the depiction of the parameters on a universal energy diagram.

The two sides of Equation (7) have special names that identify their specific roles in the fracture process. The term on the left is known as the crack driving force, or the strain energy release rate, G, expressed as a script G. The term on the right is known as the crack growth resistance, R. The R, of course is the origin of the concept of the R-curve, which for glass is constant and equal to $2\,\gamma$. Glass has a flat R-curve. Note that G contains σ_f^2 and a C term, but that R does not. Rather than plotting the familiar Griffith energy balance of U_T vs C, where the maximum specifies the fracture criterion, for crack growth it is preferable to plot a (G,R) vs ΔC diagram. Broek[14] extensively discusses the universal energy diagram, where the abscissa is the crack length and the ordinate is the energy, either the crack growth resistance R, or the strain energy release rate G. This universal energy diagram for the description of crack growth during fracture is presented below in Figure 3.

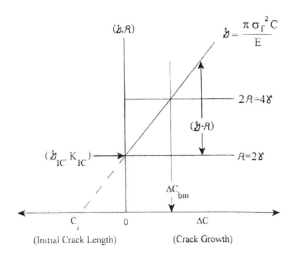

Figure 3. Fracture depicted on the (G,R) vs C, universal energy crack growth diagram. Note that the initial crack length, C_i, is left of the zero, while crack growth, ΔC, is depicted to the right of zero on the transformed ordinates.

Prior to addressing the criterion for the formation of the mirror/mist boundary on this universal energy diagram, it is desirable to point out several other positive features of this representation. First, note that $G > (R = 2\ \gamma)$ on the right of the zero point on the diagram. Since $G = R$ is the Griffith fracture criterion, this means that once the crack extends as expressed by ΔC there is more elastic strain energy released than is required to form the two fresh fracture surfaces. That energy is $(G - R)$ as shown on the diagram. It is that surplus elastic strain energy which becomes the kinetic energy associated with the moving crack system.

The strain energy release rate, G, is proportional to ΔC, so that the rate of release of the elastic strain energy becomes increasingly greater the more the crack grows. That is precisely why most cracks in glass extend completely through glass objects once they initiate. The larger the crack grows, the greater is the strain energy that is released for it to extend even more as glass has a flat R-curve. This rapidly increasing strain energy release rate is also why initiating cracks rapidly accelerate away from the fracture origin until they reach the terminal velocity for the particular glass.

It is also possible on this type of diagram to depict the minimum crack growth that is necessary for crack branching to occur. Since a branching crack forms four fracture surfaces instead of only two, the surface energy requirement for branching is $4\ \gamma$, not $2\ \gamma$. The $2\ R = 4\ \gamma$ line on the ordinate of Figure 3 requires a minimum crack growth of ΔC_{bm} to allow for sufficient energy to be released for crack forking, or branching. It is only a minimum of the crack growth ΔC, for it does not include the kinetic energy requirements of the complete crack system. It is well known that primary cracks always grow some distance before they fork or branch as has been demonstrated by Kepple and Wasylyk[15] for the case of exploding glass container bottles. Higher pressure explosions produce less primary crack growth prior to crack branching. This is to be expected from the stress squared term in the strain energy release rate, G. It produces a much steeper slope for G in the universal energy diagram and achieves the required branching energy release rate at a smaller ΔC value. These several examples illustrate the usefulness of this type of universal energy fracture diagram. Unfortunately, the diagram does not directly specify the kinetic energy for a crack system. Only the energy that is available for the kinetic energy is depicted as the $(G - R)$, which is indicated on the right of the diagram.

STRESS INTENSITIES AND MIRROR CONSTANTS FOR THE MIRROR/MIST BOUNDARY

To locate the mirror/mist boundary on a diagram such as Figure 3, it is first necessary to estimate the stress intensity, K_I, to calculate the energy, $G_{m/m}$, when the mirror/mist boundary forms. Richter[16] has provided us with an experimental measurement of that stress intensity for a soda lime silicate glass. He presented the crack growth on a classical K_I-V diagram determining the crack velocity by ultrasonic monitoring of the extending crack. Richter's results are depicted in Figure 4.

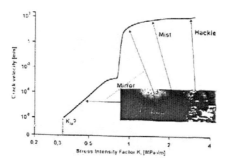

Figure 4. Depiction of the location of the mirror, mist and hackle of a soda lime silicate glass on a K_I–V diagram after Richter.[16]

Two very important aspects of Richter's results are essential to understanding the formation criterion for the mirror/mist boundary during glass fracture. One is the stress intensity factor when the mirror/mist boundary forms. Richter identifies the mirror, mist and hackle regions in Figure 4 rather than the specific boundaries between them. From his diagram it appears that the mirror/mist boundary of the soda lime silicate glass specimen is ~2 MPa·m$^{1/2}$. Shand[17] reports an $A_{m/m}$ value of 2.05 MPa·m$^{1/2}$, Orr[18] reports 2.09 MPa·m$^{1/2}$ and Kirchner and Gruver[19] list 1.9, 2.0 and 2.3 MPa·m$^{1/2}$ values. None are substantially different from Richter's experimental result. Quinn[3] summarizes numerous other measurements of soda lime silicate fracture mirror constants and they are mostly about two, but some are slightly larger and some smaller. This variation is not surprising considering Quinn's elaboration on the difficulties of accurately measuring the mirror boundary radius.

An equally important aspect of Richter's observations in Figure 4 is that the mirror/mist boundary forms after the extending crack has reached its terminal velocity of ~1.5 x 10^3 m/s (~3,500 mph) on the plateau region at the right of the K_I-V diagram. According to Richter's measurements, the crack accelerates from ~0.1 m/s at K_{IC} to its terminal velocity, ~1.5 x 10^3 m/s entirely within the mirror region. This suggests that the extending crack, having already reached its terminal velocity, has already achieved the maximum kinetic energy level that it can accommodate when the mirror/mist boundary forms. When the mirror/mist boundary forms, it is because the crack can no longer apply the vast amount of elastic strain energy that is being released by the crack growth to further increase the crack velocity. The crack is already at the terminal velocity. Thus, the continually increasing elastic strain energy that is being released must be applied to something other than further accelerating the crack. It is this excess energy that transforms the extending planar crack of the mirror with a stable planar crack front to an unstable crack front, one that develops the local perturbations that produce the appearance of droplets. This is the formation of the mist region.

Richter's experimental results also indicate that the well known and often measured mirror constant, the $A_{m/m}$ in Equation (1), is probably the stress intensity, the K_I value at that point when the mirror/mist boundary forms as the mist initiates. That is the reason why the units of fracture toughness and those of the mirror constant are identical as $MPa \cdot m^{1/2}$. Even though the concept of stress intensity was unknown when glass fracture mirrors were first observed to follow Equation (1), both quantities are stress intensities. The former was originally achieved empirically by plots of strength and fracture mirror radius data, while the latter is the product of fracture mechanics.

There is other evidence that the mirror constant is in fact the stress intensity at the mirror/mist boundary. Figure 5 below presents an internal fracture mirror in an optical fiber after Chandan, et al.[19] It is a perfect circle, not surprisingly for the fiber was under a uniform tensile stress, thus the product of the square root of the mirror radius and the stress at fracture, which has the dimensions of stress intensity, is constant around the circumference of the circular mirror/mist boundary. It could be argued that the loss of the area of the mirror will redistribute stress to the remaining ligament. That may be true, however, usually the area of the mirror relative to the total cross section of the glass object is small. Therefore, any stress redistribution can be expected to be small as well.

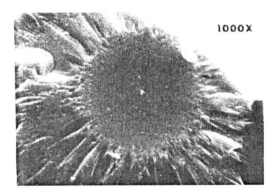

1000X

Figure 5. The fracture mirror from an internal flaw in an optical fiber broken in uniform tension, after Chandan, et al.[19] Note the uniform mirror radius about the flaw, confirming that the stress intensity is constant at the mirror/mist boundary.

These factors indicate that the criterion for the formation of the mirror/mist boundary is related to a K_I value that is equivalent to the mirror constant, $A_{m/m}$. This enables calculation of the strain energy release rate at the mirror/mist boundary, $G_{m/m}$. The physical process of the mist formation is not completely certain, but it evidently is one of crack front instability that occurs after the crack has achieved its terminal velocity. This instability results because the propagating crack is continually releasing more elastic strain energy, but the crack front cannot accommodate it as additional kinetic energy associated with an increasing velocity. This additional energy apparently creates perturbations of the crack front which appear as the mist droplets. The mist region is not the result of either the formation of microcracks or the nucleation of voids ahead of the crack front, as previously proposed.

MIRROR/MIST BOUNDARY CRITERION ON A (G,R) VS. ΔC ENERGY DIAGRAM

Presenting mirror/mist boundary formation on a (G,R) vs. ΔC universal fracture diagram not only graphically depicts the $G_{m/m}$ criterion, but it also enables an estimate of the amount of kinetic energy relative to the total crack energy that is associated with formation of the mirror/mist boundary. The Griffith criterion for fracture is at $G = R$ on the diagram as noted on Figure 3. In that figure, the angular G line crosses the ordinate at the 0 point where G_{IC} is equal to K_{IC}. All of the released elastic strain energy is converted to fracture surface energy at this point. The necessary step to find the crack extension for the formation of the mirror/mist boundary is to locate the position of the $G_{m/m}$, or alternately $A_{m/m}$ for the boundary formation on the ordinate on a line parallel to $R = 2\,\gamma$. It will be at the energy level on the ordinate commensurate with $K_I = A_{m/m}$ as previously demonstrated.

Table I lists four common silicate glasses and their pertinent mirror/mist boundary related parameters, some measured and some calculated. Different glasses will have slightly different mirror/mist boundary formation energy levels on the ordinate, so restricting the group to a few common silicates for illustrative purposes will reduce differences to a manageable level, but still present the general form of the graphical description. Other structural groups of glasses such as the chalcogenides or oxynitrides will be similar, but they are not included here for clarity of the diagram.

Table I. Mirror/Mist Boundary Parameters for Several Silicate Glasses.*

GLASS	$A_{m/m}$	K_{IC}	$A_{m/m} / K_{IC}$	E	$G_{m/m}$
fused silica	2.22	0.79	2.81	72	7.90 γ
aluminosilicate	2.17	0.81	2.68	89	7.16 γ
borosilicate	1.90	0.71	2.57	64	6.60 γ
soda lime silicate	2.05	0.76	2.70	73	7.67γ

*Units are MPa·m$^{1/2}$ for $A_{m/m}$ and K_{IC}, GPa for E, and $G_{m/m}$ is in γ, where γ is the fracture surface energy and $G_{IC} = R = 2\gamma$. This approach is used to accommodate several glasses on the energy scale of a single universal energy diagram, albeit for the same C_i.

The physical constants of the glasses in Table I are from the Quinn[3] and the Mecholsky[4] tabulations. The mirror constant to fracture toughness ratio is calculated and presented. The individual $G_{m/m}$ values for the formation of the mirror/mist boundaries of each of the glasses can then be estimated from the fracture mechanics relationships:

$$G_{IC} = (K_{IC}^2 / E) = 2\,\gamma \quad \text{and} \quad G_{m/m} \approx A_{m/m}^2 / E \qquad (8)$$

The experimental values of $A_{m/m}$ are substituted for the stress intensity, K_I at the mirror mist boundary in the usual stress intensity energy relationship as suggested by Richter's results. This enables an estimate of the strain energy release rate when the mirror/mist boundary forms, the value of $G_{m/m}$.

Presenting the energies for the four silicate glasses on the universal energy diagram in terms of their γ values is shown in Figure 6. For clarity, the diagram is presented for just one level of initial flaw size, C_i. The individual glasses are abbreviated as FS(fused silica), SLS(soda lime silica), AS(aluminosilicate) and BS(borosilicate). All the glasses have similar $A_{m/m} / K_{IC}$ values and similar levels of $G_{m/m}$ in terms of their γ values. This is not unexpected for the four are silicate glasses. Comparison of these glasses at the C_i, at similar strengths reveals a different ΔC value for the fracture mirror radius, $R_{m/m}$, for each of the four glasses. This is evidenced by the intersections of G, the strain energy release rate, with the individual glass $G_{m/m}$ values to establish the ΔC for the mirror/mist boundary formation. The ΔC crack lengths are equal to the mirror radius, $R_{m/m}$ at those intersections. They occur in the order of the experimental mirror constants that were summarized in Table I.

These were constructed for a single C_i or initial flaw size, producing only a single G line for the ease of representation and the clarity of the diagram. One could also construct diagrams for series of different flaw sizes for each glass as shown later, or for any number of other different variables. For the former, the G lines would appear as a series of lines fanning out about the G and R intersection on the ordinate. However, this is not necessary for the principle is established in Figure 6 without cluttering the figure at this point.

It is now possible to estimate the kinetic energies that are associated with cracks when they form the mirror/mist boundary. Applying the $(G - R)$ approach as illustrated in Figure 3 yields kinetic energies that are ~3-4 x the fracture surface energies of these glasses. At first this seems like a large amount of kinetic energy. However, the terminal crack velocity is ~1.5 x 10^3 m/s, a rather high velocity. It is not unreasonable. This establishes that the majority of the elastic strain energy that is released during the fracture process of glass does not go into forming the fresh fracture surface, but rather ~70% of that energy (5.90/7.90 = 75% for fused silica) is required to provide the kinetic energy that is required by the crack system moving at its terminal crack velocity.

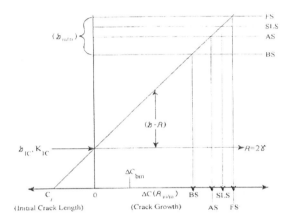

Figure 6. The universal energy diagram constructed to demonstrate the different $R_{m/m}$ values for the four silicate glasses in Table I.

The universal energy diagram in Figure 7 specifies the radius of the mirror/mist region, $R_{m/m}$, as it varies with the strength of the glass. Here the $G_{m/m}$ value is not for any particular glass, but just a convenient value to schematically illustrate the effect of the strength of the glass on the mirror/mist boundary radius, the $R_{m/m}$. Only two strength levels are presented, one a high strength glass and one a low strength. The strain energy release rate line, G, has a much steeper slope for the high strength glass for it has a much smaller C_i than does the low strength glass. Therefore the intersection with the $G_{m/m}$ line occurs at much lower ΔC values producing a much smaller $R_{m/m}$ for the high strength glass.

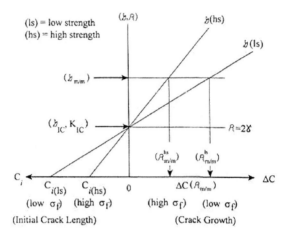

Figure 7. The universal energy diagram illustrating the different mirror radii for a high strength glass object (small C_i) and a low strength one (large C_i).

There are two other interesting fractographic observations, or features that a diagram similar to Figure 7 can explain. For very low strength glasses with very large flaw sizes, the slope of the general strain energy release rate, the G line might not intersect the individual $G_{m/m}$ before the crack propagates completely through the thickness of the glass object. In that instance no mirror/mist boundary would form and the new crack surface would be a totally flat mirror region throughout the entire specimen. This is observed for fractures of very weak glass objects. The concept can also be applied to explain the lack of mirror/mist surface features on the surface of very small silica nano-wires which has been reported by Brambilla and Payne.[20] The nano-wires have such very small diameters on the ΔC abscissa that the extending crack runs completely through the silica nano-wires before the general G line intercepts the $G_{m/m}$ line to form the mirror/mist boundary features.

SUMMARY AND CONCLUSIONS
The appearance of the fracture surface of glass is briefly described, then emphasis is addressed to the criterion for the formation of the mirror/mist boundary on the fracture surface. That boundary forms away from the fracture origin at a distance that is inversely related to the elastic strain energy in the specimen at fracture. Rearrangement of the mirror constant equation suggests a strain energy related criterion for the mirror/mist boundary formation. Experiments by Richter with the ultrasonic monitoring technique indicate that the velocity of the crack at the mirror/mist boundary is the terminal crack velocity, $\sim 1.5 \times 10^3$ m/s for a soda lime silica glass. Several reviews indicate that the mirror/mist boundary forms when the crack stress intensity is ~ 2.6-2.8 x the fracture toughness value of the glass. In the case of soda lime silicate glass, the mirror constant appears equal to the stress intensity for formation of the mirror/mist boundary. Using fracture mechanics formulae and the glass elastic modulus, the strain energy release rate can be calculated. The strain energy release rate far exceeds the energy requirement for the formation of the two new fracture surfaces. This additional energy is consumed in providing the kinetic energy of the fast moving crack system. However, once the crack reaches the terminal velocity, the ability of the crack to further accelerate is inhibited. At that point, where the increasing strain energy release rate no longer further accelerates the crack, the fracture

surface forms the mist region. Although the complete details are not understood, it appears that the mist region forms by a process of planar crack front instability after the terminal velocity is achieved.

The mirror/mist boundary formation can be represented on the universal energy diagram of (G,R) vs. ΔC by converting the mirror constant to a strain energy release rate, $G_{m/m}$ which varies for different glasses. Several silicate glasses were utilized as examples to illustrate the principles. The higher $G_{m/m}$ glasses have larger fracture mirrors, that is larger ΔC values when their mirror/mist boundary forms. The universal energy diagram is also applied to illustrate why stronger glass objects have smaller fracture mirrors.

It is concluded that the criterion for the formation of the mirror mist boundary is a strain energy release rate criterion. The mirror/mist boundary forms when the strain energy release rate can supply sufficient energy for the formation of the new fracture surfaces and also the kinetic energy required by the crack system in the glass object. The surface energy requirements appear to be only ~30% of the total strain energy released, while the kinetic energy requirement is ~70%, but these will vary slightly with individual glasses, although the kinetic energy is always expected to dominate.

ACKNOWLEDGEMENTS

The author thanks G.D. Quinn and H.C. Chandan for permission to use their photographs and also G.D. Quinn for numerous discussions on the topic. He also thanks H. Richter for Figure 4. Jeff Motz's assistance with the figures is greatly appreciated. Gratitude is also expressed to the Alton N. Scott Professorship at The University of Alabama for partial support of RCB while portions of this paper were completed.

REFERENCES

[1]E.B. Shand, "Breaking Stress of Glass Determined from Dimensions of Fracture Mirrors", *J. Amer. Cer. Soc.*, 42 (10), 474-477 (1959).
[2]J.W. Johnson and D.G. Holloway, "On the Shapes and Sizes of the Fracture Zones on Glass Fracture Surfaces", *Phil. Mag.* 14 (130), 731-743 (1966).
[3]G.D. Quinn, *"Fractography of Ceramics and Glasses"*, NIST STP 960-16, U.S. Gov. Printing Office, Washington, D.C. (2007).
[4]J.J. Mecholsky, Jr., "Quantitative Fractographic Analysis of Fracture Origins in Glass", 37-74 in *Fractography of Glass*, edit by R.C. Bradt and R.E. Tressler, Plenum Pub., NY, NY (1994).
[5]B.R. Lawn and T.R. Wilshaw, *Fracture of Brittle Solids*, 91-108, Cambridge University Press, Cambridge, UK (1975).
[6]D. Hull, *Fractography,* 121 – 129, Cambridge University Press, Cambridge, UK (1999).
[7]A.I.A. Abdel-Latif, R.C. Bradt and R.E. Tressler, "Dynamics of Fracture Mirror Boundary Formation in Glass", *Int. Jour. of Fract.*, 13, 348 – 359 (1997).
[8]W.J. Cantwell and A.C. Roulin–Moloney, "Fractography and Failure Mechanisms of Unfilled and Particulate Filled Epoxy Resins" 233-285 in *Fractography and Failure Mechanisms in Polymers and Composites*, edit by A.C. Roulin-Moloney, Elsevier Pub., NY, NY (1988).
[9]T.A. Michalske, "Quantitative Fracture Surface Analysis", 652-662 in *ASM Engineered Materials Handbook Vol. 4,* ASM International, Materials Park, Ohio (1991).
[10]*Mechanics and Physics of Energy Density,* edit by G.C. Sih and E.E. Gdoutos, Kluwer Academic Pub., Dordrecht, The Netherlands (1992).
[11]M.J. Ball, D.J. Landini and R.C. Bradt, "Fracture Mist Region in a Soda-Lime-Silica Float Glass", 110-120 in ASTM STP 827, *Fractography of Ceramic and Metal Failures,* edit J.J. Mecholsky, Jr. and D.R. Powell, Jr., pub by ASTM, Conshohocken, PA (1984).
[12]Roy W. Rice, quote in reference (11) following paper of reference (11).
[13]R.W. Hertzberg, 315-321 in *Deformation and Fracture Mechanics of Engineering Materials,* J. Wiley & Sons, NY, NY (1995).

[14]D. Broek, *Elementary Engineering Fracture Mechanics,* 128-136, Martinus Neihoff Publishers, Dordrecht, The Netherlands (1986).

[15]J.B. Kepple and J.S. Wasylyk, "The Fracture of Glass Containers" 207-252 in *Fractography of Glass* edit by R.C. Bradt and R.E. Tressler, Plenum Pub., NY, NY (1994).

[16]H.G. Richter, "Fractography of Bioceramics", 157-180 in *Fractography of Advanced Ceramics,* edit by J. Dusza, Trans.Tech. Pub., Zurich, Switzerland (2002).

[17]E.B. Shand, "Breaking Stresses of Glass Determined from Fracture Surfaces", *The Glass Industry,* 90-194, April (1967).

[18]L. Orr, "Practical Analysis of Fractures in Glass Windows", *Materials Research and Standards* 12, 1, 21-23 (1972).

[19]H.P. Kirchner and R.M. Gruver, "Fracture Mirrors in Polycrystalline Ceramics and Glass" 309-321 in *Fracture Mechanics of Ceramics* I, edit by R.C. Bradt, D.P.H. Hasselman and F.F. Lange, Plenum Pub., NY, NY (1973).

[20]H.C. Chandan, R.D. Parker and D. Kalish, "Fractography of Optical Fibers", 143-184 in *Fractography of Glass,* edit by R.C. Bradt and R.E. Tressler, Plenum Pub., NY, NY (1994).

[21]G. Brambilla and D.N. Payne, "The Ultimate Strength of Glass Silica Nanowires", *Nano-Letters* 9, 2 831-835 (2009).

SCRATCH DAMAGE IN ION-EXCHANGED ALKALI ALUMINOSILICATE GLASS: CRACK EVOLUTION AND THE DEPENDENCE OF LATERAL CRACKING THRESHOLD ON CONTACT GEOMETRY

Timothy M. Gross
Corning Incorporated
Corning, New York

ABSTRACT

The scratch behavior of ion-exchanged alkali aluminosilicate glass was investigated using sliding indentation. It is demonstrated that the evolution of crack systems originating in the subsurface, i.e. median and lateral cracks, differs for glass having a high surface compressive stress and a deep layer of compression when compared to non-strengthened glass. In non-strengthened glasses, the median crack forms prior to lateral cracking systems. In highly ion-exchanged glasses, the median crack is suppressed and lateral crack systems form preferentially. At levels of ion-exchange that suppress the median crack prior to the formation of lateral cracks, the outward propagation of lateral cracks does not appear to be influenced by the compressive stress.

The sliding indentation load required to produce visual lateral cracking damage in ion-exchanged glasses is highly dependent on the geometry of the indenter tip. The visual lateral cracking threshold is lower for tests performed using sharper indenter tips when compared to blunter indenter tips. Along with the standard Vickers indenter tip ($136°$), other custom four-sided pyramidal tips were used with angles slightly sharper ($120°$) and slightly blunter ($150°$). The scratch test was also performed with a standard Knoop indenter tip having intermediate sharpness between the Vickers and $150°$ tips. The scratch lateral cracking thresholds were 25 gf for the $120°$ tip, 100 gf for the Vickers tip, 400 gf for the Knoop tip, and >2 kgf for the $150°$ tip. Cross-sections of indents made with various tips show the change in deformation mechanism from shear to densification as the contact becomes less sharp.

INTRODUCTION

Ion-exchanged glass is increasingly being used as a high strength, high optical quality display cover in many portable electronic devices including cellular phones, MP3 players, and tablet computers. Corning® Code 2317, an alkali aluminosilicate glass sold under the product name Gorilla® Glass, is particularly well suited for cover glass applications since it is capable of ion-exchange to a surface compressive stress (CS) greater than 700 MPa and depth of compressive layer (DOL) of greater than 40 microns. It has been demonstrated that glasses with high CS and DOL have high levels of retained strength following contact damage events.[1] Besides the retained strength, another important attribute for glasses used as display covers is the resistance to visual scratch damage in the form of lateral cracks resulting in chipping at the surface. Alkali aluminosilicate cover glasses are intrinsically harder than polymeric display covers and as a result are much more resistant to scratching damage from everyday contacts, such as by car keys. Cover glass does, however, come into contact with very hard and very sharp materials, e.g. concrete surfaces. These contact events can produce sharp contact scratches having highly visible lateral cracks. It is therefore important to understand the fundamental mechanisms of sharp scratch crack formation in ion-exchanged glasses.

Sliding indentation has been used in many studies to evaluate the scratch cracking behavior of non-strengthened glasses.[2-7] The current study focuses on crack systems that initiate in the subsurface from deformation-induced flaws formed beneath the sliding indenter. These types of crack systems are the highly strength limiting median crack oriented perpendicular to the surface and highly visual lateral cracks that are oriented nearly parallel to the surface. Frictive-type cracks originating at the surface can occur during sliding indentation, but the resulting damage has less of a strength limiting effect as median cracks and less of a visual effect as lateral cracks, so will be excluded in the present study. Previous sliding indentation scratch studies describe the cracking evolution in non-strengthened glass[2-7]. At low loads, a scratch groove will be present in glass without any associated median or lateral cracking. As the scratch load is increased a median crack forms during the contact and is extended by the residual stress during removal of the load. Further increases in scratch load lead to the formation of lateral cracks that accompany the median crack. The lateral cracks are residual stress driven so that they form upon unloading. At high enough loads, the residual stress drives the lateral cracks sufficiently outward and up so that they intersect the surface of the glass and form highly visible chips. Figure 1 is a schematic showing the order in which subsurface median and lateral cracking systems form as the scratch load is increased for non-ion-exchanged glasses.

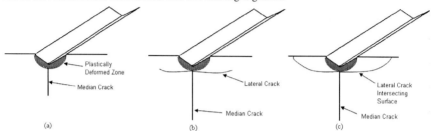

(a)   (b)   (c)

Figure 1. The formation of subsurface cracking systems as the scratch load is increased in non-ion-exchanged glass. (a) The initial crack system to form is the median crack. (b) As the load is increased lateral cracks form. (c) As the load is further increased the lateral cracks intersect the surface to form chips.

The present work describes the differences in the crack evolution for highly ion-exchanged glasses when compared to the non-strengthened glasses described in previous work.[2-7] To further understand the scratch cracking behavior in ion-exchanged glasses the role of indenter sharpness is investigated. Since the indenter sharpness alters the subsurface deformation mechanism[8], i.e. densification plays a larger role in the deformation for less sharp contacts, it can then be expected that the scratch crack resistance will be affected.

EXPERIMENTAL

Fusion drawn 1.3 mm thick specimens of Corning® Code 2317 alkali aluminosilicate glass were ion-exchanged at 410°C for 6 hours in KNO$_3$ salt to give a CS of 770 MPa and a DOL of 48 μm. All scratch tests were performed on a CETR Universal Material Tester (UMT) equipped with a 2 kgf load cell at 25°C in a 50% relative humidity environment. Constant load scratches were made at a constant rate of 0.4 mm/s for a length of 10 mm. For each scratch test, the scratch load was applied and removed at 7 gf/s. Ion-exchanged specimens were scratched with a Knoop tip at constant loads of 50 and 100 gf to study the effect of ion-exchange on the subsurface cracking behavior. The specimens were then cross-sectioned and the subsurface damage was viewed by SEM. Cross-sectioning was accomplished by cutting a groove in the back side of the glass with a diamond saw then the specimens were lightly scribed at the edge of the top surface to cause full separation. Secondary electron images were taken of the back-cut fractured surface.

The scratch test was used to measure the visual lateral cracking threshold for ion-exchanged glass specimens using a series of different indenter tips. The visual lateral cracking threshold is defined in this work as the load which produces visual lateral cracks (typically accompanied by chips) that extend on either side of the scratch track to a length of greater than twice the scratch groove width. For each scratch load used, the test was repeated five times and the threshold criteria had to be met in a minimum of three out of the five scratches. Three different four-sided pyramidal indenter tips were used having angles between opposing faces of 120°, 136° (Vickers), and 150°. The four-sided pyramidal indenter tips were aligned so that a major diagonal of the indenter impression was aligned with the scratch groove. A Knoop indenter tip was also used with the wide angle across the edges of the indenter tip (172°-30') aligned with the scratch groove.

The median/radial indentation crack threshold was also measured for ion-exchanged and non-ion-exchanged samples of 2317 with the 120°, 136°, and 150° indenter tips. Indentation threshold was measured on an Instron using 250 gf, 1 kgf and 10 kgf load cells. The indentation load was applied and removed at 0.2 mm/min with the maximum load held for 10 seconds. The indentation cracking threshold is defined in this work as the indentation load at which more than 50% of 10 indents exhibit any number of radial/median cracks emanating from the corners of the indent impression. Measurements were made 24 hrs after indentation to account for delayed crack pop-in. For each tip, a cross-section of a 500 gf indent was made using the technique described by Hagan[9] in order to show the differences in deformation mechanism. All indentation was performed at 25°C in 50% relative humidity.

RESULTS

The scratch cracking behavior is different for ion-exchanged glasses when compared to non-strengthened glasses. Figure 2 shows the SEM cross-sectional image of a 50 gf Knoop scratch in ion-exchanged alkali aluminosilicate. This image clearly indicates that the first crack system to form in the ion-exchanged 2317 glass is a lateral crack. Figure 3 shows the cross-sectional image of a scratch made at 100 gf. The lateral cracks have intersected the surface to cause chipping. Prior to sectioning, the scratches did not have any chipping, however, the act of sectioning the glass propagated the subsurface lateral cracks to intersect with the surface. Median cracks are not present in either cross-section image. Similar cross-sectioning of Vickers scratches have shown the same behavior.

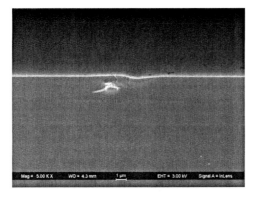

Figure 2. SEM image of the cross-section through a 50 gf Knoop scratch in ion-exchanged 2317 with CS = 770 MPa and DOL = 48 μm. A lateral crack forms prior to median crack formation.

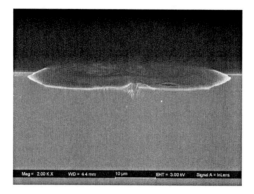

Figure 3. SEM image of the cross-section through a 100 gf Knoop scratch in ion-exchanged 2317 with CS = 770 MPa and DOL 48 μm. Sectioning propagated the lateral cracks to the surface.

Figure 4 shows a pre-threshold load scratch made at 25 gf and a post-threshold load scratch made at 50 gf with the sharpest 120° indenter tip. The less sharp 136° Vickers tip does not produce visual lateral cracking damage until higher loads are reached. Figure 5 shows a pre-threshold load scratch made at 50 gf and a post-threshold load made at 100 gf with the Vickers indenter tip. A Knoop tip has intermediate sharpness between a Vickers tip and the 150° tip as determined by the volume displaced at a given applied indentation load. Figure 6 shows a pre-threshold load scratch made at 300 gf and a post-threshold load made at 400 gf with the Knoop indenter tip. Figure 7 shows that the least sharp 150° tip requires loads exceeding 2000 gf to produce lateral cracking damage. Slight frictive damage can be seen in some of the pre-threshold scratches for the different tips, but this damage has low visual impact when viewed with the naked eye. These tests were repeated with glasses with several different ion-exchange conditions in the ranges of CS from 750 to 900 MPa and DOL from 40 to 60 μm and gave the same results.

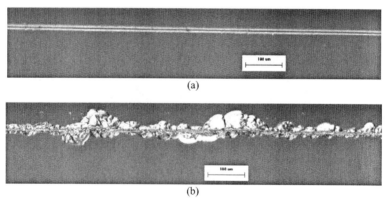

(a)

(b)

Figure 4. Scratches made on ion-exchanged 2317 with CS = 770 MPa and DOL 48 μm with the sharpest 120° 4-sided pyramidal tip at (a) 25 gf (pre-threshold) and (b) 50 gf (post-threshold).

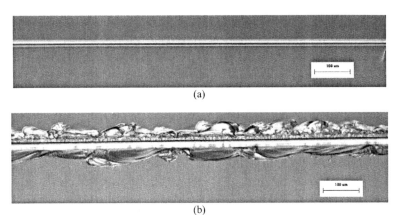

(a)

(b)

Figure 5. Scratches made on ion-exchanged 2317 with CS = 770 MPa and DOL 48 μm with the 136°
4-sided pyramidal tip (Vickers) at (a) 50 gf (pre-threshold) and (b) 100 gf (post-threshold).

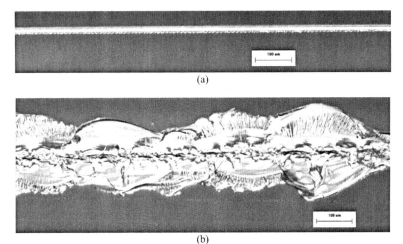

(a)

(b)

Figure 6. Scratches made on ion-exchanged 2317 with CS = 770 MPa and DOL 48 μm
with the Knoop tip at (a) 300 gf (pre-threshold) and (b) 400 gf (post-threshold).

Figure 7. Scratch made on ion-exchanged 2317 with CS = 770 MPa and DOL 48 μm with least sharp 150° 4-sided pyramidal tip at 2000 gf (pre-threshold).

The indentation median/radial crack initiation loads are given for non-ion-exchanged and ion-exchanged alkali aluminosilicate glass in Table I. Like the visual lateral cracking threshold in scratch tests performed on ion-exchanged glass, the indentation median/radial crack threshold on both non-ion-exchanged and ion-exchanged glasses is strongly dependent on the indenter tip geometry. The indentation cracking patterns at the cracking thresholds of the non-ion-exchanged glasses shown in Figure 8 demonstrate how the deformation mechanism changes towards anomalous behavior as the tip becomes less sharp. This is evident by the presence of ring cracks for the indent made at 8000 gf with the 150° tip. Images at the cracking threshold for the ion-exchanged glasses were omitted since the transition towards anomalous behavior could not observed as the threshold for the 150° tip exceeded the maximum of the 10 kgf load cell. The cross-sections of 500 gf indents made with each tip are shown in Figure 9 for the non-ion-exchanged glasses. Again, the transition from normal to anomalous behavior can be seen. The 120° tip produces the most plastic deformation as shown by the large amount of subsurface shear damage and by the long extension of cracks caused by the larger magnitude of residual stress. The 136° Vickers tip also produces significant subsurface shear damage, however, a lower residual stress is apparent by less crack extension. The 150° tip produces what appears to be deformation by densification. Subsurface shear damage is not visible by optical microscope.

Table I. Indentation Median/Radial Crack Initiation Load for Various Contact Geometries

Indenter tip	Non-ion-exchanged median/radial cracking threshold (gf)	Ion-exchanged median/radial cracking threshold (gf)
120°	15-30	50-100
136°	300-500	5000 - 7000
150°	7000 - 8000	>10,000

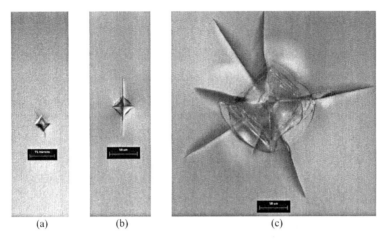

Figure 8. Cracking behavior in non-ion-exchanged 2317 at the Vicker indentation cracking threshold. (a) Normal cracking behavior at 30 gf with 120° tip. (b) Normal cracking behavior at 500 gf with 136° tip. (c) Mixed normal/anomalous cracking behavior at 8000 gf with 150° tip.

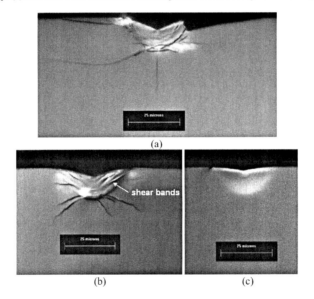

Figure 9. Cross-Section of 500gf indents in non-ion-exchanged 2317 with (a) 120° tip, (b) 136° Vickers tip, and (c) 150° tip. The transition from plastic deformation to densification is apparent.

DISCUSSION

The way in which crack systems evolve is shown to be different for glasses ion-exchanged to high compressive stress and deep compressive layer. Previous scratch experiments[2-7,10] on non-strengthened soda-lime glass have clearly shown that the median crack is the initial crack system originating in the subsurface of the glass. Images by Dick and Peter[2] and by Howes and Szameitat[10] show that the median crack is the only subsurface crack present at low scratch loads and lateral cracks will form upon increasing the load. Figures 2 and 3 clearly illustrates that the first subsurface crack systems to form in ion-exchanged 2317 glass are the lateral cracks. The crack initiates at a depth of a few microns at what appears to be the boundary of the deformation zone. At this depth the compressive stress is near its maximum and acts to prevent the initial flaw from forming a median crack. The initial flaw preferentially extends in the lateral direction where the compressive stress does not appear to influence the propagation. In glasses with low levels of compressive stress, such as thermally tempered soda-lime silicate, a median crack has been observed to form prior to lateral cracking systems. The same crack evolution has been seen in samples scratched with a Vickers indenter tip. Future work will involve viewing these cross-sections with very blunt and very sharp tips as well, but the same behavior is expected.

The scratch lateral cracking threshold is highly dependent on the geometry of indenter. The sharpest $120°$ indenter tip formed visual lateral cracks in the ion-exchanged 2317 glass at 50 gf while the least sharp $150°$ tip requires loads exceeding 2000 gf to form visual lateral cracks. The cracking threshold is highly dependent on the contact geometry since the deformation mechanism is also highly dependent on the contact geometry. The inelastic portion of indentation/scratch deformation for glasses occurs by a combination of densification and plastic flow.[11-13] Densification is not a volume conserving process and does not contribute to the volume strain around an indentation. It is produced by an interlocking of structural units during the application of a load and can be recovered by sub T_g heat-treatment.[12-14] Plastic flow is commonly thought of as a volume conserving process, however it has also been suggested that volume dilatation occurs along shear bands.[15,16] Plastic deformation results in a pile up of material at the glass surface surrounding the indent/scratch. Yoshida et. al. have demonstrated that wider angle contact geometries promote more densification whereas sharper contacts promote more plastic flow.[8] For normal glasses, a contact that results in deformation with more densification will be more resistant to crack initiation, whereas a contact that results in more plastic flow is more likely to produce crack systems originating in the subsurface at lower loads. The reason for this is two-fold. First, the deformation mechanism will determine the density of crack nuclei present. Shear deformation has been shown to produce a high density of shear micro-cracks within the deformation zone that nucleate strength limiting flaws.[9,17] Glasses that deform primarily by densification do not have such a high density of crack nuclei. The second reason that densification results in higher crack resistance is that for a given amount of deformation the residual stress is less in the region surrounding densified material as it is in the region surrounding material that has been displaced by plastic flow.[12,13,18] The work by Kato et al. clearly demonstrated that the amount of densification is related to the indentation crack resistance for many glasses.[19] At the extreme of very high levels of densification in glasses such as fused silica, the propensity to form ring cracks at moderate loads will negate the advantages of high levels of densification. In a scratch test, a high propensity to form ring cracks will be apparent in the form of "chatter marks".

The scratch visual lateral cracking resistance of the highly ion-exchanged glass increases as the indentation median/radial cracking threshold increases since the mechanism for crack initiation in the subsurface, whether by single point or sliding indentation, is similar. In Table I, it is shown that each change in indenter angle from $120°$ to $136°$ to $150°$ results in an order of magnitude change in the indentation median/radial crack threshold of non-ion-exchanged glasses. When indenting with the $150°$ tip the median/radial indentation cracking threshold could not be determined for ion-exchanged glasses due to equipment limitations. Figure 6 shows the cracking patterns observed in the non-ion-exchanged glasses for each of the indenter tip used. The cracking patterns shown for indents made with the 120° and 136° indenter tips are considered typical of "normal" glass. This means that the deformation proceeded with a high amount of shear deformation and produced median/radial cracks without any ring or cone cracking present. The cracking pattern shown for the indent made at 8000 gf with the 150° tip indicates that the deformation mechanism has at least partially changed. Median/radial cracks are still present, but are now accompanied by a cone crack. This feature is typical of an "anomalous" glass. Cone cracks form in highly densifiable glasses such as fused silica during indentation with a Vickers tip. The subsurface deformation images shown in Figure 9 also demonstrate that the deformation mechanism transitions from highly plastic to highly densifiable as the contact becomes blunter. These observations support recent works by Yoshida et al.[8] showing that wider angle contacts promote more densification and by Kato et al.[19] showing that increasing the amount of densification will increase the indentation crack resistance. Kato et al. modified his compositions to achieve greater densification and crack resistance when indenting with a single indenter geometry, while in this work, it is shown that for this particular glass-type (2317), the crack resistance is increased as blunter contacts produce more densification. However, there must be a limit to which densification is useful in providing crack resistance. Again, consider fused silica that is known to deform almost entirely by densification during relatively sharp Vickers indentation. This glass cannot be considered damage resistant since it readily forms ring cracks at low Vickers indentation loads typically less than 200 gf. A high amount of densification produces large tensile stresses at the surface surrounding the area of contact during application of the indentation load and produces these ring cracks. While increasing the amount of densification at the expense the shear deformation, it is expected that the density of the subsurface crack nuclei and the resulting residual stress will decrease, but at the same time the propensity to form ring cracks is increasing as the contact tensile stresses increase. It is then suggested that some minimum amount of shear deformation must be required to reduce the propensity towards ring crack formation for a given contact.

The scratch visual cracking threshold has been examined for ion-exchanged 2317 glass samples with a relatively wide range of high compressive stresses and deep depths of layer and results indicate that the scratch load required to initiate visual lateral cracking damage is independent of the ion-exchange level. This is only true if the ion-exchange is sufficient to prevent the formation of the median crack prior to lateral crack formation. Lateral cracks extend parallel to the surface where the ion-exchange compressive stress does not act against their extension. The lateral cracking threshold in ion-exchanged 2317 glass appears to be highly dependent on the load required to produce subsurface damage sites, i.e. starter cracks within that deformation region, that are driven to extension by the residual stress generated along the length of the scratch. The extension of the lateral cracks at low applied scratch loads is likely due to the large magnitude of residual stress generated along the length of the scratch. This is analogous to the reduced load required for indentation cracking when the deformation regions of side-by-side indents overlap.

CONCLUSION

It was demonstrated that the scratch cracking behavior for highly ion-exchanged glasses differs from non-strengthened glasses by the preferential formation of lateral cracks prior to the formation of the median crack. At ion-exchange levels that prevent median crack formation prior to the formation of lateral cracking systems, the scratch visual lateral cracking threshold appears independent of compressive stress profile. The scratch lateral cracking threshold is highly dependent on the indenter geometry and increases as the indenter tip becomes less sharp. The scratch crack resistance increases as the deformation mechanism moves towards densification and away from shear.

ACKNOWLEDGEMENT

I would like to thank John Bartoo for SEM images of scratch cross-sections.

REFERENCES

[1]J. Price, G. Glaesemann, D. Clark, T. Gross, K. Barefoot, A Mechanics Framework for Ion-Exchanged Cover Glass with a Deep Compression Layer, *SID Symposium Digest*, **40**, 1049-51 (2009).
[2]E. Dick and K. Peter, Generation of Deep Cracks in Glass, *J. Am. Ceram. Soc.*, **52**, 338-339 (1969).
[3]M. Swain, Microfracture about Scratches in Brittle Solids, *Proc. R. Soc. Lond. A.*, **366**, 575-97 (1979).
[4]V. Bulsara, S. Chandrasekar, Direct Observation of Contact Damage Around Scratches in Brittle Solids, *Proc. SPIE*, **3060**, 76-88 (1997).
[5]Y. Ahn, N. Cho, S-H. Lee, D. Lee, Lateral Crack in Abrasive Wear of Brittle Materials, *JSME International Journal Series A*, **46**, 140-44 (2003).
[6]Y. Ahn, T. Farris, S. Chandrasekar, Sliding Indentation Fracture of Brittle Materials: Role of Elastic Stress Fields, *Mechanics of Materials*, **29**, 143-52 (1998).
[7]V. Le Hourou, J-C. Sanglboeuf, S. Deriano, T. Rouxel, G. Duisit, Surface Damage in Soda-Lime-Silica Glasses : Indentation Scratch Behavior, *J. Non. Cryst. Solids*, **316**, 54-63 (2003).
[8]S. Yoshida, H. Sawasato, T. Sugawara, Y. Miura, J. Matsuoka, Effects of the Indenter Shape on the Indentation-Induced Densification of Soda-Lime Glass, *J. Mater. Res.*, **25**, 2203-11 (2010).
[9]J. Hagan, Shear Deformation Under Pyramidal Indentations in Soda-Lime Glass, *J. Mater. Sci.*, **15**, 1417-24 (1980).
[10]V. Howes, A. Szameitat, Morphology of Sub-Surface Cracks Below the Scratch Produced on Soda-Lime Glass by a Disc Cutter, *J. Mater. Sci. Letters*, **3**, 872-74 (1984).
[11]K. Peter, Densification and Flow Phenomena of Glass in Indentation Experiments, *J. Non. Cryst. Solids*, **5**, 103-15 (1970).
[12]S. Yoshida, J-C. Sangleboeuf, T. Rouxel, Quantitative Evaluation of Indentation-Induced Densification in Glass, *J. Mater. Res.*, **20**, 3404-12 (2005).
[13]S. Yoshida, J-C. Sangleboeuf, T. Rouxel, Indentation-Induced Densification of Soda-Lime Silicate Glass, *Int. J. Mater. Res.*, **98**, 360-364(2007).
[14]J. Mackenzie, High-Pressure Effects on Oxide Glasses: I, Densification in Rigid State, *J. Am. Ceram. Soc.*, **46**, 461-76 (1963).
[15]J. Gilman, Mechanical Behavior of Metallic Glass, *J. Appl. Phys.*, **46**, 1625-33 (1975).
[16]T. Gross, M. Tomozawa, Indentation-Induced Microhardness Change in Glasses: Possible Fictive Temperature Increase by Plastic Deformation, *J. Non-Cryst. Solids*, **354**, 4056-4062 (2008).
[17]J. Hagan and M. Swain, The Origin of Median and Lateral Cracks Around Plastic Indents in Brittle Materials, *J. Phys.*, **11**, 2091-102(1978).
[18]T. Rouxel, H. Shang, J.C. Sangleboeuf, The Driving Force for Indentation Cracking in Glasses, *International Congress on Fracture 11* (2005).
[19]Y. Kato, H. Yamazaki, S. Yamamoto, S. Yoshida, J. Matsuoka, Effect of Densification on Crack Initiation Under Vickers Indenter Test, *J. Non-Cryst. Solids*, **356**, 1768-73 (2010).

FRACTOGRAPHY LESSONS FROM KNAPPING

Are Tsirk
Fractography Consultant
Upper Montclair, NJ, USA

ABSTRACT
    Knapping entails the production of fractures by flaking. The crack propagation is almost always subcritical.* The great variety of fractures in knapping is not encountered in laboratory research nor in industrial applications. Brief references are made to a number of fracture markings and other fracture features observed, along with suggested explanations for their formation. The objective is to show that investigation of fractures in knapping has much to contribute to the understandings in fractography.

INTRODUCTION

    Flintknapping, or simply knapping, refers to the manufacture of artifacts from any lithic material, including flint and obsidian, by flaking.[1,2,3] Lithic analysis in archaeology deals primarily with the interpretation of stone tools – their use and manufacture by knapping – as well as other lithic artifacts.

    Applications of fractography in the study of lithic technology rely to a great extent on intuitive understanding of fractures. This understanding has been achieved by dealing with a great number of fracture specimens and from hands-on experience in knapping. Most archaeologists studying stone tools nowadays have knapped to some extent. With rare exceptions (for example, Speth[4] and Pelcin[5]), the experiments in flintknapping are not experiments in the sense of controlled laboratory experiments.[6] Nevertheless, flintknapping experiments (for example, Crabtree[7] and Pelegrin[8]) have provided much of today's understanding of knapping. Because of the complexities involved, theoretical efforts in fracture mechanics analysis of flintknapping have been extremely rare, with results that are understandably limited.[9,10,11] An intuitive understanding of fractures was obviously achieved already by prehistoric knappers. It is reasonable to say that the use of fractography dates back at least 2.5 million years.

    A number of fracture markings and other fracture features encountered in knapping are considered. Several of these are new to fractography literature. Such markings and features are discussed briefly, with suggested explanations for their formation. Some of the features are surprising and difficult to explain. The article is based on the writer's practice of knapping over a period of 40 years, and his communications with others on knapping, lithic technology and fractography.

    In my communications with Prof. Fréchette over many years, he indicated that the great variety of fractures in knapping is not encountered in laboratory research and industrial applications. He noted that the investigation of such irregular fractures supplements the understandings in fractography obtained from other research. The objective here is to show that this is indeed the case. Therefore references are briefly made also to fracture markings and other features observed in knapping that already appear in the literature on knapping.

FRACTOGRAPHY APPLICATIONS IN INDUSTRY AND LITHIC TECHNOLOGY

    Application of fractography for industry and lithic technology both rely on theoretical and experimental research as well as intuitive understandings from experience. Limited theoretical background is more evident in lithic technology applications. They rely more on vastly greater intuitive understanding. Lithic analysts often deal with over 10,000 fracture specimens from a single archaeological site, and a master knapper produces over a million fractures in his lifetime.

Fractures in knapping involve a great variety of specimen geometries, loadings and boundary conditions, as well as material properties and flaw characteristics. In contrast to industrial concerns, production rather than prevention of fractures is of primary interest in knapping (Figure 1). But prevention of fractures is also of interest, to avoid accidental breakages.[12]

Fractures in knapping are essentially the products of subcritical crack growth. Two primary reasons are seen for this. First of all, a knapper tries to use minimum forces to detach a flake. Secondly, the loading used is self-relieving. That is, it is gradually reduced as the fracture to detach a flake advances.

FRACTURE FEATURES

The cantilever curls (known as compression lips to knappers) (Figure 2c) manifested along the compressive edge, together with mist-hackle along the tensile edge, are indications that bending was associated with the break. On accidental breakage of blades in knapping, the variations in a lip length in the transverse and longitudinal directions were observed to relate to torsion and axial effects, respectively. No quantitative correlations were made.

The cantilever curls and the mist along the tensile edge, as well as other evidence, can be used as indicators of the direction for the fracturing at a break. In accidental transverse breakage of blades and other flakes, it has been observed that the fracture for the break usually progressed from the outer to the inner surface, even for straight blades. This serves as conclusive evidence that the detachment of the flakes involved double rather than single curvature bending (Figure 3).[13]

Instead of having a cantilever curl, a compression wedge (Figure 2d) is sometimes released near the compression edge of a bending break. The associated branching occurs at relatively low fracture velocity V_F, with no mist. The branching can be understood upon reflection of the divergent compressive stress trajectories ahead of the breakage crack as it approaches the compressive face. The divergence of these stress trajectories is apparent by visualizing the tensile stress trajectories curving to run around the crack tip.

The characteristic features on an accidental transverse break of a biface are shown schematically in Figure 4. Beyond the usual mist and hackle, there are often unsuccessful attempts of macroscopic branching, referred to as incipient macrobranching in the figure. Such branching attempts are unsuccessful because the branching crack extending towards the compressive side was forced to stop. Short or long bifurcation cracks can often be seen only on the tensile face of the specimen. Such cracks are easy to detect on light-colored flint but very difficult to see in black obsidian, for example. Beyond or in lieu of these incipient macrobranching features, the branching cracks sometimes do successfully extend to the compressive face, occasionally releasing a lateral wedge by the side of the specimen (Figure 4). These, of course, are associated with high V_F in contrast to the low V_F for the compression wedges. Such lateral wedges are sometimes also encountered with transverse breakage of blades and other flakes. On accidental transverse breakage of blades, fracture features resembling bowties are occasionally encountered that may consist of a compression wedge together with two lateral wedges.

In Figure 2, schematic illustrations of a compression lip (cantilever curl), a compression wedge and two other related fracture features are shown. The latter two features are most frequently encountered in blade production. The stepout fracture in part (b) resembles a transverse bending break in the presence of significant axial compression with the compression lip extended due to the compression. However, with a stepout fracture, the primary fracture usually extends beyond the stepout location. A popout fracture as in part (a) is formed when a fracture that starts out as a stepout fracture turns back toward the inner surface. Two drastic variations of popout fractures are depicted in Figure 5, and some variations observed are shown schematically in Figure 6. It is curious to note that

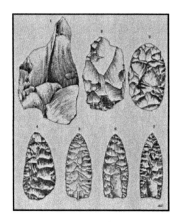

Figure 1. Reduction of a flake to a point (9.4 cm long). Heat-treated Belton chert. (From Callahan[12])

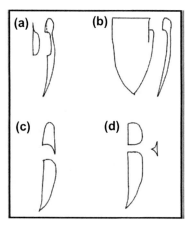

Figure 2. Popout and related fractures: (a) Popout fracture, (b) Stepout fracture, (c) Compression lip, (d) Compression wedge. Adapted from Reference[15].

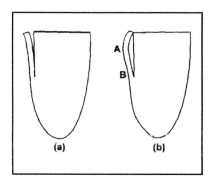

Figure 3. Blade detachment with singe (a) and double (b) curvature bending. Adapted from Reference.[13]

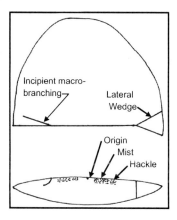

Figure 4. Characteristic features on a broken biface (schematic): Origin, mist, hackle, and lateral wedge from macro- branching. Incipient macrobranching often seen as cracks on the tensile surface and as partial penetration cracking on the section.

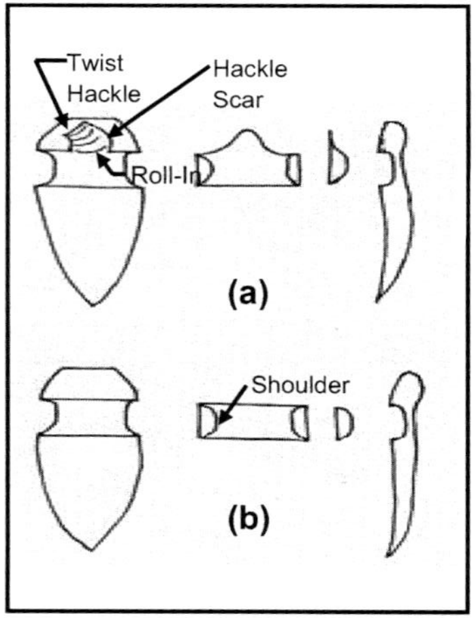

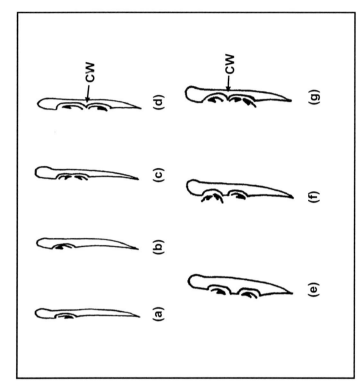

Figure 5. Regular popout fractures with (a) and without (b) roll-in from a hackle scar. Adapted from Reference.[15]

Figure 6. Schematic profiles and fracture directions for some popout fractures observed. Regular (a), reverse (b) and compound (c) popouts have cracking in the distal, proximal and both of these directions, respectively. The popouts in (d) and (g), but not in (e) and (f), were initiated at a compression wedge (CW).

the kinds of popout fractures shown in Figure 5b can be associated with "forward" or "backward" propagation of the secondary fracturing for the popout feature (Figure 6). It is of interest to compare the fracture features in Figures 2 and 6 with those in Figures 4.10 and 4.11 in Quinn.[14] The formation of popout and related fractures is considered in Tsirk.[15]

FRACTURE MARKINGS

The frequently encountered irregularities in flake geometry provide an abundant opportunity to observe the effects of geometry on crack front configuration and other fracture surface features. A variation in the flake thickness in the transverse direction is reflected on the flake inner surface by the Wallner lines and other ripples tending to lead at the thicker and lag at the thinner flake portions[16]. This is often seen near the distal parts of flakes with two or more ridges on their outer surfaces (Figure 7). Ripples concave in the fracture direction are often seen in such cases. A fracture front also tends to lead at the thicker and lag at the thinner side of a flake. The effects observed in the above cases can be understood by the fact that more of the strain energy is stored and released by the thicker portions (See also Tsirk[16]). Related to the above observation is also the manifestation of twist hackles at lagging thinner portions of a flake earlier because the lagging contributes to participation of fracture mode III.

Many other geometrical effects of thickness variations in the transverse as well as the longitudinal flake direction are reflected by the fracture markings. Thus in lithic technology it is possible to see evidence of the flake geometrical variations on the negative flake scar on a core. Another way of stating the above observations is that the geometrical irregularities of the flake outer surface can be reflected on its inner surface. The detachment of a relatively thin flake from an obsidian nodule with many irregularities at its outer surface (a.k.a. cortex) can lead to the manifestation of a rather irregular or "ruffled" inner flake surface (Figure 8) as well as the negative flake scar. These "primary ruffles" on the negative flake scar can in turn lead to "secondary ruffles" on the inner surface of a subsequent flake.

Wallner lines concave in the fracture direction can be manifested due to a number of reasons in addition to geometric effects. In a knapping context, a "Wallner wake" was often observed due to a stress pulse source upstream of a fracture front.[17] Not surprisingly, such downstream concavity of Wallner lines has also been seen in knapping contexts with localized variations in fracture toughness and residual stress. At a finer scale, partial dry and wet fracture fronts can also lead to Wallner lines concave downstream (Figure 9).

On accidental breakages in knapping obsidian, localized enhancements of mist-hackle manifestation have been observed that are analogous to Wallner lines.[18] Numerous observations of such Wallner mist-hackle configurations in knapping enabled the distinguishing of the cases where the responsible stress pulse originated at the fracture front from those where it originated behind it. Since a stress pulse can lead to mist-hackle enhancement, it is expected that an opposite effect can exist due to a highly localized obstruction to fracturing, manifested by patterns of localized suppression of mist hackle. It is thought that the manifestation of such effects was indeed observed, but with difficulties.[18]

Fracture parabolas have also been referred to as double tails, conic lines, and Kies figures. These markings in glass have usually been shown at rather high magnification, as at 890X in Figure 7.10 of Kerkhof.[19] On accidental breakages and on one controlled flake surface in knapping obsidian, very large such markings have been seen.[18] An explanation has been offered for glass that is similar to that for PMMA. That is, they are said to form when a new crack originates ahead of the fracture front.[20] Though this may perhaps sometimes be the case, for many of the large "fracture parabolas" observed in knapping obsidian, there is another mechanism. They are formed with the fracture shifting to another plane as it passes through an inclusion.[17]

Figure 7. Wavy ripples on obsidian flakes due to thickness
variation in transverse direction; the black and white bars are 1 cm long.

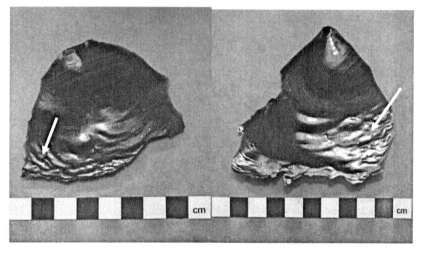

Figure 8. Ruffles (arrows) on obsidian flakes resembling ripples in a disarray.
The black and white bars are 1 cm long.

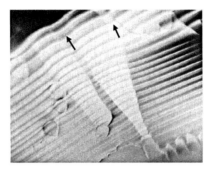

Figure 9. Various liquid-induced fracture markings,[24] due to the presence or absence of water at the partial fracture fronts, are seen on the obsidian pressure flake produced with sonic modulation at 183 Hz. The fracture is propagating upwards slowly (ca 2 to 4 cm/s). The dry parts (arrows) are lagging behind the adjacent wetted parts, as seen by the ripples concave in the fracture direction; the image width is ~2.8 mm.

Some observed contexts for mist lines, twist hackles and localized mist-hackle manifestations posed tantalizing questions. In at least one case, a twist hackle was seen to revert to a mist line or a mist-hackle line and then back again to a twist hackle. In most cases, an arm of a fracture parabola consists of a mist line or a mist-hackle line, but in some cases it consists instead of a linear series of twist hackles. Related to the above observations are the gradual changes occurring in the tail features as they get laterally closer to a general mist region. Sufficiently far from the mist region the usual tail or wake hackle resembling a twist hackle is manifested. Closer to the mist region, such a tail is seen to have a trailing mist line. Closer yet, the tail is seen to consist entirely of a mist line or a mist-hackle line. Some of these trends are seen in Figure 10. The presence of mist lines may be viewed as a precursor to general mist manifestation. All of the above observations suggest some similarity or "equivalence" in at least some aspects of the causal mechanisms for mist lines and twist hackles. In particular, a mist line can be suggested to accommodate the fracture mode III contributions piece-wise while a twist hackle (or a wake hackle) does so locally in a single step. The term "mist line" or "mist-hackle line" is used here for the linear feature of mist and hackle. The terms "streak"[21] and "hackle streak"[22,23] have also been used for it. A number of the observations and ideas discussed by Beauchamp[22,23] are consistent with those also observed in knapping, discussed above.

It is difficult or impossible to recognize many LIFMs (liquid-induced fracture markings) as such based solely on those observed in the laboratory. The many LIFMs observed in the context of knapping or use-chipping are useful for overcoming this difficulty.[24] Although LIFMs have been observed in single crystals,[14,25] it is of interest to ask whether LIFMs can be observed in polycrystalline materials. Because of their subtle nature, it is unlikely that the usual scarps manifested as ridges or "escarpments" will be observed in today's crystalline materials. Some scarps manifested as hackles, termed "hackle scarps,"[18] are not at all subtle. Observation of hackle scarps is expected in some polycrystalline materials.

Because of material irregularities, material interface markings are often seen in knapping. At a material interface oblique to the fracture direction, a ridge or hackles in the direction of the interface are often seen. These are termed here a "material interface ridge" and "material interface hackle". In terms of stresses, the formation of these is analogous to that of a scarp and a hackle scarp, respectively.[18] At an interface more or less parallel to a fracture front, a ridge termed here a "material transition ridge" can be manifested. In terms of stresses, its formation is suggested to be analogous to that of a cavitation scarp.

When a flake is split lengthwise during its detachment, fracture markings are often left on the negative flake scar[26] as evidence of the splitting. To be indicative of their respective appearance, these split marks have been referred to as a split ridge, a split step and split ripples (Figure 11).

The formation of twist hackles has been discussed by Kerkhof and Müller-Beck[27], Sommer[28] and Fréchette.[20] When a lateral breakthrough associated with their formation occurs to both sides, a sliver or a "lance" is usually released. But sometimes a breakthrough to one side is rather long. More generally, the terms "hackle scar" and "hackle flake" are used here in reference to a secondary fracture due to extended fracture propagation associated with the lateral breakthrough of a twist hackle. These features are usually associated with locally higher contours, as the bulbar scar or éraillure scar on the bulb of a flake (Whittaker[2]: Figure 2.3 and Faulkner [29]). A rather long hackle scar is shown in Figure 12. In another case, a 6.3 cm long hackle flake was seen by a twist hackle only 1.3 cm long. A fracture lance may be viewed as a very short hackle flake. When the fracture surfaces separate as the primary fracture extends, a lateral breakthrough must occur.[20] But why is there extended fracture propagation of the lateral breakthrough? How is that possible? One would expect the fracture path for the secondary fracture furthest from the unbroken ligament of the twist hackle to turn towards the free surface (that is, the inner flake surface). With the usual fracture lances, it does so. But with the longer hackle flakes the fracture path of the secondary fracture remains more or less parallel to the inner flake surface. For this to be possible, it is suggested that compression must be present in the direction of the hackle flake. It is not clear how such compression is developed in all the cases of interest.

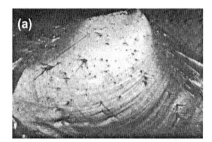 

Figure 10. Gradual transition of wake hackles to mist lines to general mist region in obsidian. Same view in (a) and (b) except for illumination. Field width 2.8 mm.

Figure 11. A "split ridge" (arrow) on a negative flake scar. Normanskill chert.

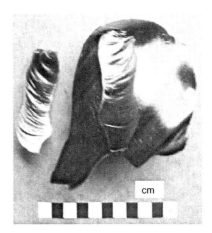

Figure 12. A long hackle flake and its scar manifested on an obsidian flake during its detachment by direct percussion.

CONCLUSION

Investigation of fractures in knapping has indeed provided much information that supplements the understandings in fractography. Future research by fractographers of this intellectually rewarding avenue is expected to yield further results useful for other applications.

FOOTNOTE

As used here, subcritical crack propagation is characterized by crack propagation at well below the terminal velocity.

ACKNOWLEDGEMENTS

I owe my greatest debt to the late Prof. V.D. Fréchette for the many enlightening and stimulating discussions. I am also grateful to Stephen Freiman for a number of discussions, helping to introduce me to fractography. I thank Jack Cresson for his comments on parts of this article, and an anonymous reviewer for useful clarifications. For my interests in the production of controlled fractures, I gratefully acknowledge my indebtedness to the late Don Crabtree, the pioneer in the art and science of knapping.

REFERENCES

[1]D.E. Crabtree, An Introduction to Flintworking, *Occasional Papers of the Idaho Museum of Natural History* 28, Pocatello, ID (1972).
[2]J.C. Whittaker, *Flintknapping: Making & Understanding Stone Tools*, University of Texas Press, Austin (1994).
[3]M.-L. Inizan, M. Reduron-Ballinger, H. Roche and J. Tixier, Technology and Terminology of Knapped Stone, *Préhistoire de la Pierre Taillée*, Volume 5, C.R.E.P., Nanterre (1999).
[4]J. D. Speth, Experimental Investigation of Hard-Hammer Percussion Flaking, *Tebiwa*, 17, 7-36 (1974).
[5]A. W. Pelcin, Controlled Experiments in the Production of Flake Attributes, Ph.D. Dissertation, Department of Anthropology, Univ. of Pennsylvania (1996).

[6]L. Lewis Johnson, A History of Flint-Knapping Experimentation, 1838-1976, *Current Anthropology*, 19 (2) 337-72.

[7]D.E. Crabtree, Mesoamerican Polyhedral Cores and Prismatic Blades, *Am. Antiquity*, 33, 446-78 (1968).

[8]J. Pelegrin, Blade-Making Techniques from the Old World: Insights and Applications to Mesoamerican Obsidian Lithic Industry, pp. 55-71 in *Mesoamerican Lithic Technology: Experimentation and Interpretation*, K.G. Hirth, ed., Univ. Utah Press, Salt Lake City (2003).

[9]J.G. Fonseca, J.D. Eshelby and C. Atkinson, The Fracture Mechanics of Flint-Knapping and Allied Processes, *Int. J. Fract.*, 7, 421-33 (1971).

[10]B.A. Bilby, Tewksbury Lecture: Putting Fracture to Work, *J. Mat. Sci.*, 15 (3) 535-56.

[11]B. Cotterell, J. Kamminga and J. Dickson, The Essential Mechanics of Conchoidal Flaking, *Int. J. Fract.*, 29, 205-221 (1985).

[12]E. Callahan, *The Basics of Biface Knapping in the Easter Fluted Point Tradition – A Manual for Flintknappers and Lithic Analysts*, 3rd ed., Piltdown Productions, Lynchburg, VA (1996).

[13]A. Tsirk, A Knapping Dilemma with Transverse Blade Breakage, *Lithic Technology*, **34**, 111-18 (2009).

[14]G.D. Quinn, *Fractography of Ceramics and Glasses*, NIST Recommended Practice Guide, Special Publication 960-16, (2007).

[15]A. Tsirk, Popouts and Related Fractures, *Lithic Technology*, 35, 149-70 (2010).

[16]A. Tsirk, On a Geometrical Effect on Crack Front Configuration, *Int. J. Fract.*, 17, R183-8 (1981).

[17]A. Tsirk, Formation and Utility of a Class of Anomalous Wallner Lines on Obsidian, pp. 57-69 in *Fractography of Glasses and Ceramics*, V.D. Fréchette and J. R. Varner, eds., Am. Ceram. Soc., Westerville, OH (1988).

[18]A. Tsirk, Hackles Revisited, pp. 447-72 in *Fractography of Glasses and Ceramics III*, J. R. Varner, V.D. Fréchette and G. D. Quinn, eds., Am. Cer. Soc., Westerville, OH (1996).

[19]F. Kerkhof, *Bruchvorgänge in Gläsern*, Deutsche Glastechnishe Gesellschaft, Frankfurt (1970).

[20]V.D. Fréchette, Failure Analysis of Brittle Materials, *Advances in Ceramics*, Volume 28, Am. Ceram. Soc., Westerville, OH (1970).

[21]E. Poncelet, The Markings on Fracture Surfaces, *Trans. Soc. Glass Technology*, 42, 279-88 (1958).

[22]E.K. Beauchamp, Crack Front Stability and Hackle Formation in High Velocity Glass Fracture, *J. Am. Ceram. Soc.*, 78 (3), 689-97, (1995).

[23]E.K. Beauchamp, Mechanisms for Hackle Formation and Crack Branching, pp. 409-45 in *Fractography of Glasses and Ceramics III*, J. R. Varner, V.D. Fréchette and G. D. Quinn, eds., Am. Ceram. Soc., Westerville, OH (1996).

[24]A. Tsirk, Liquid-Induced Fracture Markings: An Overview, pp.79-91 in *Fractography of Glasses and Ceramics V*, J. R. Varner, G. D. Quinn and M. Wightman, eds., Am. Ceram. Soc., John Wiley & Sons, Hoboken, NJ (2007).

[25]E.K. Beauchamp, Personal communication (2002).

[26]A. Tsirk, Fracture Markings from Flake Splitting, *J. Arch. Sci.*, 37, 2061-5 (2010).

[27]F. Kerkhof and H.-J. Müller-Beck, Zur Bruchmechanischen Deutung der Schlagmarken an Steingeräten, *Glast. Ber.* 42, 439-48 (1969).

[28]E. Sommer, Formation of Fracture 'Lances' in Glass, Eng. Fract. Mechs, 1,539-46 (1969).

[29]A. Faulkner, Mechanics of Eraillure Formation, *Lithic Technology* 2, 4-12 (1973).

FAILURE AND DAMAGE MECHANISMS IN CERAMIC NANOCOMPOSITES

Ján Dusza
Institute of Materials Research of SAS
Košice, Slovakia

ABSTRACT

In-situ reinforced Si_3N_4-SiC micro/nanocomposites and carbon nanofibers reinforced zirconia ceramics have been developed and investigated. In hot pressed Si_3N_4-SiC systems the influence of the rare-earth oxide additives (La_2O_3, Nd_2O_3, Y_2O_3, Yb_2O_3 and Lu_2O_3) and the addition of SiC nanoparticles on the microstructure, mechanical properties and failure and damage mechanisms have been studied. In hot pressed and spark plasma sintered zirconia/carbon nanofiber composites, prepared by addition of 2.0 and 3.3 vol.% carbon nanofibers (CNFs) the effect of the sintering route and the carbon nanofiber additions on the microstructure and fracture/mechanical properties was investigated. In both cases the properties of composites have been compared to that of the monolithic material. Fracture toughness, four-point bending strength and tribological characteristics have been investigated. Micro and macro-fractography was applied for the study of fracture and damage mechanisms.

In Si_3N_4 based ceramics the additives with smaller size rare-earth cations influenced positively the fracture toughness and wear characteristics. However, the positive influence of finer microstructure of the composites on the strength was not observed due to the present strength degrading processing flaws in their microstructures in the form of agglomerates of SiC and non-reacted carbon areas.

In ZrO_2-CNF composites potential toughening mechanisms in the form of crack deflection and pull-out have been identified, which however were not effective enough, due to the cluster formation of the CNFs. The addition of a low amount of CNFs into ZrO_2 significantly decreased the friction due to the lubrication effect by carbon phases.

INTRODUCTION

Nanoscience and nanotechnologies are new approaches to research and development involving precise manipulation and control of atoms and molecules and creating new structures with unique properties. The goal is to produce new materials, devices, and systems tailored to meet the needs of a growing range of commercial, scientific, engineering, and biomedical applications opening new markets and providing dramatic benefits in product performance.[1]

Advanced ceramic materials have a special position in the recently very attractive and intensively developing science and research of the "nano-world." During the last decades a growing need has been observed for structural and functional materials that can be used as parts and tools for different combinations of load at temperatures up to 1600°C in oxidizing and aggressive environments. Such materials should exhibit high hardness and wear resistance, high melting temperature, oxidation resistance, chemical stability, and high resistance to creep deformation, thermal shock resistance, and sufficient levels of strength, fracture toughness and multifunctional character. Advanced ceramics fulfill most of these requirements and are already used or are potential materials for many room and high temperature applications. Their industrial applications include high temperature filters, igniters, lances for liquid metals and glasses, high-temperature heat exchangers, power generator components, as well as furnace elements and components.[2,3] Automotive applications involve turbocharger rotors, glow plugs, valves and so forth. Aerospace applications include turbine engine hot-section components such as blades, combustors, seals, and others. The use of advanced ceramics can minimize, for example, turbine air-cooling, which is commonly required for super-alloys, and also reduce the number of engine parts, which simplifies the design and reduces the engine weight. Furthermore, with higher operating temperature capabilities, more efficient fuel combustion and beneficial environmental effects are expected. Although structural ceramics have a number of excellent properties, their wider application is still limited by their brittleness, low flaw tolerance, low reliability and their limited functional properties. Different approaches have been used during the last two decades with the aim to improve their room and high-temperature properties, reliability, lifetime, and to prepare multifunctional ceramics.[4]

Nanocomposite ceramics can be defined as composites of more than one solid phase where at least one of the phases has nanometric dimensions. Niihara proposed the concept of ceramic nanocomposites in 1991 on the basis of the results achieved during the study of the Si_3N_4-SiC and Al_2O_3-SiC systems.[5] During recent years these systems were developed intensively using different processing steps.[6,7] In these systems second phase nanoparticles are located inter or intragranularly and can positively influence the room and high-temperature properties. The discovery of carbon nanotubes (CNTs) has generated considerable interest thanks to their small size, high aspect ratio, low mass and excellent mechanical, electrical and thermal properties.[8] In the last few years new ceramic/carbon nanotube and ceramic/carbon nanofiber composites have been developed and a number of authors have reported improved mechanical and functional properties in the case of these composites compared to the monolithic material.[7,9-11]

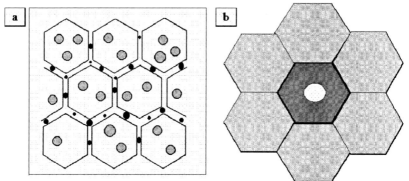

Figure 1. Schematic illustration of inter/intra particle reinforced nanocomposite
(a) and ceramic – nanofiber micro/nanocomposite (b)[5]

The aim of the present contribution is to study some mechanical properties and fracture and damage mechanisms in Si_3N_4 + SiC and ZrO_2 + carbon nanofibers micro/nanocomposites.

EXPERIMENTAL MATERIALS AND METHODS

The starting mixtures of the set of six Si_3N_4/SiC nanocomposites consisted of the following powders: α–Si_3N_4, carbon black and different rare–earth oxides RE_2O_3 (RE = La, Sm, Y, Yb, or Lu). All compositions contained the same atomic amount of RE element. The amount of SiO_2 and C was calculated to achieve 5 vol.% of SiC after in situ carbothermal reduction process. Simultaneously, the set of six reference monolithic Si_3N_4 materials with the same composition of sintering additives were prepared in order to compare the microstructure and mechanical properties of the composites and monoliths. Bulk bodies of 65 mm x 65 mm x 5 mm were then hot–pressed at 1750°C with a load of 30 MPa and 0.15 MPa pressure of nitrogen for 1 hour.

ZrO_2 + carbon nanofibers was prepared using 3 mol% Y-TZP zirconia, which is well characterized with good mechanical properties. Carbon nanofibers grade HTF150FF were used. The CNFs were dispersed by milling in Millipore water with dodecylbenzenesulfonic acid (DBSA) as a dispersing agent which was pre-mixed in the water before adding the CNFs. After adding the CNFs the mixture was ultrasonically dispersed for 10 minutes using a probe. Separately Y-TZP powder was also ball milled in Millipore water. The CNFs solution was added to the ceramic slip, and the mixture was then given a further ultrasonic treatment. This slip was continuously stirred and spray dried in a small laboratory spray dryer. Composite powders were prepared with 2.0 and 3.3 vol.% CNF content. The resultant powder granulates were die pressed into 20 mm diameter discs for hot pressing. These samples were hot pressed in an argon atmosphere at a dwell temperature of 1300°C for 30 minutes at a pressure of 41 MPa. For comparison monolithic ZrO_2 was prepared under similar conditions, e.g. hot pressed at 1300°C for 30 minutes at a pressure of 41 MPa. For SPS, discs of 20 mm and 30 mm diameter were prepared. The SPS samples were sintered at different dwell temperatures. The samples with 2.0 vol.% CNFs were sintered at 1400°C and the samples with 3.3 vol. % CNFs were sintered at 1500°C in both cases with a 5 minute dwell time and at 60 MPa.

    ZrO_2 specimens for microstructure examination were prepared by diamond cutting, grinding, polishing and thermal etching at a temperature of 1250°C in air and carbon coated before the examination. The average grain size of the zirconia matrix grains of the monolithic and composites were measured using the standard line intercept technique from SEM micrographs of thermally etched polished surfaces. For TEM, thin foils were prepared using a focused ion beam (Quanta 3D system). Microfractography was used to analyse the fracture lines and surfaces of the specimens to study the fracture micromechanisms in the monolithic material and in the composites. Analysis of microstructure has been realized after plasma and thermal etching of the polished surface of the specimens (perpendicular to the hot pressing direction) using scanning electron microscopy (SEM). Grain diameter (width of the elongated silicon nitride grains) distribution was evaluated by measuring 2300 grains of each sample. The apparent aspect ratio (AR) of the Si_3N_4 grains was obtained by the measurement of grain width and length visualized on the micrographs taken at a magnification of 20,000. Bulk ceramic samples were ground to a fine powder in the agate mortar. This powder was used for identification of crystalline phases using X-ray.

    Fracture toughness of the Si_3N_4 + SiC composites was measured by four different test methods: Chevron Notched Beam (CNB), Single Edge V-notched Beam (SEVNB), Indentation Strength (IS) and Indentation Fracture (IF) methods. Fracture toughness of the ZrO_2 + carbon nanofibers micro/nanocomposites was measured by Indentation Fracture (IF) methods. Chevron notches with angles of 90° were introduced perpendicular to the long axes of the specimens with dimensions of 3 x 4 x 45 mm.[3] The specimens were loaded in three-point bending at a cross-head speed of 0.01 mm/min. The K_{IC} values were calculated from the maximum load (F_{max}) and the corresponding minimum value of geometrical function (Y^*_{min}):[12]

$$K_{IC} = \frac{F_{MAX}}{B.W^{1/2}} Y_m \qquad (1)$$

where B and W are the width and height of the specimens, respectively. The calculation of the geometric function Y^*_{min} for CN bars was based on the Bluhm's slice model.[13]

    SEVNB method was performed on the specimens with dimensions of 3 x 4 x 45 mm[3]. The notched samples were broken under 4-point-bending with a cross-head speed of 0.5 mm/min and the fracture toughness was calculated according to:[14]

$$K_{IC} = \frac{F}{B\sqrt{W}} \cdot \frac{S_1 - S_2}{W} \cdot \frac{3\sqrt{\alpha}}{2(1-\alpha)^{1.5}} \cdot Y^* \qquad (2)$$

$$Y^* = 1.9887 - 1.326\alpha - (3.49 - 0.68\alpha + 1.35\alpha^2)\alpha(1-\alpha)(1+\alpha)^{-2} \tag{3}$$

where F is the fracture (maximum) load in four point bending, B the specimen width, W the specimen height, S_1 and S_2 the outer and inner roller span of the four point bending fixture, respectively, a is the precrack depth and α is the ratio a/W.

Each of the IS bar specimens (3 x 4 x 45 mm^3) were indented at the center of the tensile surface using the Vickers indenter with a load (P) of 98 N. The samples were broken in 4-point-flexure mode with a cross-head speed of 0.5 mm/min. The strength (σ_f) value has been determined using the maximal applied load and the fracture toughness was calculated using the following equation:[15]

$$K_{IC} = 0.88.(\sigma_f.P_i^{1/3})^{3/4} \tag{4}$$

IF method was determined by the measurement of the crack lengths created by the Vicker's indentations at a load of 98 N using the formula proposed by Anstis:[16]

$$K_{IC} = 0.016.\left(\frac{E}{H}\right)^{1/2}.\frac{P}{c^{3/2}} \tag{5}$$

where E is the Young's modulus of materials, H is the hardness of materials, P is the indentation load and c is the indentation cracks length.

Bending strength was determined using four-point bending test with inner and outer span of 20 mm and 40 mm, respectively. Wear behavior was studied by a ball-on-disc method at room temperature without any lubricant. The tests were performed in air with a relative humidity of about 60%. The applied load, sliding distance and velocity were 5 N, 500 m 0.1 m/s, respectively.

Wear testing was carried out on High Temperature Tribometer THT, by CSM, Switzerland, using ball-on-disc technique. For ZrO_2 based materials alumina balls with 6 mm diameter were used as friction partners. Sliding speed of 10 cm/s and loads of 1 N and 5 N were applied. In the case of Si_3N_4 based composites the friction partners were commercial silicon nitride bearing balls (6 mm diameter), the sliding speed was 10 cm/s and 0.5, 1.5, 2.5 and 5 N loads were applied. Testing was done on air, at room temperature in dry conditions. Coefficient of friction (COF) was measured and the damage was studied on the worn surfaces of discs and balls using optical and scanning electron microscopes. The wear tracks were measured by a stylus profilometer on three or more places, the average trough cross section area was calculated and subsequently the volume of the removed material was estimated. The specific wear rates (W) were then expressed according to the standard ISO 20808:2004[17] as the volume loss (V) per distance (L) and applied load (F_p):

$$W = \frac{V}{L.F_p} \left[\frac{mm^3}{m.N}\right] \tag{6}$$

RESULTS

Microstructure Characteristics

All investigated silicon nitride based materials exhibited bimodal character of microstructures which consist of elongated β-Si_3N_4 grains embedded in the matrix of much finer Si_3N_4 grains. Fig. 2 shows the examples of microstructures of the monolithic Si_3N_4 as well as the composite Si_3N_4-SiC sintered with the smallest one (Lu^{3+}). The composites additionally contain globular nano and submicron sized SiC particles, located intragranularly in the Si_3N_4 grains or intergranularly between the grains. The microstructures of the composites were always finer than the microstructures of monoliths because the SiC particles, formed at the grain boundaries of the Si_3N_4 grains, hinder the growth of the β-Si_3N_4 grains during the evolution of the microstructure.

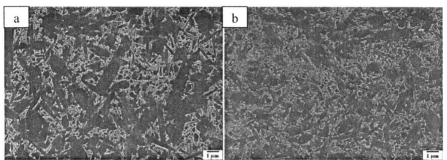

Figure 2. Characteristic microstructure of the monolithic Si_3N_4
and Si_3N_4 + SiC composite with Lu-doped sintering additive.

The aspect ratio of β-Si_3N_4 grains slightly increased with decreasing ionic radius of RE^{3+} both in nanocomposites and monolithic ceramics. These observations are in good agreement with the fact that the larger RE ions delay phase transformation[18] and that the RE ions impact the temperature at which the phase transformation initiates.[19] The presence of RE^{3+} with a large ionic radius at the interface may reduce the stability of the prismatic plane, and thereby decrease its growth rate. Non-significant influence of the various RE oxide additives on the microstructural evolution was observed by Hyuga et al.[20] On the other hand, Satet and Hoffmann[21] showed that the aspect ratio of Si_3N_4 grains increased with increasing ionic radius of RE and anisotropy of grain growth increased in the same way in the Si_3N_4 with RE_2O_3 and MgO.

TEM microstructure observation revealed that the SiC particles are situated more frequently at the grain boundaries rather than inside β-Si_3N_4 grains in the case of Lu-doped composite. On the other hand, intragranularly located SiC particles were observed more frequently in the La-doped composite. It seems that lower viscosity liquid phase (larger RE elements) tend to form intragranular SiC inclusions whereas higher viscosity (smaller RE elements) promotes formation of intergranular SiC inclusions. The grain boundary between the intragranularly located SiC particles and Si_3N_4 was without an intergranular phase, on the other hand an intergranular phase was found between the intergranularly located SiC and the Si_3N_4 grains, Fig. 3(a),(b).[22,23]

The monolithic zirconia consists of submicron/nanometer sized grains with randomly occurring defects in the form of pores with dimensions of approximately 100-200 nm.[24] The microstructure of the composite consists of a similar or an even smaller grained matrix with relatively well dispersed CNFs in the matrix. The detailed microstructure of the zirconia matrix of the composites are illustrated in Fig. 4 a. The average grain size of the monolithic material is 160 nm and in the composites it is smaller and varies from 95 nm to 135 nm. Clusters of CNFs have been found in all composites with size from 20 μm to 50 μm, Fig. 4b. The smaller grain size of the zirconia in the composites compared to the monolithic material suggests that the CNFs hinder the grain growth in the composite during sintering. The slightly smaller grained matrix in the composites sintered by SPS is explained by a shorter sintering time compared to the HP regime.

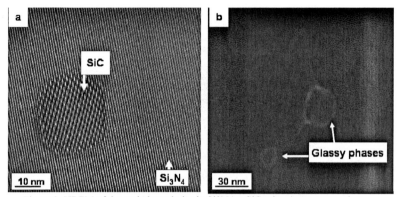

Figure 3. HREM of the grain boundaries in Si3N4 + SiC micro/nano composites.
(a) Grain boundary between SiC and Si_3N_4 is without intergranular phase;
(b) glassy grain boundary between intergranularly located SiC and Si_3N_4 grains.

In monolithic zirconia the grain boundaries were clean interphases between two ZrO_2 grains with no secondary phase. In the composites beside these grain boundaries we found two other types of boundaries illustrated in Figs. 4c and 4d. The first was a ZrO_2 / ZrO_2 boundary with a carbon nanofiber with a diameter of approximately 25 nm and the second a ZrO_2 / ZrO_2 boundary with disordered graphite, resulting from the degradation of CNFs during the powder preparation and the sintering routes.[25]

Fracture Toughness and Fracture Mechanisms

In the case of silicon nitride based ceramics the tendency in fracture toughness values for materials prepared with the different sintering additives is very similar for each testing method, Fig. 5. The fracture toughness values measured by IF is slightly overestimated in comparison to the results of other methods. The CNB method resulted in the lowest fracture toughness in all investigated materials, however statistically there are other values that match the CNB data. The composites exhibited lower fracture toughness compared to that of monolithic Si_3N_4 because of their finer microstructures. In such microstructures there are only limited possibilities for toughening mechanisms. Higher fraction of crack deflection at the elongated grains in the monoliths resulting in further toughening from mechanical and frictional interlocking of grains.

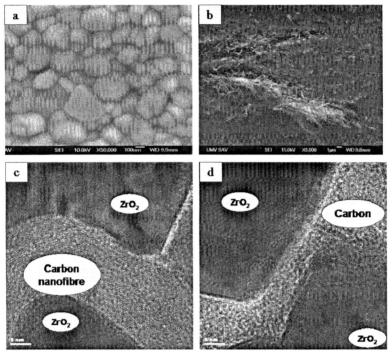

Figure 4. Microstructure of the zirconia matrix of the composites (Fig. 4a) and a cluster of SiC CNTs (Fig. 4b). ZrO_2 / ZrO_2 boundary with a carbon nanofiber (Fig. 4c) and with disordered graphite (Fig. 4d).

The materials doped with Lu exhibited the greatest aspect ratio and also the greatest fracture toughness in both monoliths and composite materials. This fact agrees with observations of the crack propagation in the materials. In the ceramics with a smaller ionic radius of RE the toughening mechanisms were observed slightly more frequently compared to that doped with a larger one, Fig. 6. Satet et al.[27] found that the fracture toughness of silicon nitride increased with increasing radius of RE^{3+} from 5.5 MPa.m$^{1/2}$ to 7.2 MPa.m$^{1/2}$ similar to the aspect ratio of Si_3N_4, which is contrary to our results. However, the materials in their study were prepared by a different processing route in comparison to the materials in this paper. Probably the difference in processing is the reason for the different observations with regards to the influence of the RE elements on the microstructures and fracture toughness of Si_3N_4 ceramics.

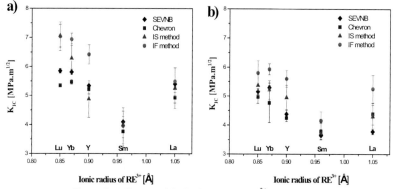

Figure 5. Influence of the ionic radius of RE^{3+} on the fracture toughness of monolithic Si_3N_4 and $Si_3N_4 + SiC$ composites[26]

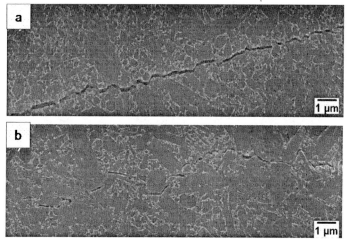

Figure 6. Crack propagation in La-doped (a) and Lu-doped (b) monolithic silicon nitride

In the ZrO_2 + CNFs system the indentation toughness decreased after addition of CNFs to the zirconia except for the SPS composite with 2.0 vol% of CNFs where a slight improvement has been found, Fig. 7.[28] The fracture in the monolithic zirconia is mainly intergranular with a very low roughness of the fracture lines/surface. No toughening mechanisms have been revealed on the fracture lines/surfaces in this system in the form of frictional or mechanical interlocking of the grains because of the very small grain size.[29] The composites prepared with both HP and SPS exhibit a slightly different behavior as the monolith with rougher fracture line/surfaces with crack deflection at the larger singular CNFs. The crack deflection is similar to the crack deflection in whisker reinforced ceramics and represents one of the toughening mechanisms in similar systems that can improve the fracture toughness of the composites. Besides crack deflection, crack bridging and CNFs pull out (Fig.8(a),(b)) was often detected on the fracture surface of the failed composite, however the pull-out length were usually short. Based on the fractographic examination it seems that the toughening mechanisms by CNFs were more effective in the SPS materials, which is probably the reason for the higher indentation toughness of SPS composites. As regards the fracture toughness, our results show that even the use of relatively coarse whisker like CNFs is not effective in toughening a zirconia ceramic matrix at the volume fractions used. On the other hand we have to note that on the fracture surface/line we frequently found different toughening mechanisms, mainly in the form of crack deflection at the CNFs and pull-out of CNFs. The reason for the relatively low indentation toughness is probably the poor dispersion and therefore the limited toughening effect of the CNFs. The carbon nanofibers used in the present investigation are coarser compared to the nanotubes in the experiment of Kothari et al.[30] The hollow or bamboo - shaped nanofibers used in the current study have similar graphitic structure as the MWCNTs used in[31] and can act as a toughening agent in a similar way. As has already been mentioned, a very important issue for toughening in fiber-reinforced composites is the nature of the interface between the fiber and the matrix, which must be of sufficiently low toughness to debond and be able to slide with friction. Xia et al reported[9] that the lack of molecular-scale perfection in nanotubes may provide some benefit in the toughening process. Ideal MWCNTs may exhibit extremely easy inter-wall sliding that prevents toughening behavior because of the easy telescoping of inner walls from outer walls. Imperfect nanotubes or nanofibers may provide more effective load transfer from outer to inner walls, resulting in enhanced toughness and strength. From the point of view of toughening there are probably two important interfaces; the outer between the CNTs/CNFs and the ceramic matrix and the internal between the individual graphite layers of the CNTs/CNFs. The first has to be designed during the processing of the composite, the second during the preparation of the CNTs and CNFs. However we have to note that the toughening mechanism connected with the internal interface can be effective in the case of cylindrical hollow CNFs only (Fig. 8b), and not in the case of bamboo-shaped CNFs, partially used in the present experiment.

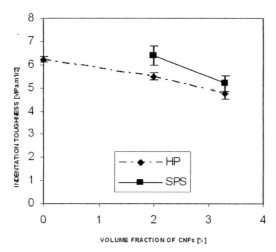

Figure 7. Influence of the volume fraction of CNFs on the indentation toughness.[28]

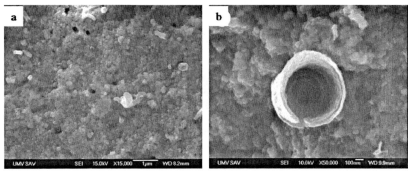

Figure 8. Pull – out of the CNFs from the ZrO_2 matrix (a)
and pull–out of the inner part of CNT from the outer part.[28]

Strength and Strength Degrading Defects

In Table 1 the mean bending strength and the wear parameters of the Si_3N_4 and Si_3N_4-SiC composites are illustrated.

Table 1. Mean Bending Strength and Wear Characteristics of all Studied Materials

Sintering additive	RE^{3+} [Å]	Si_3N_4 monoliths			Si_3N_4-SiC composites		
		Strength [MPa]	Friction coefficient	Wear rate x 10^{-5} [mm³/Nm]	Strength [MPa]	Friction coefficient	Wear rate x 10^{-5} [mm³/Nm]
La_2O_3	1.06	529 ± 21	0.739 ± 0.02	3.2 ± 0.4	532 ± 65	0.71 ± 0.04	1.9 ± 0.4
Sm_2O_3	0.98	514 ± 28	0.74 ± 0.02	1.9 ± 0.3	536 ± 35	0.69 ± 0.03	1.5 ± 0.3
Y_2O_3	0.90	759 ± 68	0.71 ± 0.02	1.8 ± 0.1	624 ± 38	0.69 ± 0.02	1.1 ± 0.1
Yb_2O_3	0.87	601 ± 32	0.71 ± 0.03	1.2 ± 0.2	688 ± 52	0.66 ± 0.04	0.9 ± 0.2
Lu_2O_3	0.85	686 ± 43	0.70 ± 0.02	1.1 ± 0.1	652 ± 46	0.64 ± 0.03	0.9 ± 0.1

According to the results, the values of mean strength slightly increased with a decreasing ionic radius of rare-earth elements except the Si_3N_4 doped with Y_2O_3 which exhibited the greatest strength value (Table 1). Fractographic observation revealed that in many cases the fracture was initiated with processing flaws, Fig. 9a. Such flaws are: pores and inclusions of impurity (Fe) in both monolithic and the composite materials. In addition, agglomerate of SiC grains and non-reacted carbon area (Fig. 9b) were often observed in the Si_3N_4-SiC composites. On the other hand, chemical inhomogenities were observed only in the case of monoliths sintered with Sm_2O_3. The monoliths as well as the composites containing La_2O_3 additives exhibited pores as the only critical defects.[23]

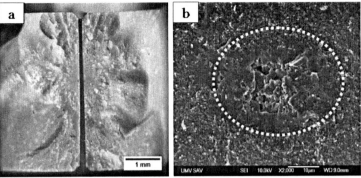

Figure 9. Macrofractography of the fracture origin in the La – doped monolith (a) and the detail of the fracture origin in the form of agglomerate of pores and carbon in Lu – doped composite, (b).

Wear Behavior and Wear Damage

Fig. 10 illustrates the influence of the size of rare-earth cations on the friction coefficient and specific wear rate of the investigated monolithic Si_3N_4 and Si_3N_4-SiC composites. In general the composites exhibit lower friction coefficient and lower specific wear rate in comparison to the monoliths. The friction coefficient of monoliths slightly decreased with decreasing ionic radius of RE^{3+} from the value of 0.74 to 0.71 and of composites from 0.71 to 0.64. Similarly, decreasing ionic radius of RE^{3+} the specific wear rate decreased in the case of monoliths from 3.2 $x10^{-5}$ to 1.2 $x10^{-5}$ [mm^3/Nm] and in the case of composites from the value of $1.91x10^{-5}$ to $0.89x10^{-5}$ for the materials sintered with La_2O_3 and Lu_2O_3, respectively.[32] Hyuga et al[20] investigated the influence of rare-earth sintering additives (Lu, Yb and Y) on the wear behavior of monolithic silicon nitride and reported increasing wear rate with increasing ionic radius of RE, which is in agreement with the result of our experiments.

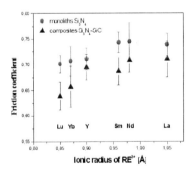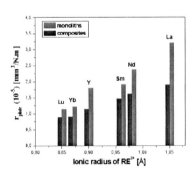

Figure 10. Influence of the ionic radius of RE^{3+} on the
frictional coefficient (a) and specific wear rate (b)[32]

Recently it was reported[34] that the bonding strength between grains and grain boundary phases of silicon nitrides increased with decreasing ionic radius of RE^{3+}. This results in the highest bonding strength in Si_3N_4 containing Lu_2O_3 as a sintering additive and restricts dropping of the individual silicon nitride grains during the wear experiment. This, together with the highest hardness and fracture toughness explain the highest wear resistance measured for Lu doped materials. According to the results of our wear tests the influence of different rare-earth additives on the high temperature friction coefficient of the investigated materials was observed only at 100°C and was the most remarkable between – La -doped and Lu - doped materials. Fractography revealed that coherent beneficial tribological film was created on the surface of Lu - doped composite after approximately 100 meters sliding resulting in significantly decreased friction coefficient during the rest of the test. Similar film was also created on the surface of further materials however the film was not coherent as in the case of Lu_2O_3. Friction coefficients measured at other temperatures varied only very slightly with changing of sintering additives, however some trends can be found especially during tests at room temperature and 700°C where friction coefficients decreased with decreasing ionic radius of rare-earth elements. The beneficial influence of additives with smaller ionic radius of RE^{3+} was negligible for tests at 300°C, 500°C and 900°C. In addition, there is no significant difference between the values of friction coefficients of the silicon nitride and the silicon nitride-silicon carbide micro/nano composites except in the already mentioned test at 100°C.

Fractography of the wear track revealed that with increasing wear resistance the worn surface became rather smooth but adherent debris can still be observed, Fig. 11. The greatest amount of coherent debris was observed for the material with the greatest resistance, i.e. for composite sintered with Lu_2O_3. The coherent debris layers act as a tribofilm, which affords some protection for the ceramic surfaces decreasing the wear coefficient of materials.[11] More fractured areas and a higher amount of silicon nitride debris have been observed in the wear tracks of the specimens having lower wear resistance. These fractured areas show a mixture of intergranular and transgranular failure modes in the worn surfaces. EDX analysis revealed that the coherent layers formed on the wear surfaces contain large amounts of oxygen, suggesting that the layers are composed mainly of the oxidation products of silicon nitride. These oxide based products form by tribochemical reaction at the wear interface and can occur at loads less than 10 N and relatively low temperatures (<400°C).[33] The wear mechanism was found very similar for all studied materials in the form of mechanical wear (micro-fracture) and tribochemical reaction. The tribochemical reaction area is characterized by a relatively smooth surface and the micro-fracture area is characterized by a rough surface and accumulated wear debris. The tribochemical reactions create a film on the tested samples and above a critical load the tribochemical film was partially removed, resulting in micro-fracture in discrete regions.

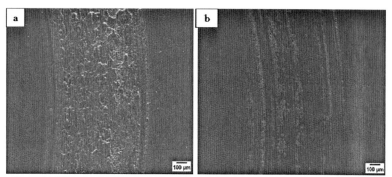

Figure 11. Wear track of the La – doped monolith (a) and Lu – doped composite.

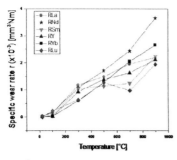 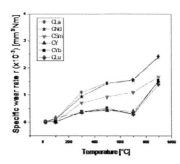

Figure 12. Influence of the temperature on the specific wear rate
of Si₃N₄ (a) and Si3N4+ SiC composite (b).

Fig. 12 illustrates the influence of the testing temperature on the specific wear rate. The wear rate generally increased with increasing testing temperature and the composites exhibit significantly lesser wear rates at all temperatures than the monoliths. In general, the wear rate decreased with decreasing ionic radius, however this trend is not so evident at the high temperatures. The composites with Lu, Yb and Y exhibited almost constant and very similar specific wear rate from room temperature up to 700°C. Above 700°C the wear rate of these materials increased very rapidly and almost reached the values characteristic for the other composite materials. This dramatic change in wear behavior visible also in other systems is probably connected with tribochemical reaction occurring at the interface between the ball and the disk. In Fig. 13 the severe damage below the wear track is illustrated.

Figure 13. TEM of the damage zone below the wear track of the Lu – doped composite.

In Fig. 14(a),(b) the results of wear testing of the ZrO₂+CNF composite and monolithic ZrO₂ is illustrated at load of 1N.[35] It can be seen that the friction coefficient of the composite did not depend on the sliding speed and remained stable during the whole test. In the case of the monolithic zirconia the friction coefficient values were greater with slightly greater scatter. By increasing the applied load the friction coefficient and the wear rate increased, Fig. 14b. The wear damage of the monolithic zirconia was low, the surface was mostly polished by low intensity abrasion. The wear of ZrO_2+CNF composite was comparatively more severe despite low values of friction coefficient. Fractography revealed many places from which clusters of CNFs or individual CNFs were pulled-out during the wear test. This suggests that the nanofibers were only weakly bonded with the matrix and could be easily loosened and removed, similar to that reported in literature.[36] This suggests that the removed CNFs present in the contact interface acted as lubricating media, not only lowering the friction but also the damage resistance.

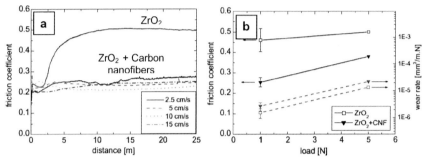

Figure 14. Influence of the sliding distance (a) and applied load (b) on the friction coefficient and wear rate.[35]

CONCLUSIONS

The microstructure characteristics, some mechanical properties and the fracture and damage mechanisms of Si_3N_4-SiC and ZrO_2 + carbon nanofibers micro/nano composites have been investigated.

- The grain growth in composites is suppressed by the presence of SiC particles and CNFs, which results in a very fine silicon nitride and zirconia matrix.
- In Si_3N_4-SiC composites the presence of SiC and fine microstructure resulted in less toughening mechanisms during the crack propagation and in less fracture toughness. However, it was found that by properly selecting a sintering additive (in the form of Lu_2O_3), the aspect ratio of the grains and the fracture toughness can be increased.
- The indentation toughness values of ZrO_2 + carbon nanofiber composites (except SPS ZrO_2 +2.0 vol% CNFs) are less than the monolithic zirconia due to the presence of CNFs clusters in their microstructure. However, a potential for toughening by CNFs has been recognized in the form of crack deflection and carbon nano-fiber pull-out.
- The positive influence of the finer microstructure of Si_3N_4-SiC composites was not observed due to the presence of processing flaws in the form of agglomerates of SiC and non-reacted carbon areas which acts as fracture origin during the bending test.
- In general, the friction coefficient decreased and the wear resistance increased with a decreasing ionic radius of RE additives and the combination of mechanical wear (micro-fracture) and tribochemical reaction was found as main wear mechanism for all studied Si_3N_4 based materials.
- In general, Si_3N_4-SiC composites exhibit significantly better tribological characteristics at room and elevated temperatures in comparison to the monolithic silicon nitride.

- The addition of low amounts of CNFs into ZrO_2 significantly decreased the frictional coefficient due to the lubricating effect arising by the help of removed CNFs during the wear test.

ACKNOWLEDGEMENT

The paper was supported by Nanosmart Centre of Excellence of SAS, by VEGA 2/0156/10, by APVV 0171-06, APVV 0034-07, by LPP 0203-07 and by HANCOC-MNT.ERA-NET. The experimental work was realized by Peter Tatarko, Viktor Puchý, Annamaria Duszová and Pavol Hvizdoš. The experimental materials were prepared with the help of Pavol Šajgalík, Jakob Kuebler and Michael Reece. For TEM analysis many thanks to Jerzy Morgiel, Ivo Vávra and the FEI Company.

REFERENCES

[1] T. Kishi, Prospects of Materials Science - History of Materials Science and Future Trends in Research, *Materials Science Outlook*, 3-5 (2005).
[2] R. N. Katz, Applications of Silicon Nitride Based Ceramics, *Industrial Ceramics*, **17**, 158-164 (1997).
[3] A. P. Bromley, D.A. Hutson, F. L. Riley, and L. S. Tovey, Clean room processing of an alumina ceramic *Ind. Ceram.*, **16**, 23-26 (1996).
[4] J.Dusza, M. Steen: Fractography and fracture mechanics property assessment of advanced structural ceramics, *International Materials Reviews*, **44**, 165-216 (1999).
[5] K. Niihara, New design concept of structural ceramics-ceramic nanocomposites, *J. Ceram. Soc. Jpn.* **99**, 974-1061 (1991).
[6] M. Sternitzke, Structural ceramic nanocomposites, *J. Eur. Ceram. Soc.*, **17**, 1061-1082 (1997).
[7] J.Dusza, P.Šajgalík, Silicon nitride and alumina-based nanocomposites, *Handbook of Nanoceramics and their Based Nanodevices, Stevenson Ranch, California: American Sci.Publ.*, **30**, 253-283 (2009).
[8] S. Iijima, Helical microtubules of graphitic carbon, *Nature*, **354**, 56-58 (1991).
[9] Z. Xia, L. Riester, W. A. Curtin, H. Li, B. W. Sheldon, J. Liang, B. Chang, J. M. Xu, Direct observation of toughening mechanisms in carbon nanotube ceramic matrix composites, *Acta Mat.*, **52**, 931-944, (2004).
[10] X. Wang, N. P. Padture, H.Tanaka, Contact damage-resistant ceramic/single wall carbon nanotubes and ceramic/graphite composites, *Nat. Mat.*, **3**, 539-544 (2004).
[11] A. Peigney, S. Rul, F. Lefevre-Schick, C. Laurent, Densification during hot-pressing of carbon nanotube-metal-magnesium aluminate spinel nanocomposites, *J. Europ. Ceram. Soc*, 27, 2183- 2193 (2007).
[12] ASTM C1421, Standard Test Method for Determination of Fracture Toughness of Advanced Ceramics at Ambient Temperature (2002).
[13] J.I. Bluhm, Slice Synthesis of a Three Dimensional Work of Fracture Specimen, *Eng. Fract. Mech.*, 7, 593-604 (1975).
[14] J. Kuebler, Fracture toughness of ceramics using the SEVNB method; round robin, *VAMAS-Report, EMPA*, Dübendorf **37**, (1999).
[15] P. Chantikul, G.R. Anstis, B.R. Lawn, D.B. Marshall, A critical evaluation of indentation techniques for measuring fracture toughness: II. strength method *J. Am. Ceram. Soc.*, **64**, 539-544 (1981).
[16] G.R. Anstis, P. Chantikul, B.R. Lawn and D.B. Marshall, A critical evaluation of indentation techniques for measuring fracture toughness: I. direct crack measurement, *J. Am. Ceram. Soc.*, **64**, 533-538 (1981).
[17] ISO 20808:2004(E), Fine ceramics – Determination of friction and wear characteristics of monolithic ceramics by ball-on-disc method (2004).
[18] P.F. Becher, G.S. Painter, N. Shibata, S.B.Waters, H.T. Lin, Effects of Rare-Earth (RE) Intergranular Adsorption on the Phase Transformation, Microstructure Evolution, and Mechanical Properties in

Silicon Nitride with RE_2O_3+MgO Additives: RE=La, Gd, and Lu, *J. Am. Ceram. Soc.*, **91**, 2328-2336 (2008).

[19]M. Kitayama, K. Hirao, S. Kanzaki, Effect of Rare-Earth Oxide Additives on the Phase Transformation Rates of Si_3N_4, *J. Am. Ceram. Soc.,* **89**, 2612-2618 (2006).

[20]H. Hyuga, M.I. Jones, K. Hirao, Y. Yamauchi, Influence of Rare-Earth Additives on Wear Properties of Hot-Pressed Silicon Nitride Ceramics Under Dry Sliding Conditions, *J. Am. Ceram. Soc.*, **87**, 1683-1686 (2004).

[21]R.L. Satet, M.J. Hoffmann, R.M. Cannon, Experimental evidence of the impact of rare-earth elements on particle growth and mechanical behaviour of silicon nitride, *Mater. Sci. Eng.,* **422**, 66-76 (2006).

[22]Š. Lojanova, P. Tatarko, Z. Chlup, M. Hnatko, J. Dusza, Z.Lences, P. Šajgalik, Rare-earth element doped Si_3N_4/SiC micro/nano-composites-RT and HT mechanical properties, *Journal of the European Ceramic Society*, **30**, 1931-1944 (2010).

[23]P. Tatarko, Š. Lojanova, J. Dusza, P.Šajgalik, Influence of various rare-earth oxide additives on microstructure and mechanical properties of silicon nitride based nanocomposites, *Materials Science and Engineering A*, **527**, 4771-4778 (2010).

[24]A. Duszova, J. Dusza, K. Tomasek, G.S. Blugan, J. Kuebler, Microstructure and Properties of Carbon Nanotube/Zirconia Composite, *Journal of the European Ceramic Society*, **28**, 1023-1027 (2008).

[25]J. Dusza, J. Morgiel, P. Tatarko, V. Puchy, Characterization of interfaces in ZrO_2 - carbon nanofiber composite. *Scripta Materialia*, **61**, 253-256 (2009).

[26]P. Tatako, Š. Lojanova, J. Dusza, P. Šajgalik, Fracture Toughness of $Si3N_4$ Based Ceramics with Rare-Earth Oxide Sintering Additives, *Key Engineering Materials*, **409**, 377-381 (2009).

[27]R.L. Satet, M.J. Hoffmann, Influence of the Rare-Earth Element on the Mechanical Properties of RE-Mg-Bearing Silicon Nitride, *J. Am. Ceram. Soc.*, **88**, 2485-2490 (2005).

[28]J. Dusza, G.S. Blugan, J.Morgiel, J. Kübler, F. Inam, T. Peijs, M.J. Reece, V. Puchý, Hot pressed and spark plasma sintered zirconia/carbon nanofiber composites, *Journal of the European Ceramic Society*, **29**, 3177-3184 (2009).

[29]A. Duszová, J.Dusza, K. Tomášek, J. Morgiel, G.S.Blugan, J. Kübler, Zirconia/Carbon Nanofiber Composite, *Scripta Materialia*, **58**, 520-523 (2008).

[30]A. K. Kothari, K. Jian, J. Rankin, B. W. Sheldon, Comparison Between Carbon Nanotube and Carbon Nanofiber Reinforcements in Amorphous Silicon Nitride Coatings, *J. Am. Ceram. Soc.*, **91**, 2743-2746 (2008).

[31]S. Pasupuleti, R. Peddetti, S. Santhanam, Jen Kei-Peng, Z.N. Wing, M. Hecht, J.P. Halloran, Toughening behavior in a carbon nanotube reinforced silicon nitride composite, *Mat. Sci. and Eng.*, **A491**, 224-229 (2008).

[32]P.Tatarko, M. Kašiarová, J.Dusza, J.Morgiel, P.Šajgalík, P.Hvizdoš, Wear resistance of hot-pressed Si_3N_4/SiC micro/nanocomposites sintered with rare-earth oxide additives, *Wear*, **269**, 867-874 (2010).

[33]S.S. Kim, Y.H. Chae, D.J. Kim, Tribological characteristics of silicon nitride at elevated temperatures, *Tribol. Lett.*, **9**, 227-232 (2000).

[34]P.F. Becher, N. Shibata, G.S. Painter, F. Averill, K. van Benthem, H.T. Lin, S.B. Waters, Observations on the Influence of Secondary Me Oxide Additives (Me=Si, Al, Mg) on the Microstructural Evolution and Mechanical Behavior of Silicon Nitride Ceramics Containing RE_2O_3 (RE=La, Gd, Lu*), J. Am. Ceram. Soc.*, **93**, 570-580 (2010).

[35]P.Hvizdoš, V. Puchý, A. Duszová, J. Dusza, Tribological behavior of carbon nanofiber-zirconia composite, *Scripta Materialia,* **63**, 254-257 (2010).

[36]D.S. Lim, D.H. You, H.J. Choi, S.H. Lim, H. Jang, Effect of CNT distribution on tribological behavior of alumina–CNT composites, *Wear*, **259**, 539-544 (2005).

FAILURE ANALYSIS OF CERAMIC LAMINATES USING QUANTITATIVE FRACTOGRAPHY AND FRACTAL GEOMETRY

J. J. Mecholsky, Jr.
Department of Materials Science and Engineering
College of Engineering, University of Florida
Gainesville FL 32611-6400 USA

ABSTRACT

Ballistic impact of laminated structures often results in comminution. In order to understand the method to better design protective materials, it is important to understand the failure mechanism(s). Some of the primary tools that can be used to reconstruct the failure process are quantitative fractography and fractal analysis. Quantitative fractography is the identification of fracture markings on fractured pieces that can be used to determine the magnitude of the stress state and direction of crack propagation. Fractal analysis is based on fractal geometry and is a mathematical tool that can be used to identify the toughness of materials as well as the creation of order out of seemingly complex fracture patterns. Comminution is in fact a localized crack branching that occurs due to localized high stresses. Fractal geometry has been successfully used to analyze crack branching processes in complicated stress fields. We present the methodology by which comminuted laminated structures can be analyzed to identify the stress at fracture and the fracture process that led to failure. Previous work has shown that the initiation of fracture is related to the toughness of the material. The toughness of the material, γ, is directly related to the fractal dimensional increment (D^*), i.e., $\gamma \propto D^*$. In addition, the crack branching pattern and stress at fracture is directly related to the fractal dimension of the pattern, known as the crack branching coefficient (CBC). Thus, it will be shown how the combination of quantitative fractography, fractal geometry and crack branching analysis provide the tools for failure analysis of laminated, ceramic and glass structures.

INTRODUCTION

In many applications for ceramic materials, lamination is used to enhance the structural reliability of the component such as with tempered laminated glass, dental crowns and bridges, armor materials and electronic substrates. In developing these materials it is important to understand the fracture processes to make sure the fracture resistance is optimized. Certainly, quantitative fractography is a very useful tool for understanding the fracture processes and for knowing the weakest link for improving fabrication techniques. An additional tool that can be used to enhance the fracture forensics is the application of fractal geometry. This paper addresses the application of quantitative fractography and fractal geometry to fracture forensics of laminated materials.

BACKGROUND

Fractography

Quantitative fractography has been used for decades to analyze the failure of industrial and laboratory components.[1-5] The principles of fracture mechanics have been shown to be very powerful tools that complement the observations during fractographic analysis.[6] It is suggested that a non-Euclidean geometry, i.e., fractal geometry, can be applied to fracture surfaces to provide a mathematical tool for describing the irregular surfaces produced during fracture. Using this approach, we discover relationships between the fracture features and the energy required to fracture the sample. The implications of fractal geometry are that there exists a scaling of energy and geometry at all length scales so that the fracture process is fractal in nature.

149

The observations of the mirror, mist and hackle regions on the fracture surface have been known for many decades (Figure 1). As early as the 1920's, Preston[1] observed that the mirror region contained perturbations from the plane of fracture, and thus was not a "smooth" surface. If we examine the mist and hackle regions of the fracture surface, we notice that if we magnify the mist region, it appears similar to the hackle region.[7] Thus, the process that produces mist evidently is the same as that which produces hackle. This similarity with different magnification is called scale invariance. I suggest that the perturbations in the mirror region have the same source as mist and hackle.[7] Examination of the fracture surfaces show that they are self-similar, i.e., one region appears statistically the same as another region at the same radial distance from the origin.

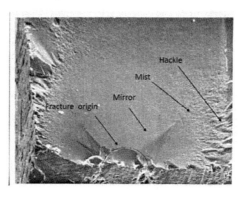

Figure 1. SEM image of soda-lime silica glass fracture surface indicating the fracture origin and the mirror, mist and hackle regions. The solid curve line is just outside the fracture initiating crack.

Almost all of the mechanically induced cracks can be idealized as a semi-elliptical, sharp crack of depth, a, and half-width, 2b.[8,9] The relationship between the stress at fracture, or strength, σ_f, and the fracture toughness, K_{IC}, is:[10]

$$K_{IC} = Y\,\sigma_f\,(c^{1/2}) = \sqrt{(2E'\gamma_c)} \tag{1}$$

where K_{IC} is the critical stress intensity factor (fracture toughness) and Y is a geometric factor which accounts for loading geometry and the shape of the fracture-initiating crack, E' is E, the elastic (Young's) modulus for plane stress conditions, and $E' = E / (1-v^2)$ for plane strain conditions where v is Poisson's ratio, and γ_c is the critical fracture energy, i.e., all the energy involved in fracture including the creation of new surface. The quantity Y in Eqn. (1) depends on the ratio a/b. However, the crack size can be approximated by an equivalent semi-circular crack size, c [$c = (a \cdot b)^{1/2}$].[8-10] This approximation allows many irregular crack shapes to be analyzed and avoids the complications of calculating a geometric factor for each crack.[10] For surface cracks without local residual stress and those that are small relative to the thickness of the sample, Y ~ 1.26.

We observe that from the fracture origin, the crack propagates in a relatively smooth plane (the mirror region) to the boundary, r_1, and progressively gets rougher by deviating slightly out of plane in a region that resembles mist (the mist region), between r_1 and r_2. Finally, the crack deviates locally from the main plane of fracture (the hackle region), between r_2 and r_3, getting very rough and finally branching into two or more cracks at r_3. Beyond macroscopic crack branching, the process can repeat itself on each branch of the propagating crack. These regions are all related to the applied (far field) stress at fracture, σ_f:

$$\sigma_f \, r_j^{\,1/2} = \text{constant} = K_{Bj} \,/\, Y'(\Theta) = \sqrt{(2E'\gamma_{Bj})} \,/\, Y'(\Theta) \qquad (2)$$

where $r_j = r_1$, r_2 or r_3 correspond to the different regions in an analogous equation to Eqn. (1).[11,12] K_{Bj} is the crack branching stress intensity and $Y'(\Theta)$ is a crack border correction factor where Θ is the angle from the surface to the interior, i.e. $\Theta = 0$ to $90°$. Note that K_{Bj} is proportional to K_{IC}, i.e. $K_{Bj} = \lambda \, K_{IC}$ where λ is 2-4 (for j=1) for most ceramic materials.[13] γ_{Bj} is the branching energy associated with the different regions, respectively, and E' is the same as above. Figure 2 shows the strength of silica glass versus the fracture mirror radius, r_1, for fibers loaded in tension and bending, for disks loaded in biaxial tension and for bars loaded in three-point flexure.[14] The straight line represents equation 2. The fact that the relationship represented by Eq. (2) is valid for several stress states during failure provides increasing confidence that this equation is useful in analyzing in-situ field failures. Also, notice the large range of stress values and of sizes of fracture mirror radii.

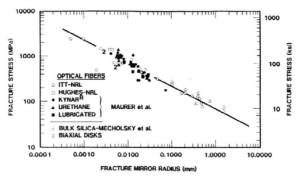

Figure 2. Fracture stress versus fracture mirror radius for silica glass. The optical fibers were failed in tension. The bulk silica bars were failed in three point flexure. The biaxial disks were failed in piston on three balls.

Fractal Geometry

Fractal objects are characterized by their fractal dimension, D, which is the dimension in which the proper measurement of a fractal object is made.[15] For example, if we had a plane square, then the only dimension in which to make a meaningful measurement is 2, i.e., the area of the square. The length or volume of a square is meaningless. This same concept is generalized for fractals that are allowed to have non-integer dimensions. Thus, a plane square with "bumps" out of the plane would have dimension, 2.D*, where D* is the fractional part of the fractal dimension and represents the amount of tortuosity out of the plane. An object with a fractal dimension of 2.1 (D* = 0.1) would be relatively flat and an object with D=2.9 (D* = 0.9) would almost be a volume-filling object. If the same relatively flat "bumpy" plane (with D = 2.1) was measured by a contour line, then the fractal dimension would be D = 1.1, but D* would still be 0.1.

The fractal dimension can be measured in a number of ways. The author prefers the slit island analysis because it offers the most reliable values, if sufficient care is taken to protect the surface during contour polishing and the sections are taken approximately perpendicular to the original fracture surface.[16,17] The length of part, or all, of the contour of an "island" that appears from polishing of an embedded fracture surface is measured as shown in Figure 3. The total length, L, is graphed versus its step (ruler) length, S, on a log-log graph, and fitted with a straight line:

$$L = k\,S^{\,1-D} \tag{3}$$

where the slope of logarithmic graph of Eq. 3 is equal to $1-D$, the fractal dimensional increment, and k is a proportionality constant. This measurement is made repeatedly using different starting measurement rulers. The highest point of the fracture surface is measured first, and subsequent (lower) areas are measured on subsequent horizontal contours (Fig. 3). Generally, this is a random section on the fracture surface. Thus, an average representation of the most tortuous portion on the fracture surface is usually obtained. The details of the measurements can be obtained from the literature.[18, 19]

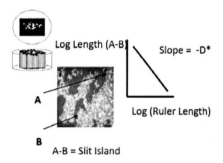

Figure 3. Schematic of Slit Island Analysis Process.

Fractal dimensions for fracture surfaces of glasses, ceramics, glass-ceramics, single crystals, intermetallics and polymers have been measured.[20] Atomic force microscopy studies have shown that the fracture surface is not smooth at the atomic level for several materials[21,22,23] and has the same tortuosity at the atomic level as at the macroscopic level. Thus, the crack front is neither smooth nor continuous.

Fracture Mechanics and Fractal Geometry
    A fundamental relationship between basic material constants, fractal geometry, and fracture parameters has been shown to exist for brittle materials:

$$\gamma_c = \gamma_s + \gamma_t + \gamma_i = \gamma_o + \frac{a_0\,D^*E}{2} \tag{4a}$$

$$K_c = E\,a_0^{\,1/2}\,D^{*1/2} \tag{4b}$$

where γ_C is the critical fracture energy (toughness), γ_S is the surface energy created during fracture,

γ_t is the thermal energy during fracture, γ_i is the summation of all other energy contributions during fracture, γ_0 is the fracture energy for a flat surface ($\gamma_0 \sim 0$), D* is the fractal dimensional increment of the fracture surface, E is Young's modulus, a_0 is a material structure constant and K_c is the fracture toughness. The value of γ_C in Eqn. (4a) can be thought of as the toughness, inclusive of all contributing terms. The variables D*, E, and γ_C are independently measurable; a_0 can be calculated from these three variables without introducing any adjustable parameters.[24,25] Eq. (4) can be considered representative of the toughness-tortuosity relation for glass, Ocala chert,[26] intermetallics,[27] ceramics and glass ceramics,[28] i.e., the toughness increases as the tortuosity. Note that a crack propagating through a perfectly homogeneous material with thermal vibrations will not result in a perfectly smooth surface.[29,30] Thus, we should not expect any fracture surface that involves fracture of primary bonds during no-equilibrium crack propagation to be smooth if the fracture occurs above absolute zero!

The quantity a_0 is a representation of a length on the scale of the structure of the material. It is most likely the expected average, $<a_0>$, of all of the configurations of bonds at fracture related to a linear unit, similar to the crack "size" represented as a length. The value of a_0, at this time, cannot be measured. The calculation of a_0 based on Eq. (3) above involves relatively macroscopic measurements, i.e., length scales on the order of microns. The estimation of the magnitude of a_0 can also be made from Molecular Orbital (MO) calculations. Both of these approaches result in similar estimates of the values for a_0. For example, for silica glass it is ~ 14 Å and for silicon single crystal it is ~ 8 Å [25]. It can be shown that:[31]

$$[Y'(\theta)^2/Y(\theta)^2]r_j/c=1/D* \qquad (5)$$

The relationship in Eq. 3 is valid for all values of j, i.e., j=1,2,3, but the r_j/c will be numerically different as will the $(Y_j'(\theta)^2 / Y(\theta)^2)$ proportionality constants. Thus, the length scale of fracture surface features in the fracture plane is directly related to the perturbations out of the plane.

Fractal geometry can be used to describe branching structures such as diffusion limited aggregation (DLA),[32] dense branching growth in electrochemical deposition,[33] and crack branching in fracture.[34-36] Most, if not all, of these studies have been analytical in nature after observations of physical phenomena. Many of the studies of crack branching related to fractal geometry have been analytical in nature without the benefit of experimental measurements. One exception was the work of Sakai et al.[37] in which the fractal dimension of chemically strengthened glass bus stop shields was determined by extensive measurements of the branching patterns. Their work is important because it experimentally demonstrated that the fractal dimension could be used to quantify crack branching patterns. The fractal dimension determined by these investigators did not agree with previous measurements on glass fracture surfaces for fractal dimensions from other studies.[7,38] The other studies focused on fracture surface measurements which were essentially perpendicular to the crack branching pattern. At the time, there was not really a good explanation for the differences. This paradox will be discussed later. However, the important point to note is that the fractal patterns of branching that develop are self-similar. A grid dimension technique was used to determine the fractal dimension of the disks.[34,35] In this technique, a transparent grid with a particular box size is overlaid on an image of the reassembled, fractured disk displaying the crack branching pattern. The number of intersected boxes containing branched cracks is compared to the total number of boxes in the grid. This technique is repeated for different box sizes. The fractal dimension characterizes the covering of an object by uniform boxes of linear size L. If the entire object can be contained within one box of size, L_{max}, then each of the $N = 1/r^D$ portions will fall into one box of size $L = r L_{max}$, where r is the reduction or magnification factor. Thus, the number of boxes, N(L), of size, L, needed to cover the object is given by

$$N(L) = (L_{max}/L)^D \qquad (6a)$$

or

$$D = [\log(N(L)]/ \log(L_{max}/L) \qquad (6b)$$

The crack branching fractal dimension (with a value between 1 and 2) can be determined from the above equation. For the purposes of this study it is assumed that the grid dimension is a close approximation to the fractal dimension. To avoid confusion of the fractal dimension as a measure of the crack branching pattern and the fractal dimension of the fracture surface (which has values between 2 and 3,[34] the fractal dimension determined was termed the crack branching coefficient (CBC). The fracture surface is a three-dimensional structure, albeit the structure in the vertical direction is small compared to the other directions, and thus would have a fractal dimension between 2 and 3. For a thin disk or plate with uniform thickness, the crack branching pattern is essentially two dimensional and thus will have a fractal dimension between 1 and 2. In both cases, it is the mantissa that provides the structural geometric information. For spherical objects which fracture and branch, a crack branching fractal dimension between 2 and 3 would be appropriate. Thus, the crack branching fractal dimension should be reported per unit volume.

The crack branching coefficient is distinct from the fractal dimension obtained from the fracture surface. The fractal dimension of the fracture surface is, in most cases for brittle materials, a constant and related to the crack tip stress field. It should not be surprising, then, that there is a correlation between the fracture stress and the number of pieces. In addition, the strain energy in an isotropic material is proportional to $\sigma^2/2E$ where E is the elastic modulus. Thus, the number of pieces will be proportional to the strain energy released in the material at fracture for primary branching. The fractal dimension is a quantitative measure of repeatable patterns, especially of branching patterns. The correlation between crack branching patterns is based on two properties: the angle of branching and the distance to branching. The angle of branching is based on the far field stress.[39-43] The distance to branching is based on the energy released at the time of fracture and the critical crack size.[44-46] The crack branching fractal dimension is a function of the stress at fracture and related to the far-field stress distribution, or in other words, related to the type and magnitude of loading.

APPLICATIONS

Dental Restorations

All-ceramic fixed-partial dentures (FPDs) are becoming more and more popular because of their excellent aesthetics and biocompatibility.[47] However, they are more susceptible to brittle fracture and fatigue failures, especially when being used as dentures.[48,49] Marquardt et al.[50] reported that the survival rate of IPS Empress II all-ceramic crowns after 5 years was 100% but only 70% for FPDs. They found that framework fracture, biologic failure and irreparable partial veneer fracture were failures that occurred exclusively in the three-unit FPDs.

Fractographic analysis can be a critical tool in analyzing failure mechanisms, identifying fracture initiation sites, and determining the probable causes of the failures in retrieved clinical specimens. Thus, fractographic analysis should be one of the key elements in the design and development of dental structural materials.[51-54] In addition, fractal geometry has been applied to dental restorations to identify the fractal dimensional increment on lithia disilicate glass ceramics, alumina-zirconia ceramics and glass veneers.

Many of the restorations require a veneer on the core ceramic. As demonstrated in the past, most failures occur due to failure in the veneer first and then propagation into the core material. One of the uses of fractal geometry is to quantitatively monitor the crack propagation fracture surface to identify crack paths. For example, a lithia disilicate glass ceramic bar with a glass veneer was fabricated to study the fracture of the laminate bar. The purpose was to identify the character of the fracture surface when the composite bar failed in the glass veneer. Intentional (indentation) cracks were introduced into the veneer side of the laminate bar and the bar was fractured with the veneer side in tension. The fractal dimensional increment was then measured in the veneer fracture surface and the glass ceramic fracture surface. Separately monolithic bars of the glass veneer and glass ceramic were fabricated and fractured. The fractal dimension of these bars were also measured. As can be observed in Table I, the fractal dimension of the veneer fracture surface and the lithia disilicate glass ceramic in the laminate bar were statistically the same as opposed to the values when fractured separately. These results imply that, for brittle materials with low fracture toughness, the fracture process and surface topography are controlled by the initiation of fracture and not the material microstructure. Thus, for dental applications it is the toughness of the veneer toughness properties that is more important to prevent failure then the core toughness.

Table I. Fractal Dimensional Increment, D* as a Function of Testing Condition.

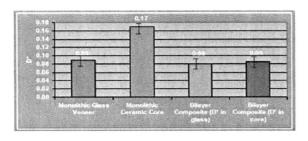

Glass Panels
    Glass and ceramics are brittle materials that, in most cases, fail catastrophically at stresses well below their theoretical strengths. They also have very low reliability and produce fragments that are large and have sharp edges. Characterizing and understanding the fragmentation behavior of glass or ceramics is critical in several different applications. For instance, during the Oklahoma City bombing in 1995, the resulting glass shards injured more than 400 people.[55] Fragmentation phenomena are also encountered in the fracture of glass and ceramic armor materials under ballistic impact.
    In the laboratory, simulating complex loading conditions is not easily accomplished; however, valuable insight into fragmentation behavior may be gained by using tempered glass as a model material. There are various tempering treatments that dramatically change the strength of tempered glass so that many different fracture patterns can be developed. An additional benefit of this approach is that tempered glasses are used in a variety of applications, and the knowledge gained may be directly applicable. Tempering processes are divided into thermal and chemical (ion-exchange) tempering. In thermal tempering the glass is rapidly cooled through the glass transition temperature using air or liquid jets.[56] This process generates compressive surface stress, and can approximately double the glass's base strength (from ~50-80 MPa to >100 MPa). The fragments produced range in sizes from 3-10 mm, depending on the original thickness of the glass plate.[57]
    Research using fractal analysis to characterize the macroscopic surface crack branching patterns for brittle materials is limited. Mecholsky et al.[34] found a direct relationship between the fractal dimension of the crack branching pattern and the strength of annealed magnesium fluoride disks and proposed that the fractal dimension is proportional to the stress at fracture. Quinn et al. created a computer program that models crack propagation in biaxially-stressed glass disks using fractal

geometry.[45] Sakai et al. have shown that it is possible to quantify and simulate crack branching patterns on the surfaces of thermally-tempered glass using fractal geometry.[35,37] The data of Sakai et al.[35,37] which measures the branching pattern for compressively strengthened glass, are easily explained because of the loading and residual stress. The value of fractal dimension for branching that they determined, i.e., 1.58, will be variable depending on overall stress levels and quite different from the fracture surface fractal dimension for that glass, i.e., a constant value of approximately 2.1. The findings in this work have strong implications for many commercial processes such as comminution, attrition, grinding, and basic studies in crack branching.[57] Thus, it is imperative that further studies be made utilizing the fractal geometric approach for studying crack branching.

Ion Exchanged Glasses

The chemical tempering process involves replacing small ions near the glass surface with larger ions.[59,60] This replacement produces a residual near-surface compression that is balanced by a residual internal tension. The exchange depth, the concentration of ions that can be replaced by larger ions, and any stress or structural relaxation phenomena that may occur determine the magnitudes of the compressive and tensile stresses. The exchange depth is controlled by the exchange time and temperature. Using chemical ion-exchange processes, it is possible to achieve very high residual compression, leading to strength values up to 1000 MPa and fragments that are sub-millimeter in size.[61,62]

From prior work, it is clear that the CBC is a useful description of the extent of crack branching observed. However if the CBC could be related, in a simple manner, to the stress distribution in the material prior to fracture, it would become an invaluable tool in post-mortem analyses of fracture. To investigate this possibility, Kooi et al. measured the CBC of identically prepared, chemically-tempered glasses varies for samples fractured via indentation and via biaxial flexure.[36] The CBC is determined not only by the far-field stresses but also by the details of the residual stress distribution. Examining the fracture surface features and analyzing the stresses at fracture, they concluded that for materials with residual stresses, the CBC is not simply related to the strength, or to stress at fracture or even the total stored tensile strain energy. Since the CBC appears to be related, in a complex manner, to the distribution of stresses in the body, determination of stress-state prior to fracture may not be possible. Figure 4 demonstrates that the same ion exchanged material can result in totally different branching patterns depending on the loading. Of course, this can be explained by the effect of the far-field stress on the final crack branching pattern.

A

B

C

Figure 4 – Different crack branching fracture patterns depending on loading.
A. Ring-on-ring biaxial flexure
B. Point loading with diamond indent
C. Piston=on-three-ball biaxial flexure.

Figure 5 is a graph of the crack branching coefficient, (left ordinate, CBC) and the number of fragments/cm^2 (right ordinate, NOF) as a function of exchange time for both indentation and biaxial flexure samples from the work of Kooi et al. The CBC for unexchanged samples, fractured in biaxial flexure, is included. For both sets of samples, the CBC and the NOF increase with exchange time. Therefore, there appears to be a good correlation between the CBC and the NOF. For all exchange conditions, the biaxial samples had both a greater number of fragments and a greater CBC than the indented samples. Additionally, for biaxial samples, the CBC appears to be approaching a limiting value ($\approx$1.6) as indicated by the asymptote of the CBC curve for longer exchange times. The CBC for the indentation samples, however, appears to still be increasing. For a very large number of fragments, the crack branching pattern must effectively cover the entire surface, yielding a CBC value of 2. Therefore, we expect the CBC to asymptotically approach 2 if the number of fragments continues to increase.

Figure 5. Crack Branching Coefficinet (CBC) and Number of Fragments (NOF) vs. Exchange Time for Chemically Strengthened Glass.

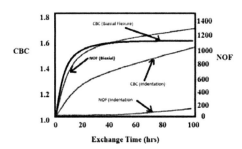

As we have observed, the CBC and NOF for the indention samples increases with exchange time, and these increases follow the increased change in the stored tensile energy. So the CBC and the tensile stored energy appear to be correlated. However, for the biaxial samples, the rise in CBC and NOF as a function of exchange time, does not correlate with the observations in the bending strain energy (decreasing) or the total tensile strain energy ($\approx$ constant). Here the CBC and number of fragments are dictated by distribution of stresses, both residual and applied. The evolution of fracture and creation of a fragment have already been described. Therefore, for a general fracture process in a brittle material, CBC (or NOF) may not be related simply to the strength, stress at failure or to the stored tensile energy, and depends instead on the distribution of stress in the material.

CONCLUSIONS
When materials fail in complicated stress states the fragments that remain are numerous and seemingly complex. However, by separating the fracture process into initiation and propagation, failure analysis techniques can be applied to separate the contributions to the failure process. Quantitative fractography, fractal analysis and fracture mechanics can be combined to aid in the failure analysis of complex stress field fractures. The fractal dimension of the fracture surface features provides a mathematical tool to quantitatively describe the energy required for failure. The fractal dimensional increment, D*, is directly related to the fracture energy, γ, at fracture [$\gamma = E\ a_0\ D* / 2$]. D* is not a function of the strength. When the material has enough energy to crack branch, then the pattern of crack branching can be modeled as a fractal process and the Crack Branching Coefficient (CBC) can be used to identify the pattern of branching. In relatively simple stress fields, the CBC can be related to the strength of the material and the number of fragments. In more complex stress fields,

such as with thermal tempering and ion exchanged glasses, the relationship between the CBC and strength is much more complex and needs additional investigation. However, the combination of fracture mechanics, quantitative fractography and fractal analysis offers a unique combination of tools for analyzing the fracture of materials that fail in a brittle manner.

REFERENCES
[1]F.W. Preston: J. Soc. of Glass Tech. 10 (1926) 234-269.
[2]E.B. Shand: J. Am. Ceram. Soc., 42(10) 474-477(1959).
[3]H. Wallner: Z. Phys., 114 (1939) 368-78.
[4]Ch. De Freminville: Rev. Met, 11 (1914) 971-1056.
[5]R.W. Rice : *Ceramic Fracture Features, Observations, Mechanisms and Uses*, in Fractography of Ceramic and Metal Failures, edited by J.J. Mecholsky, Jr. and S.R. Powell, Jr., ASTM STP 827, Philadelphia PA (1984) 5-102.
[6]S.W. Freiman, J.J. Mecholsky, Jr., and P.F. Becher: *Fractography: A Quantitative Measure of the Fracture Process* in Ceramic Trans. 17, Am. Ceram. Soc.,(1991) 55-78.
[7]J.J. Mecholsky, Jr., Application of Fractal Geometry To Fracture Forensics, Research and Production, Advances in Science and Technology Vol. 45 pp. 1646-1651(2006).
[8]G.K. Bansal, "Effect of Flaw Shape on Strength of Ceramics." J. Am. Ceram. Soc. 59 [1-21 87-88 (1976).
[9]J.J. Mecholsky, Jr.: S.W. Freiman: and Roy W. Rice, Effect of Grinding on Flaw Geometry and Fracture of Glass, V. 60, NO. 3-4, 114-117 (1977).
[10]P.N. Randall; pp, 88-126 in Plane Strain Crack Toughness Testing of High- Strength Metallic Materials, Am. Soc. Test. Mater., Spec. Tech Pub., No. 410, (1967).
[11]J.J. Mecholsky, Quantitative Fractography: An Assessment, Ceram. Trans. 17, in Fractography of Glasses and Ceramics II, eds. Frechette & Varner, Am. Ceram. Soc. (1991).
[12]H.P. Kirchner and J.W. Kirchner, Fracture Mechanics of Fracture Mirrors, J. Am. Cer. Soc. 62, 3-4, 198-202 (1979)
[13]H.P. Kirchner, Brittleness Dependence of Crack Branching in Ceramics, J. Am. Ceram. Soc.., 69 [4] 339-42 (1986).
[14]J.J. Mecholsky, Jr., Fractography of Optical Fibers, in ASM Engineered Materials Handbook, 4, Ceramics and Glasses, Section 9: Failure Analysis (1992).
[15] B.B. Mandelbrot, Fractal Geometry of Nature, Freeman Press, NY (1982).
[16]T.J. Hill, A. Della Bona, and J.J. Mecholsky, Jr., "Establishing a Protocol for Optical Measurements of Fractal Dimensions in Brittle Materials, J. Materials Science 36,2651-2657(2001).
[17]A. Della Bona, T.J. Hill, J.J. Mecholsky, Jr., "The Effect of Contour Angle on Fractal Dimension Measurements for Ceramic Materials," J. Materials Science 36,2645-2650 (2001).
[18] J.C. Russ: Fractal Surfaces. Plenum Press, NY (1996).
[19]J.J. Mecholsky, D.E. Passoja and K.S. Feinberg-Ringel : J. Am. Ceram. Soc. 72[1] (1989).
[20]J.J. Mecholsky, Jr., Fractography, Fracture Mechanics and Fractal Geometry: An Integration, Ceram. Trans. 64, in Fractography of Glasses and Ceramics III, eds. J. P. Varner, V.D. Frechette, & G. D. Quinn, Am. Ceram. Soc., pp. 385-393 (1996).
[21]D.M. Kulawansa, L.C. Jensen, S.C. Langford and J.T. Dickinson, J. Mater. Res. 9 [2] 341-370 (1994).
[22]E. Guilloteau, H. Arribart, and F. Creuzet, "Fractography of Glass at the Nanometer Scale," Mater. Res. Soc. Symp. Proc. V., 409 , 365-370 (1996).
[23]J. Cuneo and J.J. Mecholsky "Evaluation of Fractal Analysis Techniques using Atomic Force Microscopy," Ceramic Engineering & Science (Peer Reviewed) Proceedings 20 (2), The American Ceramic Society, (1999).

[24]J.K. West, J.J. Mecholsky, Jr, and L.L. Hench, "The Quantum and Fractal Geometry of Brittle Fracture", J. Non-Crystalline Solids 260 (1999) 99-108.

[25]J.J. Mecholsky, Jr.,J.K. West and D.E. Passoja , Fractal Dimension As A Characterization of Free Volume Created During Fracture in Brittle Materials, Phil. Mag A **82** [17/18] Nov. (2002).

[26]T.J. Mackin and J.J. Mecholsky, Fractal Analysis of Fracture in Ocala Chert, J. Mater. Sci. Ltrs. 7:1145-47; Nov. (1988).

[27]Y. Fahmy, J.C. Russ and C.C. Koch: J. Mater. Sci. 6, 1856-1861(1991).

[28]J.J. Mecholsky, Jr., "Application of Fractal Geometry To Fracture Forensics, Research and Production", Advances in Science & Technology **45** 1646-51 (2006).

[29]A.J Markworth, J.K. McCoy, and R.W. Rollins, J. Metals Res. **3**[4] 675 (1988).

[30]T.P. Swiler, T. Varghese, and J.H. Simmons, J.Non-Crystalline.Sol., **181**, 238(1995).

[31]J.J. Mecholsky, Jr., and S.W. Freiman, Relationship Between Fractal Geometry and Fractography, J. Am. Cer. Soc. 74 (12) 3136-38 (1991).

[32]e.g., T. A. Witten and L. M. Sander, Diffusion-limited aggregation, PHYS. Rev. B, V 27, N. 91 (1983).

[33]e.g., I. Mogi, M. Kamiko and S. Okubo, Magnetic field effects on fractal morphology in electrochemical deposition, Physica B: Condensed Matter, V. 211, 1-4, p. 319-322 (1995).

[34]J.J. Mecholsky, Jr., R. Linhart, B.D. Kwitkin, and R. W. Rice, "On the Fractal Nature of Crack Branching in MgF$_2$", *J. Mater. Sci.* November, , **13**, [11], pp. 3153-3159 (1998).

[35]T. Sakai, M. Ramulu, A. Ghosh, and R.C. Bradt, "A Fractal Approach To Crack Branching (Bifurcation) In Glass", pp. 131-146 in *Ceramic Transactions*, **17**, Fractography of Glasses and Ceramics II, Edited by V. D. Fréchette and J. R. Varner, The American Ceramic Society, (1991).

[36]J.E. Kooi, R. Tandon, S.J. Glass and J.J. Mecholsky, Jr., Analysis of Macroscopic Crack Branching Patterns in Chemically-Strengthened Glass, J. Mater. Res., Vol. 23, No. 1, (2008) 214-225.

[37]T. Sakai, M. Ramulu, A. Ghosh and R.C. Bradt, Cascadating fracture in a laminated tempered safety glass panel, Intern. J. of Fracture V 48,[1] 49-69 (1991).

[38]J.J. Mecholsky, D. E. Passoja and K.S. Feinberg-Ringel, Quantitative Analysis of Brittle Fracture Surfaces Using Fractal Geometry, J. Am. Ceram. Soc. 72(1) (1989).

[39]H. P. Rossmanith, University of Maryland Report [NSF Grant #DAR-77-05171] (1980).

[40]R. W. Rice, in Adv. in Ceramics 22, in Fractography of Glasses and Ceramics, edited by Varner and V. Frechette (The American Ceramic Society, Inc., Westerville, OH, pp. 3–56 (1988).

[41]V. D. Frechette, Fractography of Glass and Ceramics, The American Ceramic Society, Westerville, OH, (1990).

[42]J. van Vroonhoven, Ph.D. Thesis, Eindhoven Univ. of Tech. (1996).

[44]E.K. Beauchamp, Sandia National Lab. Research Report SC-RR- 70-766, Albuquerque, NM (1971).

[45]H.P. Kirchner and J.C. Conway, Jr., in Fractography of Glass and Ceramics, Advances in Ceramics, Vol. 22 (The American Ceramics Society, Westerville, OH, pp. 187–213 (1988).

[46]G.W. Quinn, J.B. Quinn, J.J. Mecholsky, Jr. and G. D. Quinn, "Computer Modeling of Crack Propagation Using Fractal Geometry," Ceramic Engineering and Science Proceedings. V. 26,[2] pp. 77-84 (2005).

[47]Von Steyern PV, Carlson P, Nilner K. All-ceramic fixed partial dentures designed according to the DC-Zirkon (R) technique. A 2-year clinical study. J Oral Rehabil, 32:180–7(2005).

[48]Taskonak B, Mecholsky JJ, Jr., Anusavice KJ. Fracture surface analysis of clinically failed fixed partial dentures. J Dent Res 2006;85 (3):277-81.

[49]Quinn JB, Quinn GD, Kelly JR, Scherrer SS. Fractographic analyses of three ceramic whole crown restoration failures. Dent Mater 2005;21(10):920-9.

[50]Marquardt P, Strub JR. Survival rates of IPS empress 2 all-ceramic crowns and fixed partial dentures: results of a 5-year prospective clinical study. Quintessence Int 2006;37(4):253-9.

[51]J.R. Kelly, S.D. Campbell, H.K. Bowen, Fracture-surface analysis of dental ceramics. J Prosthet Dent, 62(5):536-41(1989).

[52]K.R. Kelly, R. Giordano, R. Pober, M.J. Cima, Fracture surface analysis of dental ceramics: clinically failed restorations. Int. J Prosthodont, 3(5):430-40(1990).

[53]J.J. Mecholsky, Jr., Fractography: determining the sites of fracture initiation. Dent Mater,11(2):113-6(1995).

[54]S.S. Scherrer, J.B Quinn, G.D. Quinn, J.R. Kelly, Failure analysis of ceramic clinical cases using qualitative fractography. Int J Prosthodont,19(2):185-92 (2006).

[55]H.S. Norville, J.L. Swofford, M.L. Smith, and K.W. King: Survey of window glass broken by the Oklahoma City Bomb on April 19, 1995. Revised, March 1996.

[56]R. Gardon: Thermal tempering of glass. Glass: Science and Technology V. Academic Press, New York (1980).

[57]K. Akeyoshi and E. Kanai: Mechanical properties of tempered glass. Rept. Res. Lab 17, 1, pp. 23-36 (1967).

[58]J.C. Conway and H.P. Kirchner, "Crack Branching as a Mechanism of Crushing During Grinding" J. Am. Ceram. Soc., 69, 603(1986).

[59]S. Kistler: Stresses in glass produced by nonuniform exchange of monovalent ions. J. Am. Ceram. Soc. 45, 2, pp. 59-68 (1962).

[60]M.E. Nordberg, E.L. Mochel, H.M. Garfinkel, and J.S. Olcott: Strengthening by ion exchange. J. Am. Ceram. Soc. 47, 5, pp. 215-219 (1964).

[61]A.L. Zijlstra and A.J. Burggraaf: Fracture phenomena and strength properties of chemically and physically strengthened glass. J. Non-Cryst. Solids 1, 1, pp. 49-68 (1968).

[62]R. Tandon and S.J. Glass: Controlling the fragmentation behavior of stressed glass, in Fracture Mechanics of Ceramics, Vol. 14, Active Materials, Nanoscale Materials, Composites, Glass, and Fundamentals, edited by R. C. Bradt, D. Munz, M. Sakai, and K. W. White, Springer, 2005 (8th International Symposium on Fracture Mechanics of Ceramics, Houston, TX, February 2003), pp. 77-91.

FRACTOGRAPHIC ANALYSIS OF BROKEN CERAMIC DENTAL RESTORATIONS

George D. Quinn, Kathleen Hoffman
American Dental Association Foundation, Paffenbarger Research Center
Gaithersburg, MD, USA

S. Scherrer
University of Geneva, School of Dental Medicine, Department of Prosthodontics-Biomaterials
Geneva, CH

U. Lohbauer, G. Amberger
University of Erlangen-Nurnberg, Dental Clinic 1 - Operative Dentistry and Periodontology
Erlangen, Germany

M. Karl
University of Erlangen-Nurnberg, Dental Clinic 2 – Prosthetic Dentistry
Erlangen, Germany

J.R. Kelly
University of Connecticut, School of Dental Medicine, Department of Prosthodontics
Farmington, CT, USA

ABSTRACT
Three *in vivo* fractured multiunit all-ceramic restorations were analyzed as part of an ongoing project to characterize fracture mechanisms in dental restorations. The overall breakage patterns were evaluated and stereoptical and scanning electron microscopy used to find the origins of fracture. The causes of fracture in the alumina and zirconia restorations were different in each case.

INTRODUCTION
An increasing number of all-ceramic multiple tooth restorations are now being used in prosthetic dentistry. These are termed fixed partial dentures (FPDs). Since some are breaking in service it is important to identify the causes of fracture so that manufacturers, clinicians, and laboratory technicians can minimize the incidence of such breakage. It is also important to identify the mechanisms of fracture so that researchers can develop relevant testing procedures. Questions are often raised such as: Why did it break? Where is the origin? Did it break from an unexpected cause? Was there a problem with the material or was the restoration simply overloaded? Was it misused? Was it fabricated properly? Was there a fitting problem? Did a laboratory technician or a clinician damage the restoration when making an adjustment? These are common practical questions and the fractographer can often give straightforward definitive answers.

Analysis of ceramic restorations can be difficult. Rarely are fracture mirrors centered on an origin. Fractographic analysis is a cumulative learning experience and our goal is to analyze as many fractured restorations as we can to increase our experience base and to identify all the likely causes of fracture. This is the latest in a series of papers[1,2,3,4,5,6,7,8] on this topic. Two of these were presented at the previous conference in this series held in Rochester in 2006.[1,2] Most of the earlier papers featured single-unit crowns. This paper presents results for three multiunit restorations. Two of the cases have not been published previously. The third, which has been previously documented,[8] is added for comparison.

161

EXPERIMENTAL PROCEDURE

All three restorations had only one primary fracture. In two of the three cases, both fractured halves were available and we could examine the matching fracture halves. In one case, only one half was retrieved from the patient. Conventional fractographic analysis and fractographic techniques with stereoptical and scanning electron microscopy were performed in accordance with the Guide to Fractography.[9] We used a systematic approach to evaluate each fracture.[5] Overall maps of the local directions of crack propagation across the fracture surfaces were made utilizing classical fracture markings such as wake hackle from pore/bubbles in the veneer and twist hackle in the framework or veneer ceramics. Using these markings we were able to back track to an origin site. Usually, the entire process of analyzing a fracture involves examining dozens or even a hundred photographs, but due to space limitations, only the primary results are shown in this manuscript.

CASE 1: A Three-Unit Alumina Bridge

This was a three-unit anterior maxillary bridge, shown in Figure 1, which fractured at the right central incisor in mid 2009 after approximately four years in the patient's mouth. The bridge was cemented to abutments (teeth that have been reduced to serve as supports) on the end two units #8 and #10. The pontic (middle unit #9) was not supported by the jaw, but was attached by lateral connectors to the two end units. The gingival margin end of it blends in and is concealed by the gum line. The clinician who donated the broken FPD detected a crack in the incisor (#8) during a routine checkup and monitored its progress with time. When fracture occurred, he was able to retrieve the #8 and #9 units on either side of the fracture plane. (This is not always the case since extraction often damages a restoration. In such cases, it is often possible to make a replica of the restoration for fractographic analysis.[6]) The clinician informed us that the restoration was a zirconia bridge, but as we learned later, it was actually glass-infused alumina. The following analysis and all illustrations were done by two of our team (GQ and KH).

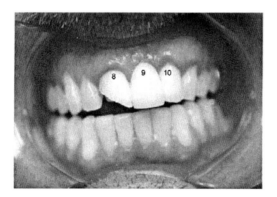

Figure 1. The three-unit anterior maxillary bridge just after fracture of the incisor #8.

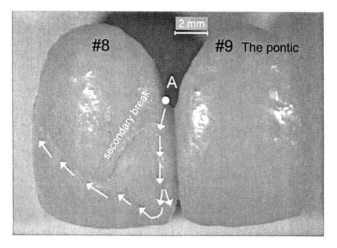

Figure 2. Close-up of the fracture after the bridge was extracted. Fracture started at the margin site A, near to, but not at the narrow connector between units #8 and #9. Crack propagation occurred as shown by the arrows. The triangular broken fragment was reattached for this photo.

Figure 2 shows a close-up of teeth #8 and #9 with arrows marking the direction of crack propagation (dcp) away from the origin site A, which was on the margin (the thin end of the restoration near the gum line where the restoration is affixed to the supporting tooth abutment. The crack went down the side wall of #8 to the incisal edge. It then went around the restoration and caused a large triangular piece to break off. The entire exposed fracture surface was examined in order to interpret the direction of fracture. Our attention was drawn to the origin region shown in Figures 3 and 4. Figures 5 and 6 are close-ups showing there was no severe material irregularity or gross flaw at the margin. The lack of a fracture mirror indicated the stress at fracture was low, or alternatively, there was a severe stress gradient such that peak stress was right at the margin, but it rapidly dropped off up the side wall. The margin was well-prepared and had a smooth rounded edge. Nevertheless, there is a small crack at the bottom, and combined with hoop stresses acting on the rim of the unit (presumably created by occlusal biting forces), the crack propagated up the thin wall of the unit #8 and split the restoration. This is a failure mode that we have observed before in a number of single crowns, especially if they are very thin or if the margin is damaged during manufacture or installation. As a consequence, we strongly advise against the use of "knife-edge" or "feather" margins, which some clinicians like to use for aesthetic reasons. Such restoration margins are simply too vulnerable, even if a strong ceramic such as zirconia is used.

Figure 7 shows an SEM image of a polished portion of a fragment from the specimen. It is a two-phase material with residual porosity. The energy-dispersive-spectroscopy analysis revealed the primary constituents were aluminum, calcium, lanthanum, and gold (from the coating). The microstructure and the chemistry are not that of a zirconia restoration, but most likely that of a glass-infused alumina framework material.[a,b]

---

[a] VITA In-Ceram® Alumina, VITA Zahnfabrik, Bad Säckingen, Germany.

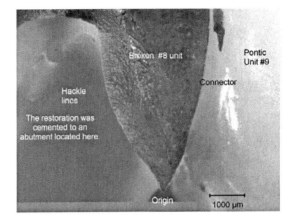

Figure 3. Optical image of the green-dye-stained fracture surface of the fracture origin region in unit #8. The thin tapered portion is the margin which blends into the gum line and is the site of the fracture origin. This origin site is near to, but not at, the connector to the next unit, #9.

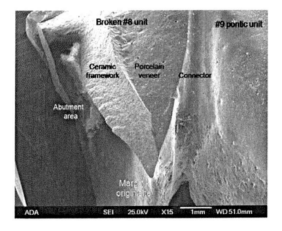

Figure 4. SEM image same area as Figure 3. In this view, the layered structure of a ceramic restoration is revealed. The interior core portion of the bridge is a solid ceramic. It is covered by a glassy-porcelain veneer for cosmetic purposes.

---

[b] Commercial products and equipment are identified only to specify adequately experimental procedures and does not imply endorsement by the authors, institutions or organizations supporting this work, nor does it imply that they are necessarily the best for the purpose.

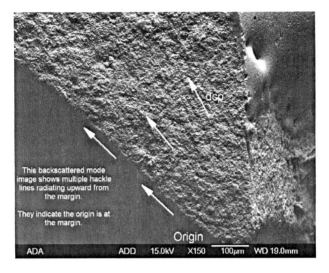

Figure 5. SEM close-up of the margin region showing hackle lines that indicate the dcp.

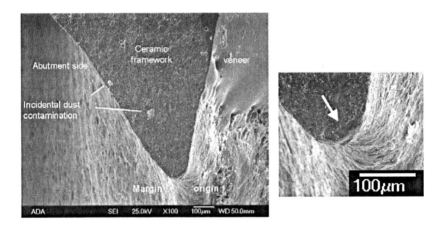

Figure 6. Close-up of the origin which is right at the tip of the margin. The margin is smooth and rounded and has no obvious faults, with the exception of a possible tiny crack (insert).

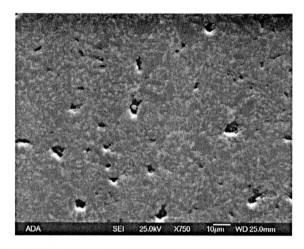

Figure 7. SEM image of the microstructure of the Case 1 framework material.

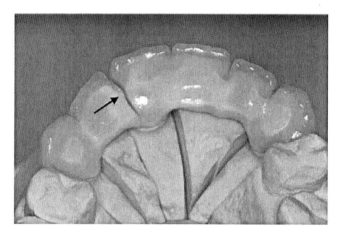

Figure 8. The Case 2 six-unit zirconia anterior maxillary bridge after removal and as mounted on a working stone model. The arrow shows the fracture origin.

CASE 2: A Six-Unit Zirconia Bridge

This was a six-unit anterior maxillary zirconia bridge with three abutments as shown in Figure 8. It was prepared by a technician trained in the use of a CAD/CAM system for the Cercon® system. The failure analysis of this case has been documented previously,[8] and it is included here for comparison to the other two cases. The framework consisted of a yttria-stabilized tetragonal zirconia polycrystal (Y-TZP) sintered at $1350°$ C. It was veneered with a feldspathic porcelain. It broke 24 h after installation with a temporary cement for a planned try-out period of 1 week. There was no abnormal loading by the patient. The break was a simple split perpendicular to the dental arch at a connector between the central right incisor tooth (#11) and the pontic tooth (#12). Since the cementation was only temporary, it was possible to extract both halves and examine both fracture surfaces as shown in Figure 9. The following analysis was done by several of our team (UL, GA, SS, GQ) and all micrographs and pictures were by UL and GA.

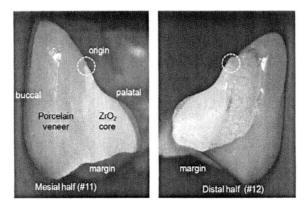

Figure 9. The exposed fracture surfaces. The core-foundation ceramic was not shaped well.

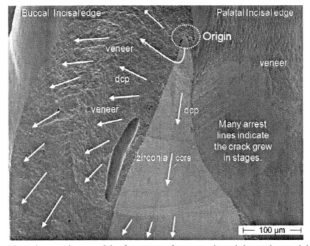

Figure 10. SEM close-up image of the fracture surface near the origin on the mesial half, #11. Fracture started at the core material tip and propagated in stages down into the core material.

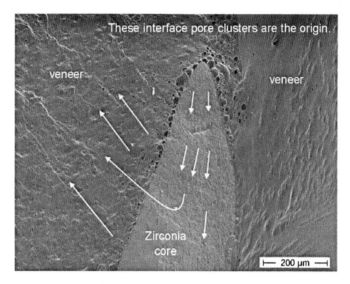

Figure 11. Close-up of the tip of the core material revealing a
cluster of interfaces pores that acted as the fracture origin.

Figures 10 and 11 show SEM close-ups of the mesial half, unit #11. Similar markings were observed on the matching distal half, #12. Although there was no fracture mirror, unequivocal markings such as wake hackle from pore/bubbles, twist hackle, and many arrest lines aided our interpretation. Fracture started at the tip of the framework ceramic and propagated down into the remainder of the framework and through the veneer. Loads from biting forces flexed the bridge and caused fracture in bending. It is possible that the three abutments did not support the FPD evenly.

Figure 11 shows a close-up of the origin site which is a cluster of pores at the interface of the framework tip and the veneer. These bubbles and the tiny cusp ligaments between them are especially vulnerable sites in a brittle ceramic, and not very much tensile stress would be needed to pop in a crack. Interestingly, fracture seems to have propagated down first in the framework in a series of steps (see the arrest lines), and then crossed the interface and swept back up into the veneer. This suggests that the framework was in tension and there may have been some residual compression stresses in the veneer. This is common in dental restorations whereby the thermal expansions of the veneer and core materials are slightly different. The thermal expansion coefficient for the veneer is slightly less than that of the framework, and consequently the weaker porcelain veneer is put into a state of residual compression and the stronger framework ceramic is able to carry some residual tensile stress.

The key issue here is that the restoration was poorly designed in this region. The framework should not have tapered to a sharp tip and should have been much thicker and rounded at the top. The veneer is much too thick and is unbalanced. The veneer is very thick on the buccal (outer) side and very thin on the palatal side. The sharp framework tip would have made manual application of the veneering ceramic very difficult, and it is not surprising there are problems at the interface.

An additional consequence of the poor design is that the massive size and unbalanced shape of the veneer probably put the thin zirconia into moderate residual tension stresses due to the thermal expansion mismatches. Calculations for the residual stresses are often based on assumptions that the veneer and framework have balanced thicknesses. In this instance, the contraction of the thin framework material at the tip would have been controlled by the far more massive veneer. It was estimated that 66 MPa tension stress in the framework would have resulted and it could have been concentrated to much greater values by the tip shape and the local holes.[8] This residual stress would have combined with the flexing stresses from biting forces along the arch.

The huge elongated pore in the veneer shown on the left of Figure 10 was harmless. Although it was enormous in size, it was very smooth and not sharp. Secondly, it apparently was not exposed to significant tensile stresses. So when fracture began elsewhere, the crack simply propagated around it.

A close examination of the right side of Figure 10 shows evidence (grinding marks) of reshaping of the framework on the palatal side in the laboratory, probably due to an interference with the occlusion. Such reshaping could have been prevented by more careful computer design and possibly a wax-up of the frame.

CASE 3: A Five-Unit Zirconia Telescoping Denture

This was a five-unit telescoping denture made of a single large piece of veneered zirconia as shown in Figure 12. It was from the upper left lateral incisor to the upper left first molar and was attached to two screw implants that anchored the part at the incisor and second premolar teeth. The piece was one of three that was used to replace the entire upper dental arch including substantial missing hard and soft tissues. The zirconia[c] was CAD/CAM machined from presintered zirconia blocks and then partially reduced in esthetically relevant areas for the application of the porcelain veneer. The restoration was colored and then sintered. The veneer was then applied. All processing steps were carried out in accordance with manufacturers' guidelines. Although this particular zirconia ceramic may be used as a monolithic body without veneering, in this instance a veneer was applied to obtain improved aesthetics.

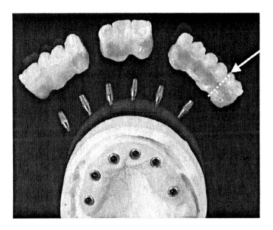

Figure 12. Laboratory view of the three segments comprising the maxillary dental arch showing the location of the supporting screw implants. Fracture occurred at the arrow.

---

[c] Zirconzahn, Gmbh, Gais Italy.

The clinical aspects of this case have been documented by Karl and Bauernschmidt[10] and a preliminary fractographic assessment by Karl.[11] The description and images that follow are the first full analysis of this fracture and were done by GQ, MK, and JRK of our team.

The cantilevered end unit broke off the restoration only 48 h after cementation as shown in Figures 13 and 14. The patient did not overload the restoration. The cantilevered upper first molar broke away from the anterior part of the prosthesis which remained in place. This fracture was very surprising since the last molar was connected to the rest of the restoration by a very large 10 mm x 12 mm cross sectional body of zirconia, a rather strong ceramic. It is hard to imagine that a patient could apply sufficient biting forces to snap off such a short, thick, stubby cantilever.

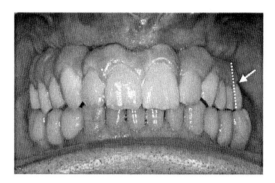

Figure 13. Clinical view showing the installed telescoping bridge segments and the eventual location of fracture (dashed line and arrow) on the left upper 5 unit piece.

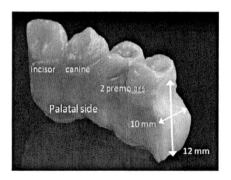
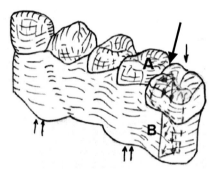

Figure 14. The five-unit segment after the end molar (on the right side in this view) broke off. The schematic on the right shows that there were two fracture planes, A and B. Plane A was a thermal crack that popped in during processing and ran halfway through the piece. It severely weakened the restoration. Plane B occurred when the end molar snapped off in the patient's mouth after installation. The small arrows depict the supporting implants on the bottom and occlusal loading on the end molar. The large slanted arrow shows the starting point of the first fracture plane A.

The fracture surface of the broken-off last molar is shown in Figure 15. With reference to Figure 14, there were two main intersecting crack systems, labeled A and B. B started at a point on plane A, although plane A continued deeper into the piece and undercut plane B (not shown). Plane B was a simpler fracture surface to analyze since it had numerous arrest lines and twist hackle. In contrast, plane A was almost featureless and was a low-energy, wavy surface. Numerous close-up

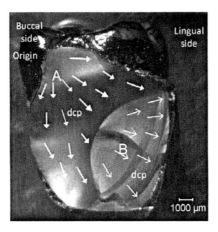

Figure 15. Stereoptical microscope image of the gold-coated molar unit that broke off the end. The initial starting point of fracture is in the upper left in this view. A thermal crack (fracture plane A) popped downward more than halfway through the piece as shown by the first set of arrows Final fracture (fracture plane B) started from the middle of the piece and radiated in steps outwards towards the lingual side as shown by the second set of arrows.

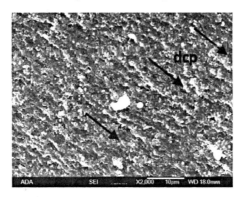

Figure 16. High magnification SEM image showing microhackle lines which indicate the direction of crack propagation (dcp). The microstructure appears normal for a 3Y-TZP in this region.

views of the microstructure in region A, such as the example shown in Figure 16, revealed tiny microhackle lines that indicated the local dcp. From these it was possible to map the overall dcp across the fracture surface and trace fracture back to the origin site, which was on the buccal side of the restoration, near to the occlusal surface. Fracture plane A was an incomplete fracture that was probably already in the restoration when it was installed in the patient, or it popped in during installation when a small fitting adjustment was made to the occlusal surface as described below. With such a large crack in the piece, it was only a matter of time before normal biting forces would break off the end molar. Indeed, this final fracture plane had a series of arrest lines, corresponding to incremental crack growth during mastication.

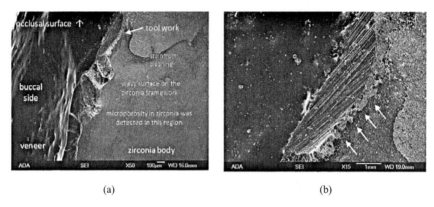

(a)   (b)

Figure 17. Two SEM views of the origin site which was near the top occlusal surface on the buccal side. (a) shows that the framework material had some waves in the surface at this location which was subsequently covered by the veneer. Some grinding marks from an adjustment are also visible and are shown as a close-up in (b). Microcracking into the zirconia (arrows) acted as the staring flaw.

The origin zone shown in Figure 17 had a number of interesting features. Firstly, the foundation zirconia material had an atypical wavy surface, presumably due to the reduction steps in preparation for the application of a veneer. Secondly, there was evidence of a grinding adjustment in the veneer that penetrated down into the zirconia framework. Microcracks were detected beneath the grinding adjustment and these are believed to have been the ultimate starting point for fracture plane A. Thirdly, it was observed that the microstructure in this region was not fully dense and that microporosity and incomplete zirconia-to-zirconia grain necks were observed as shown in Figure 18.

Very smooth, almost featureless, wavy fracture planes are commonly generated by thermal stress fractures.[9] These can occur from sudden temperature changes, or from severe thermal gradients set up by non-uniform heating or cooling. Fracture surfaces like these can also arise from pieces that break from internal residual stresses set up during firing. It is believed that the latter explanation applies in the present case as discussed below.

Thermal fractures can occur during the firing of ceramics, especially with complex shaped parts that are cooled too quickly. If the cracks form during the early stages of sintering, the fracture surfaces may be very rough and granular since the particles separate and cannot sinter together. On the

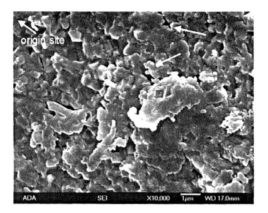

Figure 18. SEM close-up view of the zirconia microstructure in the vicinity of the origin.

other hand, cracking during cool down creates very smooth fractures that cleave through the fully dense material. It cannot be ascertained with certainty whether the A crack formed during the cool down from the furnace or immediately after removal from the furnace. It is possible that the veneering step covered over a cracked foundation piece. Alternatively, the crack may have occurred during the firing of the veneer.

One other likely scenario is that the restoration was fired in a manner that it had locked-in residual stresses. If so, crack A could have popped in after firing. Fracture plane A seems to have started from small microcracks from a grinding adjustment in the veneer. Such shaping is one of the final steps in processing. The microcracks penetrated into the framework zirconia (Fig. 17b), and residual internal stresses may have propagated the crack to form the A fracture plane.

In summary, this was a multistep fracture event. Thermal stresses or residual stresses from processing led to the formation of a large crack A that propagated from grinding damage underneath an adjustment. The fracture penetrated deep into the restoration but did not sever it. Final fracture along plane B occurred when the patient bit on the cracked and severely weakened restoration.

CONCLUSIONS

Three different *in vivo* broken multiunit ceramic restorations were analyzed. The breakages were from very different causes. A three-unit glass-infused alumina anterior bridge broke from hoop stresses acting on a thin margin. The margin was well prepared, but a very small crack acted as a fracture initiating flaw.

A six-unit CAD/CAM zirconia anterior bridge broke from bending along the dental arch and localized residual stresses. The poor design of the framework led to the fracture. The origin was located near the incisal end of the bridge and the fracture initiating flaw was a cluster of interface bubble/pores around the sharp framework tip.

A five-unit telescoping denture zirconia piece had a large crack in it from thermal or residual stresses from processing. The end molar required only a modest force to break when the patient bit down on it.

These three cases illustrate that all-ceramic restorations are susceptible to fracture from a variety of causes. Proper design and fabrication procedures must be established and followed to eliminate *in vivo* fractures.

ACKNOWLEDGEMENTS

The authors acknowledge the support of the National Institute of Standards and Technology, the American Dental Association Foundation, and the National Institutes of Health with NIH Grant R01-DE17983.

REFERENCES

[1]J.B. Quinn, S.S. Scherrer, and G.D. Quinn, "The Increasing Role of Fractography in the Dental Community," pp. 253 – 270 in *Fractography of Glasses and Ceramics,* Vol. 5, eds, J.R. Varner, G.D. Quinn, M. Wightman, American Ceramic Society, Westerville, OH, 2007.

[2]S.S. Scherrer, J.B. Quinn, G.D. Quinn, and H.W.A. Wiskott, "Descriptive Fractography on All Ceramic Dental Crown Failures," pp. 339 – 352 in *Fractography of Glasses and Ceramics,* Vol. 5, eds., J.R. Varner, G.D. Quinn, M. Wightman, American Ceramic Society, Westerville, OH, 2007.

[3]J.B. Quinn, G.D. Quinn, J.R. Kelly, and S.S. Scherrer, "Fractographic Analyses of Three Ceramic Whole Crown Restoration Failures," *Dental Materials,* **21**, 920 – 929 (2005).

[4]S.S. Scherrer, J.B. Quinn, G.D. Quinn, and J.R. Kelly, "Failure Analysis of Ceramic Clinical Cases Using Qualitative Fractography," *Int. J. Prosthodont.,* **19** [2], 151 – 158 (2006).

[5]S.S. Scherrer, G.D. Quinn, and J.B. Quinn, "Fractographic Failure Analysis of a Procera AllCeram Crown Using Stereo and Scanning Electron Microscopy," *Dental Materials,* **24** (2008) 1107 – 1113.

[6]S.S. Scherrer, J.B. Quinn, G.D. Quinn, and H.W.A. Wiscott, "Fractographic Ceramic Failure Analysis Using the Replica Technique: Two Case Studies," *Dental Materials,* **23** [11] (2006) 1397-1404.

[7]S.S. Scherrer, J.B. Quinn, and G.D. Quinn, "Fractography of Dental Restorations," pp. 72- 81 in *Fractography of Advanced Ceramics, III,* ed. J. Dusza, R. Danzer, R. Morrell and G. Quinn, TransTech Publ., Zurich, 2009.

[8]U. Lohbauer, G. Amberger, G.D. Quinn, S.S. Scherrer, "Fractographic Analysis of a Dental Zirconia Framework: A Case Study on Design Issues, *J. Mech. Behav. Biom. Mat.,* **3** (2010) 623-629.

[9]G.D. Quinn, *Guide to Practice for Fractography of Ceramics and Glasses,* NIST Special Publication SP 960-16, May 2007.

[10]M. Karl and B. Bauernschmidt, "Erste Erfahrungen mit Telescopierendem Zahnersatz aus 100% Zirconiumdioxid," (First Experiences with 100% Zirconia Telescoping Dentures) Qunitessenz *Zahntech* **36** [1] 86-94 (2010).

[11]M. Karl, "Fraktur Einer Kopiergefrästen Implantatbrücke aus Zirkonoxid-Keramik," (Fracture of a Copy-Milled Implant-Supported Fixed Dental Prosthesis Made from Zirconia Ceramic) *Das Deutsche Zahnärzteblatt,* **119** [6] 296-298 (2010).

FRACTOGRAPHY OF CERAMIC DENTAL IMPLANTS

Stephan Hecht-Mijic
CeramTec GmbH, Central Development.
Plochingen, Germany

INTRODUCTION

Fractography is a decisive tool in material development, in product development and in the development of test procedures (Richter,[1] Hecht-Mijic[2]). The present paper is focused on the fractographic analysis of ceramic dental implants which were fractured in the development steps mentioned above. This objective deals with identification of fracture origins, as described in standards like EN 843-6[3] and ASTM C 1322[4] and with orientation of fracture surfaces caused by certain stress distribution in the component. The relation between fracture surface features and special component features, such as surface roughness or stress concentration factors, is discussed by use of examples.

This paper deals with examples of fractographic features that one finds during analysis on both test specimens with simplified geometries and real components, i.e., dental implants. These examples stem from material development, product development and, finally, from test development. In the stage of material and process development by means of fractographic analysis the necessity of hot isostatic pressing in connection with special surface treatment of ceramic dental implants is shown on bending bars. The check of the correctness of the stress distribution in the component at the moment of fracture and the indication of weak regions; i.e., regions of dangerous stress concentrations on ceramic components, will be demonstrated. Finally, it will be shown how fractography supports the development of appropriate mechanical tests; e.g., proof-test in final inspection.

MATERIAL AND PROCESS DEVELOPMENT

Strength Experiments According to Valid Standards and Application of Fractographic Analysis

The application of fractography in material development starts after strength characterization on samples of standardized geometries, mostly on bending bars. 4-point-bending experiments with appropriate sample preparation are described in numerous standards (EN 843-1[5] and ASTM C 1161[6]). Fractographic analysis on bending bars should be carried out very consistently whenever deviations from a regular Weibull distribution appear. This holds for control of regular batch-control processes in a production process as well as for development batches of new ceramic compositions. An example of a deviation from expected Weibull distribution which resulted in a fractographic analysis of fragments of bending bars is shown in Figure 1.

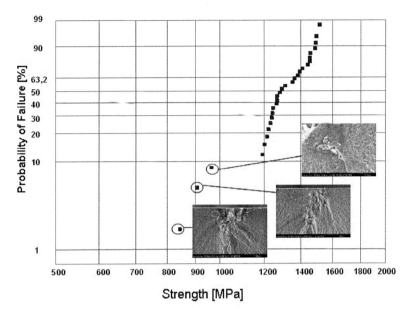

Figure 1. Deviation of single-strength values from a unimodal Weibull distribution and corresponding figures of the fracture origin from subsequent fractographic analysis. The flaw of the high-strength specimens stems from regular grinding of the specimens; e.g., the origin of the high-strength specimens is at the specimen surface. The flaw of the three low-strength specimens is a porous region and occurred only in an early stage of the development of the ceramic.

Strength experiments may not only be carried out on bend bars but also on disks by means of biaxial strength testing (ASTM C 1499[7]). Biaxial strength testing might be preferred if the later component design is similar to disk geometry, for example ceramic substrates for electronic applications, or if the load on the later component is a biaxial load (Pittayachawan[8]). However, in daily application the use of fragments of bend bars for fractographic analysis is easier and more space-saving with respect to fragment positioning, handling and archiving.

Size and Geometry Dependance of Strength
Strength values obtained on standardized sample geometries are valid together with the specific sample dimensions and test-setup dimensions. When strength values obtained on standard samples must be related to the strength of components then size and special geometric details of the components must be considered. This means that size of the area under maximum load should be calculated for the standard samples and estimated for the component. An example of such estimation is shown in

Table I.

Table I. Calculated and estimated area under maximum load of bend bar and dental implant, respectively, and corresponding relative strength.

Sample / Component	Max. loaded area [mm^2]	Strength [%]
Bending bar	88	100
Implant under bending	0.7	162
Implant under rotating bending	9	126

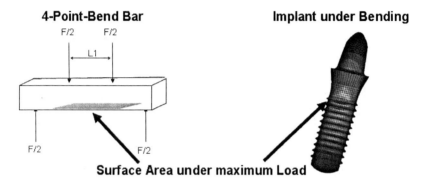

Figure 2. Schematic illustration of the area under maximum load of a bend bar and of a dental implant.

Hot Isostatic Pressing and Etching of Implant Surface

The process step of hot isostatic pressing is normally applied to achieve maximum density in ceramic material (Rieger[9]). In other words, without such a process step, a small part of residual porosity remains in the microstructure, as can be seen in the example of a ceramic microstructure produced without and with hot isostatic pressing in Figure 3.

On dental implants a specific etching step or even a series of etching steps is necessary in order to produce a roughness of the ceramic surface and to enhance growth of bone cells on the implant surface after implantation. However, etching procedures preferentially attack the edge of small open pores at the implant surface. This is exemplified and illustrated in Figure 4. Consequently, the process of hot isostatic pressing is indispensable for ceramic dental implants, due to the fact that open pores which are enlarged due to an etching process after hot isostatic pressing would lower the implant strength.

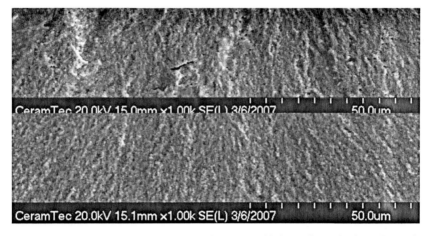

Figure 3. Ceramic microstructure without (upper figure) and with (lower figure) hot isostatic pressing.

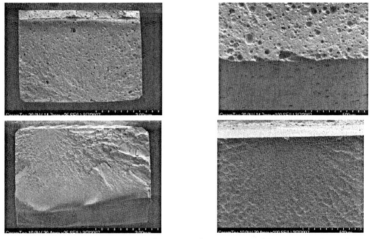

Figure 4. Upper figures – etched fracture surface and part of the outer surface of a bend bar produced without hot isostatic pressing (hip), lower figures – the same but with hip.

PRODUCT DEVELOPMENT

Fractographic Features Resulting From Fracture Under Bending According to ISO 14801[10]

The major objectives of a fractographic analysis of components are the localization of the fracture origin, then of course the analysis of the microscopic region surrounding the fracture origin, and the analysis of the primary fracture surface orientation, together with comparison of the expected stress distribution. Secondary features, such as metal transfer from test set-up or surface roughness (local or distributed) or stress concentrating characteristics of the component's geometry, must be documented in detail and compared with fractographic features.

The position of fracture origins may be described considering the specific geometry characteristics of the components. In the case of the dental implants the regular position of the fracture origin caused in a bend test according to ISO 14801[10] is in the 1st, 2nd or 3rd course of the thread, meaning that the 1st course is the one near to the so-called head and shoulder of the implant. This is illustrated in Figure 6.

Fracture origins are located at the core diameter of the thread, and the direction of crack propagation is perpendicular to the implant axis through the complete cross section of the implant. On the side of the implant which is opposite to the position of the fracture origin, the crack propagation changes its direction into parallel to the implant axis. This feature is quite similar to the change of direction of crack propagation in a bar fractured by bending. The characteristics described in this paragraph are shown in Figure 5 and Figure 7.

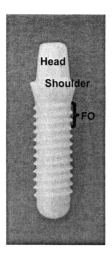

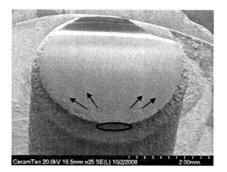

Figure 6. Ceramic dental implant with typical region of fracture origin (FO).

Figure 5. Fracture origin located at the core diameter of the thread, and direction of crack propagation perpendicular to the implant axis through the complete cross section of the implant.

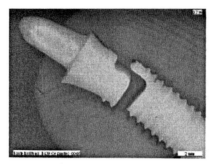

Figure 7. Crack propagation through the implant, and twofold
change of direction of crack propagation by about 90°.

These findings are summarized here and combined with these considerations:

Position of the fracture origin is always
- at the implant surface. The state of the surface and the state of the near-surface region decisively affect the implant strength.
- at the root diameter of the thread. The highest stresses must be expected to occur at the root diameter of the thread due to stress concentration.
- in the 1st, 2nd or 3rd course of the thread.

Surface Roughness Effect on 3-Point-Bend Strength of Round Bend Bars

In the development of dental implants, investigation of different implant surfaces, especially in the region of the thread, is crucial (von Chamier,[11] Hisbergues[12]). Such investigations were made, first, on samples with simplified geometries, namely on round bend bars. Roughness values are not specified here, but roughness characteristics are named only from R1 to R4, where R1 is the smoothest and R4 is the coarsest surface roughness. In Table II one finds that 3-point-bend strength has a maximum for R2 and R3. The corresponding fracture surfaces and outer surfaces of the bend bars are shown in Figure 8.

Table II. 3-Point-Bend Strength Resulting From Surface Roughness States From R1 to R4.

	Roughness1	R. 2	R. 3	R. 4
3-Point-Bend Strength [MPa]				
Mean Value	869	936	937	648
Stand. Deviation	103.6	148.1	115.1	31.5
Rel. Stand. Deviation [%]	12	16	12	5
R_a-Value [μm]	0.4–0.5	0.6–0.7	1.1–1.3	3.5–3.7

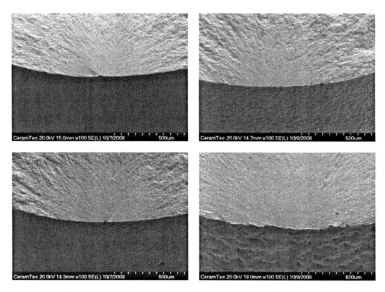

Figure 8. Fracture surfaces and outer surfaces with roughness
from R1 to R4 of round bend bars, according to Table II.

Stress-Concentration-Factor Effect on Implant Strength
    In one investigation the radius at the core diameter of the thread was the only parameter varied during implant production. In
    Table III strength results of two groups of implants are presented. Measurement of the radius at the core diameter of the thread resulted in radius mean values of 155 μm and 95 μm, see the light microscopy images in Figure 9.

Table III. Strength Results of Two Groups of Implants.

	Implant Group 1	Implant Group 2
Strength Mean Value [N]	971	791
Strength Stand. Dev. [N]	36	59
Relative Stand. Dev. [%]	4	7

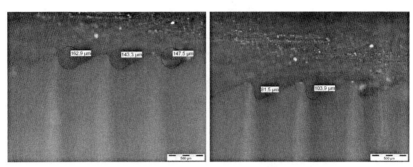

Figure 9. Measurement of the radius at the core diameter
of the thread of implants of group 1 and group 2.

The maximum stress in an dental implant under bending according to ISO14801[10] can be calculated with equation (1).

$$\sigma_{max} = \alpha_c \cdot \frac{F \cdot \sin(30°) \cdot l}{\dfrac{\pi \cdot d^3}{32}} \tag{1}$$

Under the assumption that a (defect size), Y (geometry factor) and K_{Ic} (fracture toughness) in (2) as well as l (length of the lever) and d (core diameter of thread) of Implant Group 1 and 2 are constant, then the fracture force $F_{fracture}$ is a function of the stress concentration factor α, and (3) is valid.

$$\sigma = \frac{K_{Ic}}{\sqrt{a} \cdot Y} \tag{2}$$

$$\frac{\alpha_{c_1}}{\alpha_{c_2}} = \frac{F_2}{F_1} \tag{3}$$

The results of strength measurements performed on two sets of implants with different values of the radius at the core diameter are listed in

Table IV. The results are in conformance with equation (3).
An appropriate change of the concentration factor α, i.e., appropriate change of the radius at the core diameter of the thread, leads to an increase of the implant strength. Parts of two dental implant threads from which one was improved with respect to surface roughness and stress concentration factor α are shown in Figure 10.

Table IV. Calculation of the Ratios in (3).

Implant Group	Radius at Thread Core Diameter [μm]	Stress Conc. Factor α_c (Bending)	Mean value of Fracture Force [N]	a_{K1} / a_{K2}	F_2 / F_1
1	155	2.7	971		
2	95	3.3	791		
				82	82

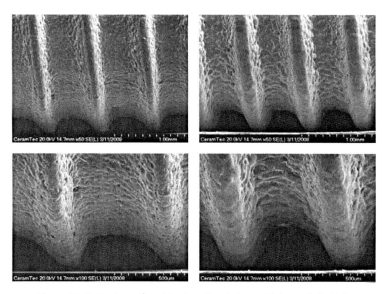

Figure 10. Parts of two dental implant threads from which one (left side) was improved with respect to surface roughness and stress concentration factor α.

TEST DEVELOPMENT

Bending Test and Rotating Bending Prooftest

A schematic drawing of the built-in situation of a dental implant is shown in Figure 11. From this figure the following characteristics of the loading situation may be deduced:

- There is a certain angle of inclination of the implant with respect to the force direction
- The load on the implant is applied via the opposite tooth

- There is a certain length of implant outside the bone; i.e., there is a lever with a certain length
- a fixation of the implant in the bone is necessary
- The resulting loading situation is bending.

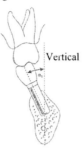

Figure 11. Schematic drawing of a dental implant in place.

The test set-up deduced from this in-vivo situation is standardized in ISO 14801[10]. It is shown in Figure 12. Part of ISO 14801[10] is also a quantitative definition of a few of the parameters mentioned above. These are:
- 3 mm bone atrophy
- 8 mm distance between bone level after surgery and intersection point of implant axis and force direction
- 30° inclination

From the worst-case assumption of 3 mm bone atrophy and from the 8 mm distance an overall lever length of 11 mm results.

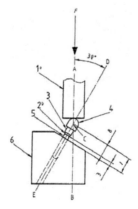

Figure 12. Test-setup deduced from the in-vivo situation, as described in ISO 14801.[10]

The requirement that weak implants should be sorted out results in the need to develop a proof test (Evans and Wiederhorn,[13] Fischer,[14] Kelly[15]). A specific requirement for test development results from the fact that the implant is screwed into the bone. The necessity of screwing it in results in the fact that any rotational angle of the implant after surgery is possible. This means that implant strength along the entire circumference of the implant has to be guaranteed. In other words no large defects along the entire circumference are allowed; i.e., implants with such large defects have to be sorted out in the proof test procedure.

It turned out in bending experiments with subsequent fractographic analysis that large defects at an angle of about 45° (see Figure 13 for definition of this angle) were not "found" by a simple bend test; i.e., the fracture origin is at another position. This fractographic – experimental result is easy to understand if one keeps in mind that the maximum tensile stress in the bend test is localized in only a very small part of the implant circumference, as is shown in the result of the finite-element calculation shown in the right half of Figure 2. Finally, from these facts the necessity of a rotating bending proof test was deduced. A schematic drawing of this test is shown in Figure 14, and the prototype test set-up is shown in Figure 15. Evidence that the proof-test set-up works properly is shown with the truncated strength distribution in Figure 16. Finally, fractographically the conformity of fractographic features resulting from a bend test according to ISO 14801[10] and from proof testing was demonstrated. The relevant comparison of fractographic features is shown in Figure 17.

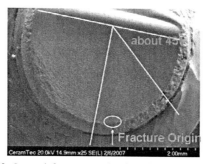

Figure 13. Large defect at an angle of 45° did not lead to fracture.

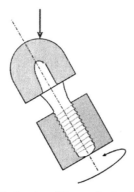

Figure 14. Schematic drawing of the rotating bending proof test.

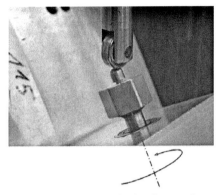

Figure 15. Prototype of the rotating bending proof-test set-up.

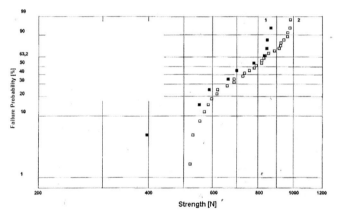

Figure 16: Truncated strength distribution obtained on implants which had survived the proof-test procedure.

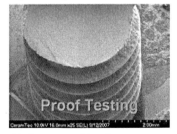

Figure 17. Conformity of fractographic features resulting from a bend test according to ISO 14801[10] and from proof testing.

Torque Load Test

During surgery the implant must be screwed in with a certain torque to achieve a primary stability of the implant in the bone. This means that a certain strength against torque load is necessary. Further consequences of the results of the investigations reported here are that the dimensions of implant and of screw-in tool must fit within small tolerances to each other, and that the material of the screw-in tool must be chosen carefully.

In the very first screw-in experiments the fracture origin was found at the implant head. In the example shown in Figure 18 three fragments originated from the head of the implant. The number of these small fragments varies from one experiment to another, but in almost every screw-in experiment the number of fragments is higher than the number of fragments from bending experiments.

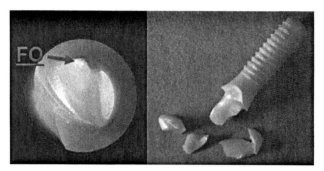

Figure 18: Fracture origin from screw-in experiments was found at the implant head. There is a higher number of fragments than from bending experiments.

Grooves, as a detail in the implant design in the head region, together with the features mentioned above (i.e., small geometries misfit between tool geometry and implant head geometry, as well as inappropriate tool material) may lead to point-like or line-like contact situations and, hence, to tensile-stress concentrations in the ceramic and, finally, to fracture events below the fracture-strength potential of the material. The effect of stress concentration at the edge of a groove in the implant head can be seen in the example shown in the right half of Figure 19

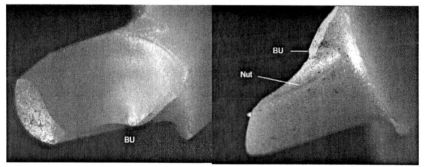

Figure 19. Fracture origin at the edge of a groove ("Nut") in implant head, note the point-like metal transfer at the position of the fracture origin ("BU").

Under torsion the direction of tension stress is inclined 45° with respect to the implant axis. The direction of the primary fracture surface in the region of the fracture origin is perpendicular to the direction of tension stress. Due to these facts the orientation of the primary fracture surface in the region of the fracture origin must be expected to be oriented at an inclination angle of 45° with respect to the implant axis, too. Evidence is shown in Figure 20 and in Figure 21 for a fracture origin in the head region of the implant as well as for a fracture origin in the thread region of the implant. The implant in Figure 21 did not fracture completely into separate fragments. The direction of illumination used in light microscopy has to be chosen carefully in order to make visible the crack which occurred due to the torsion load in the experiment.

The torsion load was applied by means of the test set-up shown in Figure 22, and the force was applied in a universal material testing machine. The implant is almost completely built-in into the test set-up. Evidence of an appropriately working proof-test procedure is shown by the truncated torque strength distribution in Figure 23.

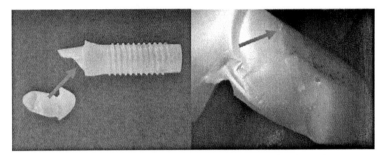

Figure 20. 45° degree inclination with respect to the implant axis, example of a fracture origin in the head region of the implant.

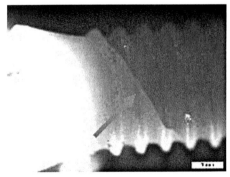

Figure 21. 45° degree inclination with respect to the implant axis, example of a fracture origin in the thread region of the implant.

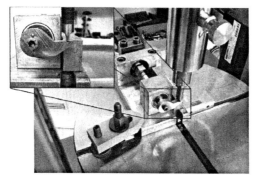

Figure 22. Test set-up for the application of torsion (torque) load.

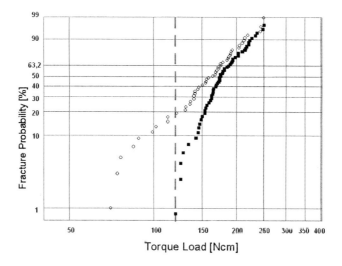

Figure 23. Truncated distribution of torque load after proof-test procedure.

In a few cases cone-like fracture surfaces were found. In these cases tracing back to the fracture origin was not possible, due to the complicated shape of the fracture surface. However, one can assume that fracture started at the outer side of the implant. It is obvious that the direction of the crack propagation changed continuously during the crack propagation. Hence, the direction of tensile stress changed its direction, too. Examples are shown in Figure 24, in Figure 25, and, finally, in Figure 26.

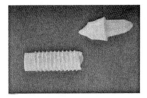

Figure 24. Cone-like fractured implant.

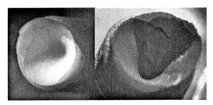

Figure 25. Cone-like fracture surface. The same fracture surface is shown twice. Left: without gold layer, right: with gold layer on the fracture surface. It seems that there are layers of several cracks.

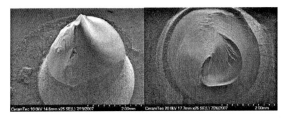

Figure 26. Cone-like fracture surface from counterpart fragments of the fragment shown in Figure 25.

SUMMARY
Fractography is a well established tool for the development of ceramic materials, of processing of ceramic materials and components, and for the development of strength-related tests, especially tests specific of a ceramic component and proof-testing. In this paper the application of fractography is presented as applied to a ceramic dental implant. In detail, the effects of fractographic results on choosing surface finishing, on the design of a dental implant and on an appropriate proof-test set-up is discussed. In addition, analysis of the orientation of fracture surfaces can provide information on the stress distribution at the moment of fracture.

ACKNOWLEDGMENTS
Explicitly the author thanks the industry partners for their cooperation in the field. Further he thanks the following CeramTec-Contributors for fruitful discussions, experimental work or accurate fractographic analysis: Bartolome, Berger, Kuntz, Rank, Richter, Rosenfelder, Sprügel, Wimmer and Ziegler.

REFERENCES

[1]Richter, H.G. [2001]: Fractography of bioceramic components for total hip joint replacement, Fractography of Glass and Ceramics, Konf., Stara Lesna. Key Eng. Mater. Vol. 223 (2002), p. 157.

[2]Hecht-Mijic, S. [2006]: Fracture mirror constants of bioceramics for hip joint replacement: determination and application. Proceedings: Fractography of Glasses and Ceramics V (2006), Rochester, USA.

[3]EN 843-6 [2010]: Advanced technical ceramics – Mechanical properties of monolithic ceramics at room temperature – Part 6: Guidance for fractographic investigation.

[4]ASTM C 1322 [2010]: Standard practice for fractography and characterization of fracture origins in advanced ceramics, American Society for Testing and Materials (2010).

[5]EN 843-1 [2006]: Advanced technical ceramics – Mechanical properties of monolithic ceramics at room temperature – Part 1: Determination of flexural strength.

[6]ASTM C 1161 [2008]: Standard Test Method for Flexural Strength of Advanced Ceramics at Ambient Temperature.

[7]ASTM C 1499 [2009]: Standard Test Method for Monotonic Equibiaxial Flexural Strength of Advanced Ceramics at Ambient Temperature.

[8]Pittayachawan, P. [2008]: Flexural Strength, Fatigue Life, and Stress-Induced Phase Transformation Study of Y-TZP Dental Ceramic. J of Biomed. Material Res., Part B: Appl. Biomat. Publ. online 13 March 2008: www.interscience.wiley.com. DOI: 10.1002/jbm.b.31064.

[9]Rieger, W; Köbel, S.; Weber, W. [2007]: Herstellung und Bearbeitung von Zirkonoxid-Keramiken für dentale Anwendungen. DIGITAL_DENTALNEWS, 1.Jahrgang, Juni 2007.

[10]ISO 14801[2003]: Fatigue test for endosseous dental implants.

[11]von Chamier, W.; Willmann, G. [1993]: Wie kann die Osteointegration bei Dentalimplantaten getestet werden? Die Zahnarztpraxis, DFZ 8/93, pp. 15-21.

[12]Hisbergues, M; Vendeville, S; Vendeville, P. [2008]: Zirconia: Established Facts and Perspectives for a Biomaterial in Dental Implantology. J of Biomed. Mat. Res., Part B: Appl. Biomat. Publ. online 17 June 2008: www.interscience.wiley.com, DOI: 10.1002/jbm.b.31147.

[13]Evans, A.G. and Wiederhorn, S.M. [1972]: Proof-testing of ceramic materials – an analytical basis for failure prediction. Intern. J. of Fracture, vol 10, no. 3 (1972) 379-392.

[14]Fischer, H.; Rentzsch, W.; Marx, R. [2002]: Elimination of low-quality ceramic posts by proof testing. Dental Materials 18 (2002), pp. 570-575.

[15]Kelly, J.R. [1997]: Ceramics in restorative and prosthetic Dentistry. Annu. Rev. Mater. Sci. 1997, 27, pp. 443-68.

MODES OF FAILURE OF BONDING INTERFACES IN DENTISTRY

Alvaro Della Bona, DDS, MMedSci, PhD, FADM
Dental School, University of Passo Fundo
Campus I, BR 285, Km 171, POBox 611
Passo Fundo, RS, Brasil 99001-970

John J. Mecholsky Jr.
Materials Science & Engineering Department
University of Florida
Gainesville, FL 32611-6400

ABSTRACT
The most common and effective dental bonding systems use resin-based adhesives. To understand the failure of bonding interfaces in dentistry it is necessary to comprehend the bonding mechanisms applied in clinical dentistry.

Failure of bonding structures is one of the most relevant problems in restorative dentistry. The mode of failure is an important aspect of bond strength tests, but it is not commonly reported in the dental literature. A detailed inspection of the fractured surfaces can indicate the failure mode of a bonded assembly. The fracture surface characterization combined with analyses of fracture mechanics parameters are of great importance to understand and predict bonded interface reliability and also to reduce the risk for data misinterpretation such as the inference that the bond strength must exceed the cohesive strength of the ceramic when the fracture initiates within the ceramic material and away from the bonding interface. Therefore, a classification of the modes of failure based on principles of fractography should assist researchers to correctly interpret the fracture phenomena.

A thorough SEM examination of the fracture surfaces using the principles of quantitative fractography, and confirmation of surface composition through the use of X-ray elemental map analyses, produce a more consistent and complete description of the fracture process. These analyses would avoid simplistic comments such as the "mixed mode of failure" that often follows the "adhesive and-or cohesive" unscientific observations, Thus, when fractography is correctly used to determine the fracture origin, a proper scientific statement on the mode of fracture can be formulated.

INTRODUCTION
Materials and procedures used either to repair fractured ceramic restorations with a resin composite or to bond indirect ceramic restorations using a resin cement are based on the results of bond strength tests, which exhibit wide variability in test data and resulting fracture surface characteristics.[1-11]

In the search for a method that produces a uniform stress distribution across the interface, investigators have evaluated similar adhesive systems under different bond test configurations.[3,12,13] These studies suggest that a tensile bond strength test is more appropriate to evaluate the bond strength of adhesive interfaces because of more uniform interfacial stresses. However, tensile tests require careful alignment of the specimens to minimize the risk of additional flexure.[4,14]

The physical contribution to the adhesion process is dependent on the surface topography of the substrate and can be characterized by its surface energy.[15,16] Dynamic contact angle (DCA) analysis has been used to evaluate the surface energy of treated ceramic surfaces and their work of adhesion (W_A) to resin. In principle, the work of adhesion can be related to the apparent interface toughness (K_A).[15]

Despite variable fracture surface morphology, the dental ceramic-adhesive interface seems to be the weakest location for the ceramic bonded tooth system. Adhesive failure is not a common outcome either clinically or during bond strength testing. This mode of failure may occur during specimen cutting procedures, i.e. during specimen preparation, suggesting a weak bond. Conditioned and resin-bonded ceramic structures usually fail cohesively from the ceramic-adhesive interface and the crack either propagates through the adhesive, or reaches the adhesive-composite interface.

Fracture mechanics allows quantification of the relationships between material properties such as toughness, and the stress level, the presence of crack-producing flaws, and crack propagation mechanisms. Another way to assess the integrity of the bond is to estimate the apparent interfacial fracture toughness (K_A) of the adhesion zone by promoting crack initiation within this zone. The K_A value reflects the ability of a material to resist unstable crack propagation.[17-20]

The objective of this work is to present the most popular bonding mechanisms used in dentistry and to correlate them to the fracture behavior of the associated adhesive interfaces.

EFFECT OF SURFACE ENERGY ON THE WORK OF ADHESION

To understand the failure of bonding interfaces in dentistry it is necessary to comprehend the bonding mechanisms. The clinical success of either a repaired ceramic restoration or a resin-cemented ceramic restoration will depend on the quality and durability of the bond between the ceramic and the resin.[14] The quality of the resin bond to dental materials and tooth tissues depends upon the bonding mechanisms that are controlled in part by the specific surface treatment used to promote micromechanical and/or chemical retention with the substrate.[4] Structural and surface analyses of treated (e.g., etched) substrates have shown that different topographical patterns are created according to the substrate microstructure and composition, and to the concentration, application time and type of surface conditioner (e.g., acid etchant).[4,21-25]

Considering dental ceramics, surface defects, often located in the glassy matrix, and phase boundaries of heterogeneous materials are preferably etched by the acids. Alteration of the surface topography by etching will result in changes in the surface area and in the wetting behavior of the substrate.[15,16,23,26] This may also change the material surface energy and its adhesive potential to resin.[14,15,27] Therefore, differences in ceramic microstructure and ceramic composition are controlling factors in the development of micromechanical retention produced by etching. Thus, the ceramic microstructure, composition, and morphologic patterns after surface conditioning should yield potentially useful information on the clinical success of the bonding procedures for indirect ceramic restorations and ceramic repairs.[14,23]

The surface changes produced by fluorine containing etchants such as hydrofluoric acid (HF), ammonium bifluoride (ABF) and acidulated phosphate fluoride (APF) have been reported.[23] HF etching produces the most aggressive effect, resulting in a substantial, consistent, surface roughness on acid-sensitive ceramics, mainly because of its action on defects and phase boundaries (Fig. 1).[4,14,21-23] These results have a positive correlation with the results from contact angle measurements between ceramic and resin.[14,15]

The etched ceramic topography can be explained by the chemical reactivity of the crystals of single-phase materials that depends on crystallographic orientation. In polycrystalline materials, etching characteristics vary among crystal types. Atoms along the crystal boundaries are more chemically active and dissolve at a greater rate than those within the crystals, resulting in the formation of small grooves or linear defects after etching.[14,23,28]

Dental ceramics with high crystalline content (alumina and/or zirconia), also called acid-resistant ceramics, have demonstrated better clinical performance than feldspar-, leucite-, and lithium disilicate-based ceramics, known as acid-sensitive ceramics. However, an increase in mechanical strength, by increasing the crystalline content and decreasing the glass content, results in an acid-resistant ceramic whereby any type of acid treatment produces insufficient surface changes for adequate bonding to resin.[4,7-12,14,23,29-33] For these acid-resistant ceramics, a silica coating process (silicatization) has been suggested to maximize the bond to resin.[7-11,14,32,34,35] Silica coating systems (e.g., Rocatec and Cojet from 3M-ESPE) create a silica layer on the ceramic surface because of the high-speed surface impact of the alumina particles modified by silica. It has been reported that the airborne particles can penetrate up to 15 μm into the ceramic and metal substrates.[35] This tribochemical effect of the silica coating systems may be explained by two bonding mechanisms: (1) the creation of a topographic pattern via airborne-particle abrasion allowing for micromechanical bonding to resin; and (2) the chemical bond of the silica coated ceramic surface, i.e., the silane agent, and the resin material. Therefore, a silica-silane chemical bond can occur with acid-resistant ceramics if a silica coating of the ceramic surface is used (Fig 2).[7-11,14,29]

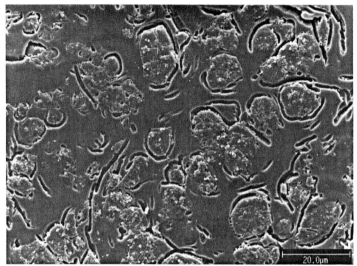

Figure 1. Light etched feldspathic ceramic. Note the etching on defects and phase boundaries (Magnification: x1000). From Della Bona, 2009.[14]

Silane has been used to enhance bonding between organic adhesives and ceramics or metals in various industries since the 1940's.[36] Silane coupling agents bond to Si-OH on ceramic surfaces by condensation reactions and methyl methacrylate double bonds provide bonding to the adhesive. The technology of organosilane coating of inorganic filler particles has improved their bonding to matrix resins.[37] This technology also improves the chemical adhesion of ceramic bonded restorations and resin-bonded ceramic repairs.[4-11,14,21,24] However, the long-term stability of the adhesive bonding using silane coupling agents has been questioned.[38-40]

Therefore, the adhesion between dental ceramics and resin-based composites is the result of a physical-chemical interaction across the interface between the resin (adhesive) and the ceramic (substrate). The physical contribution to the adhesion process is dependent on the surface topography of the substrate and can be characterized by its surface energy. Alteration of the surface topography, e.g., etching and airborne-particle abrasion, will result in changes on the surface area and on the wettability of the substrate.[14-16] This may also change the surface energy and the adhesive potential.[14,15,27,41-43]

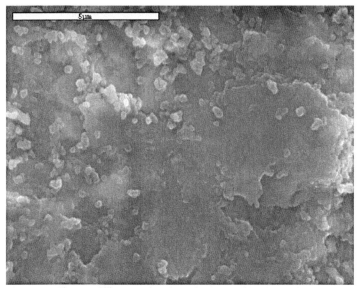

Figure 2. Silica coated surface of a zirconia-based ceramic
(Magnification: x5000). From Della Bona, 2009.[14]

The wetting behavior (wettability) of the resin on the treated ceramic substrate can be characterized using contact angle measurements and surface energy calculations. The wettability of a solid surface by a liquid (e.g., adhesive) can be characterized by Young's equation:

$$\gamma_{SL} = \gamma_{SV} - \gamma_{LV} \cos\theta \tag{1}$$

where γ_{SV} is the free energy per unit area of the solid surface in equilibrium with vapor, γ_{LV} is the surface tension of liquid balanced with its vapor tension, γ_{SL} is the interfacial energy, and θ is the contact angle. The work of adhesion (W_A) of the liquid drop on a substrate can be expressed by Dupré's equation:

$$W_A = \gamma_{SV} + \gamma_{LV} - \gamma_{SL} \tag{2}$$

Combining equations (1) and (2) yields the following Young-Dupré equation:

$$W_A = \gamma_{LV} (1 + \cos\theta) \tag{3}$$

An increase in the ceramic surface energy can improve the bond strength between ceramic and resin composite. Contact angle values can be used as an indicator of total surface energy and wettability.[14,43] It was found that chemical and mechanical treatment of ceramic surfaces yield increased total surface areas and increased total surface energies. According to equation (3), W_A is dependent on the surface tension of the liquid (*e.g.*, an adhesive) and its contact angle on the substrate. The W_A of a resin-based adhesive on ceramics was examined and the results showed that untreated ceramics displayed a larger mean contact angle (θ) than the etched ceramics. This improved ceramic wettability by low viscosity resins is a resultant of the increase in surface area, which allows a solid to draw more medium onto its surface and exerts greater interfacial force on the specimen. Consequently, clean, roughened surfaces display smaller contact angles and a greater W_A. A good correlation can be observed between the amount of surface disruption and the resulting contact angle. Thus, the greater the surface disruption, the lower the contact angle and the greater the W_A.[14,15]

BOND STRENGTH, MODES OF FAILURE AND INTERFACIAL FRACTURE TOUGHNESS

Bond strength tests have been used to predict the clinical performance of bonded structures, even though, most of these tests exhibit a wide variability in fracture patterns and bond strength values. The commonly used shear bond test often produces fracture at a distance from the adhesion zone that may lead to erroneous conclusions on bond quality. Such failures of the substrate prevent measurement of interfacial bond strength and limit further improvements in bonding systems.[2,3,22,44]

To test the integrity of bonded interfaces one can subject a bonded assembly to a variety of loading conditions to control the crack path along the interface or within the interfacial region. Analyses of bond tests have revealed several problems associated with most common test arrangements and suggest a lack of reliability of such measurements in assessing the adhesive behavior of bonded dental materials. Several studies have identified non-uniform stress distributions along the bonded interfaces. These variable stress patterns suggest that a standardized research protocol may address only part of the problem.[3,44-48]

In the search for a method that produces a uniform stress distribution across the interface, investigators have evaluated similar adhesive systems under different bond test configurations.[3,13,14] These studies suggest that a tensile bond strength test is more appropriate to evaluate the bond strength of adhesive interfaces because of more uniform interfacial stresses.

The microtensile test, a tensile bond test with reduced testing area, was developed as an attempt to eliminate the non-uniform stress distribution at the adhesive interface and to minimize the influence of interfacial defects.[14] The reduction in the number and size of defects in the adhesive zone is thought to decrease bulk cohesive failures and increase the tensile bond strengths, regardless of the cross-sectional shape. This test has been used to measure the bond strength of composite to dental tissues[13,49,50] and to ceramics.[4,6,14]

The evaluation of the structural integrity of the adhesion zone by Weibull analysis is also an important component for an integral analysis of the bonding interface.[4,6] The strength values reported using the microtensile test are considered reliable indicators of the interfacial bond quality when the fracture starts within the adhesion zone. Several modes of failure were identified for this type of test (Fig 3).[4,6,12,14] Adhesive failures (mode 1) are not common and normally happened during specimen cutting procedures. Considering resin bonded to acid-sensitive ceramics, the HF-treated ceramic specimens usually produce failures that start at the ceramic-adhesive interface and propagate through the adhesive (mode 4), then either reach the adhesive-composite interface (mode 5, see Figure 5C) or return to the ceramic-adhesive interface (mode 2, e.g. Figure 5A).[4,6,12,14]

Therefore, optical microscopy observation is often not enough to determine the mode of failure of bonding interfaces. A thorough SEM examination of the fracture surfaces following the principles of fractography and confirmation of surface composition through the use of X-ray elemental map analysis (Figures 4 and 5) produce a more consistent and complete description of the fracture process and the modes of failure.[6,7,14] These analyses would avoid simplistic comments such as the "mixed mode of failure" that often follows the "adhesive and/or cohesive" unscientific observations. Thus, when fractography is correctly used to determine the fracture origin, a proper scientific statement on the mode of fracture can be formulated.

Although the mode of failure is an important aspect of bond strength tests, this parameter is not commonly reported. A detailed inspection of the fracture surface origin can indicate the failure mode of a bonded assembly. The fracture behavior of adhesive interfaces will depend on stress level, the flaw distribution, material properties, and environmental effects. Therefore, fracture surface characterization combined with analyses of fracture mechanics parameters are of great importance to understand and predict bonded interface reliability.[11]

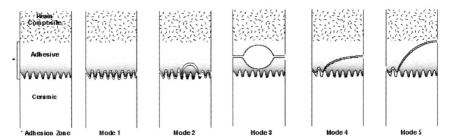

Figure 3. Schematic representation of the modes of failure for the microtensile bond strength test of ceramic bonded to resin composite, based on crack initiation and principles of fractography. Mode 1: adhesive separation at the ceramic-adhesive resin interface (adhesive failure). Mode 2: failure starts at the ceramic-adhesive interface, goes into the adhesive resin and returns to the interface (e.g. Figure 5A). Mode 3: failure from internal flaw (penny-shape internal crack). Mode 4: failure starts at the ceramic-adhesive interface propagates through the adhesive resin. Mode 5: failure starts at the ceramic-adhesive interface, goes through the adhesive resin to reach the composite-adhesive interface. Modes 2-5 are examples of cohesive failures. The adhesive-ceramic interface seems to be the weakest link of the system (e.g. Figure 5C). From Della Bona, 2009.[14]

A careful interpretation of the failure mode is required to prevent inappropriate conclusions about the utility of the microtensile test and the adhesion zone phenomena. Several dentin bond strength studies using microtensile test have reported the modes of failure based on SEM observations.[51-53] These studies based the failure classification on the substrate location where the fracture occurred. Examining the information provided in these studies one concludes that most of the failures occurred within the adhesion zone, as defined by Della Bona et al., 2003.[6] Yet, an understanding of the fracture mechanics concepts and the analysis of fracture events on the basis of fractography will reduce the risk for data misinterpretation such as the inference that the bond strength must exceed the cohesive strength of the ceramic when the fracture initiates away from the interface. Therefore, the classification of the modes of failure based on principles of fractography (Fig. 3) should assist researchers to correctly interpret the fracture phenomena.[6]

As demonstrated by finite element stress analyses (FEA), the non-uniformity of the interfacial stress distribution generated during conventional tensile and shear bond strength testing may lead to fracture initiation from flaws at the interface or within the substrate at areas of high localized stress.[3,48,54] To promote crack initiation within the interfacial zone, an interfacial toughness test can be used,[6,20] and fractography can identify the initial critical flaw, suggesting the interfacial fracture toughness. Therefore, a more meaningful property than bond strength could be assessed employing fracture mechanics principles. The fracture toughness (K_C) reflects the ability of a material to resist unstable crack propagation.[14,17,55] Following this rationale, another way to assess the strength of the bond is to estimate the apparent interfacial fracture toughness (K_A) of the adhesion zone by promoting crack initiation within the bonding interface.[20] Strictly speaking, measurement of the toughness at the interface using K_{IC} is undefined. However, tensile tests can be performed in which a crack or defect is the source of failure. Therefore, the apparent fracture toughness of the interface can be calculated from the size of the defect and the strength with the appropriate geometric factor. Thus, the apparent fracture toughness value (K_A) reflects the ability of a material to resist unstable crack propagation at the interface.[18,19]

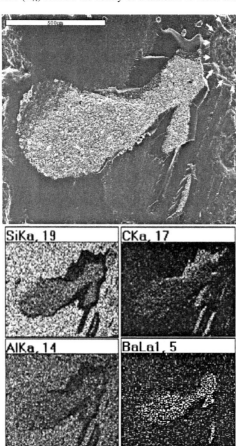

Figure 4. A fracture bonding interface surface examined using SEM to determine the mode of failure, which was confirmed using X-ray dot mapping. This figure shows a SEM image (top) and X-ray elemental maps of fracture surface of a feldspathic ceramic (IPS Empress, Ivoclar) bonded to a resin-based composite (Z100, 3M-Espe). The label at the top of X-ray maps indicates the elements and their intensity. The critical flaw is indicated by the white arrows at the top right corner of the SEM micrograph (x100). The fracture starts along the ceramic-adhesive interface, propagates through the adhesive resin to reach the composite-adhesive interface (Failure Mode 5). From Della Bona, 2009.[14]

Usually, in order to maintain equal compliances of the specimen halves, *i.e.* for the two halves to have equal strain energy, most of the interfacial fracture toughness tests require:

$$E_1 I_1 = E_2 I \qquad (4)$$

Where E_1 is the elastic modulus for the first specimen half (ceramic), I_1 is the moment of inertia of the ceramic; E_2 is the elastic modulus of the composite half, and I_2 is the moment of inertia of the composite.[6,18,19]

As the microtensile test is a uniaxial tensile test, there is no need for preparing specimens with balanced compliance for the two materials to measure K_A. Yet, the fracture toughness in the form of the stress intensity factor is really a pseudocritical stress-intensity factor, *i.e.* apparent toughness (K_A), considering the fact that it is difficult to define a stress intensity at an interface and one must determine an effective modulus for the two systems.[18] To determine the apparent interfacial fracture toughness, it is assumed that the initiation of fracture occurs at the ceramic-adhesive interface from a crack or defect whose size can be measured. The fundamental fracture mechanics expression is then applied to this condition with these assumptions:

$$K_A = Y \sigma (c)^{1/2} \qquad (5)$$

where Y is a geometric factor that accounts for crack geometry shape and location as well as the geometry of loading, σ is the applied stress at failure and c is the equivalent semi-circular crack size, i.e., a is the depth and 2b is the width of a semi-elliptical crack [$c = (a b)^{1/2}$], e.g. see Figure 5.

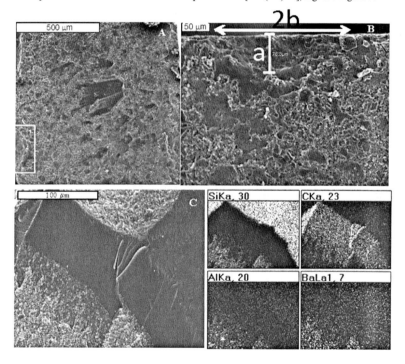

Figure 5. Representative SEM micrographs of fracture surfaces. (A) Semi-circular flaw produced by indentation is the crack origin (white box); the adhesive resin island in the middle of fracture surface has some fracture markings and represents a failure mode 2 (x80). (B) Enlargement of white box area of (A) showing the size of the crack semi-minor axis (a = 76.3 μm) (x300). (C) SEM image and x-ray elemental maps (lower right) of fracture surface representing fracture mode 5. The label at the top of x-ray maps indicates the elements and their intensity. From Della Bona et al., 2006.[20]

Another appropriate way to assess the interfacial bond is to analyze the energy per unit crack surface area, G_{IC}, that is required for a crack to advance in the bond plane. The toughness relative to the strain energy release rate (G_{IC}) is an accurate measure of the resistance of the bond to fracture since G_{IC} represents the relative energy required to create the new surfaces.[18]

In conclusion, understanding the bonding mechanisms of adhesive interfaces and a careful microscopic analysis of the fracture surfaces can produce a more consistent and complete description of the fracture process and interpretation of the modes of failure, offering important information leading to predictions of clinical performance limits, which is the ultimate test of any adhesive system.

ACKNOWLEDGEMENTS
This paper contains information from projects supported by CNPq-Brazil, grants 300659/03-2, 300748/06-0, 302364/09-9.

REFERENCES
[1]M.B. Blatz, A. Sadan, and M. Kern, Resin-Ceramic Bonding: a Review of the Literature, *J. Prosthet. Dent.*, **89**, 268-74 (2003).
[2]R.G. Chadwick, A.G. Mason, and W. Sharp, Attempted Evaluation of Three Porcelain Repair Systems- What Are We Really Testing?, *J. Oral Rehabil.*, **25**, 610-5 (1998).
[3]A. Della Bona, and R. van Noort, Shear Versus Tensile Bond Strength of Resin Composite Bonded to Ceramic, *J. Dent. Res.*, **74**, 1591-6 (1995).
[4]A. Della Bona, K.J. Anusavice, and C. Shen, Microtensile Strength of Composite Bonded to Hot-Pressed Ceramics, *J Adhesive Dent.*, **2**, 305-13 (2000).
[5]A. Della Bona, K.J. Anusavice, and J.A.A. Hood, Effect of Ceramic Surface Treatment on Tensile Bond Strength to a Resin Cement, *Int. J. Prosthodont.*, **15**, 248-53 (2002).
[6]A. Della Bona, K.J. Anusavice, and J.J. Mecholsky Jr., Failure Analysis of Resin Composite Bonded to Ceramic, *Dent. Mater.*, **19**, 693-9 (2003).
[7]T.A. Donassollo, F.F. Demarco, and A. Della Bona, Resin Bond Strength to a Zirconia-Reinforced Ceramic After Different Surface Treatments, *Gen. Dent.*, **57**, 374-9 (2009).
[8]N. Boscato, A. Della Bona, and A.A. Del Bel Cury, Influence of Ceramic Pre-Treatments on Tensile Bond Strength and Mode of Failure of Resin Bonded to Ceramics, *Am. J. Dent.*, **20**, 103-8 (2007).
[9]A. Della Bona, M. Borba, P. Benetti, and D. Cecchetti, Effect of Surface Treatments on the Bond Strength of a Zirconia-Reinforced Ceramic to Composite Resin, *Braz. Oral Res.*, **21**, 10-5 (2007).
[10]L.F. Valandro, A. Della Bona, M.A. Bottino, and M.P. Neisser, The Effect of Ceramic Surface Treatment on Bonding to Densely Sintered Alumina Ceramic, *J. Prosthet. Dent.*, **93**, 253-9 (2005).
[11]L.F. Valandro, A. Mallmann, A. Della Bona, and M.A. Bottino, Bonding to Densely Sintered Alumina- and Glass Infiltrated Aluminum / Zirconium-Based Ceramics, *J. Appl. Oral Sci.*, **13**, 47-52 (2005).
[12]A. Della Bona, Characterizing Ceramics and the Interfacial Adhesion to Resin: II- The Relationship of Surface Treatment, Bond Strength, Interfacial Toughness and Fractography, *J. Appl. Oral Sci.*, **13**, 101-9 (2005).
[13]R.F. Schreiner, R.P. Chappell, A.G. Glaros, and J.D. Eick, Microtensile Testing of Dentin Adhesives, *Dent. Mater.*, **14**, 194-201 (1998).

[14]A. Della Bona, Bonding to Ceramics: Scientific Evidences for Clinical Dentistry, Sao Paulo: Artes Medicas (2009).

[15]A. Della Bona, C. Shen, and K.J. Anusavice, Work of Adhesion of Resin on Treated Lithia Disilicate-Based Ceramic, *Dent Mater.*, **20**, 338-44 (2004).

[16]R.D.Phoenix, and C. Shen, Characterization of Treated Porcelain Surfaces Via Dynamic Contact Angle Analysis, *Int. J. Prosthodont.*, **8**, 187-94 (1995).

[17]A. Della Bona, K.J. Anusavice, and J.J. Mecholsky Jr., Fracture Behavior of Lithia Disilicate- and Leucite-Based Ceramics, *Dent. Mater.*, **20**, 956-62 (2004).

[18]J.J. Mecholsky Jr., and L.M. Barker, A Chevron-Notched Specimen for Fracture Toughness Measurements of Ceramic-Metal Interfaces, *ASTM STP*, 855: 324-36 (1984).

[19]L.E. Tam, and R.M. Pilliar, Fracture Toughness of Dentin/Resin-Composite Adhesive Interfaces. *J. Dent. Res.*, **72**, 953-9 (1993).

[20]A. Della Bona, K.J. Anusavice, and J.J. Mecholsky Jr, Apparent Interfacial Fracture Toughness of Resin/Ceramic Systems, *J. Dent. Res.*, **85**, 1037-41 (2006).

[21]J.H. Chen, H. Matsumura, and M. Atsuta, Effect of Etchant, Etching Period, and Silane Priming on Bond Strength to Porcelain of Composite Resin, *Oper. Dent.*, **23**, 250-257 (1998).

[22]A. Della Bona, and R. van Noort, Ceramic Surface Preparations for Resin Bonding: a SEM Study, *Am. J. Dent.*, **11**, 276-80 (1998).

[23]A. Della Bona, and K.J. Anusavice, Microstructure, Composition, and Etching Topography of Dental Ceramics. *Int. J. Prosthodont.*, **15**, 159-67 (2002).

[24]V. Jardel, M. Degrange, B. Picard, and G. Derrien, Correlation of Topography to Bond Strength of Etched Ceramic. *Int. J. Prosthodont.*, **12**, 59-64 (1999).

[25]A. Della Bona, T.A. Donassollo, F.F. Demarco, A.A. Barrett, and J.J. Mecholsky Jr, Characterization and Surface Treatment Effects on Topography of a Glass-Infiltrated Alumina/Zirconia-Reinforced Ceramic, *Dent. Mater.*, **23**, 769-75 (2007).

[26]A. Della Bona, Characterizing Ceramics and the Interfacial Adhesion to Resin: I - The Relationship of Microstructure, Composition, Properties and Fractography, *J. Appl. Oral Sci.*, **13**, 1-9 (2005).

[27]V. Jardel, M. Degrange, B. Picard, and G. Derrien, Surface Energy of Etched Ceramic, *Int. J. Prosthodont.*, **12**, 415-8 (1999).

[28]W.D. Callister Jr., Materials Science and Engineering: an Introduction. 7th ed. New York: John Wiley & Sons, Inc. (2007).

[29]P. Dérand, and T. Dérand, Bond Strength of Luting Cements to Zirconium Oxide Ceramics. *Int. J. Prosthodont.*, **13**, 131-5 (2000).

[30]M. Fradeani, and M. Redemagni, An 11-year Clinical Evaluation of Leucite-Reinforced Glass-Ceramic Crowns: a Retrospective Study, *Quintessence Int.*, **33**, 503-10 (2002).

[31]M. Hayashi, Y. Tsuchitani, Y. Kawamura, M. Miura, F. Takeshige, and S. Ebisu, Eight-year Clinical Evaluation of Fired Ceramic Inlays, *Oper. Dent.*, **25**, 473-81 (2000).

[32]M. Madani, F.C.S. Chu, A.V. McDonald, and R.J. Smales, Effects of Surface Treatments on Shear Bond Strengths Between a Resin Cement and an Alumina Core, *J. Prosthet. Dent.*, **83**, 644-7 (2000).

[33]J. Tinschert, G. Natt, W. Mautsch, M. Augthun, and H. Spiekermann. Fracture Resistance of Lithium Disilicate-, Alumina-, and Zirconia-Based Three-Unit Fixed Partial Dentures: a Laboratory Study, *Int. J. Prosthodont.*, **4**, 231-8 (2001).

[34]N.M. Jedynakiewicz, and N. Martin, The Effect of Surface Coating on the Bond Strength of Machinable Ceramics, *Biomaterials*, **22**, 749-52 (2001).

[35]R. Sun, N. Suansuwan, N. Kilpatrick, and M. Swain, Characterization of Tribochemically Assisted Bonding of Composite Resin to Porcelain and Metal. *J. Dent.*, **28**, 441-5 (2000).

[36]E.P. Plueddemann, Silane Coupling Agents, New York: Plenum Press (1991).

[37]R.L. Bowen, Properties of Silica-Reinforced Polymer for Dental Restorations, *J. Am. Dent. Assoc.*, **66**, 57-64 (1963).

[38]T. Berry, N. Barghi, and K. Chung, Effect of Water Storage on the Silanization in Porcelain Repair Strength. *J. Oral Rehabil.*, **26**, 459-63 (1999).

[39]S. Eikenberg, and J. Shurtleff, Effect of Hydration on Bond Strength of a Silane-Bonded Composite to Porcelain After Seven Months, *Gen. Dent.*, **44**, 58-61 (1996).

[40]R.C. Pratt, J.O. Burgess, R.S. Schwartz, and J.H. Smith, Evaluation of Bond Strength of Six Porcelain Repair Systems, *J. Prosthet. Dent.*, **62**, 11-13 (1989).

[41]J.P. Attal, V. Edard, and M. Degrange, Factors Modifying the Accuracy of Free Energy Surface Measured by Contact Angle Method, *J. Biomater. Dent.*, **5**, 143-55 (1990).

[42]M. Panighi, and C. G'Sell, Physico-chemical Study of Dental Surfaces and Mechanisms of Composite Adhesion, *J. Biomater. Dent.*, **8**, 61-70 (1993).

[43]P. Benetti, A. Della Bona, and J.R. Kelly, Evaluation of Thermal Compatibility Between Core and Veneer Dental Ceramics Using Shear Bond Strength Test and Contact Angle Measurement, *Dent. Mater.*, **26**, 743-50 (2010).

[44]A. Versluis, D. Tantbirojn, and W.H. Douglas, Why do Shear Bond Tests Pull Out Dentin? *J. Dent. Res.*, **76**, 1298-307 (1997).

[45]S. Ban, and K.J. Anusavice, Influence of Test Method on Failure Stress of Brittle Dental Materials, *J. Dent. Res.*, **69**, 1791-9 (1990).

[46]P.H. DeHoff, K.J Anusavice, and Z. Wang, Three-dimensional Finite Element Analysis of the Shear Bond Test, *Dent. Mater.*, **11**, 126-31 (1995).

[47]S. Sudsangiam, and R. van Noort, Do Dentin Bond Strength Tests Serve a Useful Purpose? *J. Adhesive Dent.*, **1**, 57-67 (1999).

[48]R. van Noort, S. Noroozi, I.C. Howard, and G. Cardew, A Critique of Bond Strength Measurements. *J. Dent.*, **17**, 61-7 (1989).

[49]D.H. Pashley, H. Sano, B. Ciucchi, M. Yoshiyama, R.M. Carvalho, Adhesion Testing of Dentin Bonding Agents: a Review, *Dent. Mater.*, **11**, 117-25 (1995).

[50]D.H. Pashley, R.M. Carvalho, H. Sano, M. Nakajima, M. Yoshiyama, Y. Shono, The Microtensile Bond Test: a Review, *J. Adhes. Dent.*, **1**, 299-309 (1999).

[51]S.R. Armstrong, D.B. Boyer, and J.C. Keller, Microtensile Bond Strength Testing and Failure Analysis of Two Dentin Adhesives, *Dent. Mater.*, **14**, 44-50 (1998).

[52]M. Tanumiharja, M.F. Burrow, and M.J. Tyas. Microtensile Bond Strengths of Seven Dentin Adhesive Systems, *Dent. Mater.*, **16**, 180-7 (2000).

[53]T. Yoshikawa, H. Sano, M.F. Burrow, J. Tagami, and D.H. Pashley. Effects of Dentin Depth and Cavity Configuration on Bond Strength, *J. Dent. Res.*, **78**, 898-905 (1999).

[54]R. van Noort, G. Cardew, I.C. Howard, and S. Noroozi, The Effect of Local Interfacial Geometry on the Measurement of the Tensile Bond Strength to Dentin, *J. Dent. Res.*, **70**, 889-93 (1991).

[55]R.W. Hertzberg, Deformation and Fracture Mechanics of Engineering Materials. 4ª ed. New York: J. Wiley & Sons, (1996).

QUALITATIVE AND QUANTITATIVE FRACTOGRAPHIC ANALYSIS OF ALL-CERAMIC
FIXED PARTIAL DENTURES

Márcia Borba
University of Passo Fundo
Passo Fundo, RS, Brazil

Humberto N. Yoshimura
University of ABC
São Carlos, SP, Brazil

Jason A. Griggs
University of Mississippi Medical Center
Jackson, MS, USA

Paulo F. Cesar
University of São Paulo
São Paulo, SP, Brazil

Álvaro Della Bona
University of Passo Fundo
Passo Fundo, RS, Brazil

ABSTRACT
Objective: to evaluate the fracture load of all-ceramic fixed partial dentures (FPDs) and to perform a qualitative and quantitative fractographic analysis.

Methods: Five ceramics were used: yttria partially stabilized tetragonal zirconia (YZ – In-Ceram YZ), glass-infiltrated alumina/zirconia (IZ – In-Ceram Zirconia), polycrystalline alumina (AL – In-Ceram AL), two porcelains (VM7; VM9) (Vita). Steel models simulating abutment teeth were constructed to design the FPDs, and CEREC InLab was used to produce the frameworks (16 mm^2 connector). FPDs frameworks were veneered with porcelain and divided into three groups (n=10): (YZ) YZ/VM9; (AL) AL/VM7; (IZ) IZ/VM7. FPDs were tested until failure with the load being applied in the center of the pontic at 0.5 mm/min (37°C distilled water). Fracture surfaces were examined using a stereomicroscope and SEM. Fractography was used to estimate the critical flaw (c) and the fracture stress (σ_f). Data were analyzed using ANOVA and Tukey's test (α=0.05).

Results: YZ showed the highest fracture load (4540N) and σ_f (1073MPa), and the lowest c-value (25μm). AL and IZ had similar values of fracture load, σ_f and c. Macroscopically, three different sites of fracture initiation were observed: connector, pontic and crown. The critical flaw was located either on the porcelain surface or on the porcelain-framework interface. Chi-square test showed that the framework material and the critical flaw location were significantly related (p=0.017, β=0.86).

Conclusion: The fracture load and failure behavior of the FPDs were influenced by the framework material. The fracture behavior observed in vitro was similar to the behavior reported for clinically failed zirconia and alumina-based FPDs.

INTRODUCTION
The mechanical behavior of ceramic materials is usually evaluated using flexural strength tests (biaxial, three- or four-point flexure). However, the specimens used in these tests, disks and bar-shaped specimens, fail to reproduce the clinical situation since the influence of the restoration geometry in the

stress distribution is neglected. In addition, different processing steps are used to produce these specimens, such as grinding and polishing, resulting in different flaw populations and failure modes that do not correspond to the clinical situation. Therefore, a more reliable prediction of the mechanical behavior of ceramic materials is obtained through the in vitro evaluation of specimens that reproduce the shape of dental restorations, such as crowns and fixed partial dentures (FPDs), and the surface properties as received from representative grinding and polishing.[1,2]

The shape of a FPD is a complex combination of multiple convexities and concavities, depending on the geometry of the replaced teeth and their alignment in the oral cavity. A change in the contour of the prosthesis components influences the stress distribution, and this effect may be more significant in posterior FPDs, where the connector height is usually limited by short clinical molar crowns and due to higher stresses produced in the posterior areas. In addition, with the objective to improve esthetics, dental technicians tend to produce sharp embrasures in the connector area. Thus, the connector represents a thin and stress concentrating section that may reach its critical stress intensity factor before the thicker portions of the FPD (i.e, pontic).[1,3]

Metal-ceramic systems are considered to be a gold standard for prosthetic restorations. However, in-vitro studies with yttria partially stabilized tetragonal zirconia (Y-TZP) all-ceramic FPDs found relatively high fracture load values, between 2237 and 3480 N.[4-6] These results are in the same range of the fracture load values reported for metal-ceramic FPDs (2800 to 3500 N).[7] A study observed that the fracture load values for inlay-retained metal-ceramic FPDs (1318 N) were similar to the results obtained with the Y-TZP system (1248 N).[8] In addition, clinical evaluation of zirconia-based all-ceramic FPDs (In-Ceram Zirconia and Y-TZP systems) reported low failure rates, between 2 to 6%. However, the follow-up period for these clinical studies is still relatively short, varying from 2 to 5 years.[9-12] For all-ceramic FPDs, the probability of failure is related to the properties of the ceramic material, to the size and shape of the connectors, to the span of the pontic, to the position in the oral cavity and to the present defects which are due to the surface machining by grinding and polishing.[13,14]

Analyzing the fracture surface through fractography is one of the key elements to understand the mechanical behavior of materials. Fractography includes the examination of fracture surfaces that contain features resulting from the interaction of the advancing crack with the microstructure of the material and the stress fields.[15-17] Depending on the fractured sample under investigation, quantitative fractography might be helpful to obtain information about one or more of the following features that influenced the immediate fracture event and the subsequent crack propagation: the size and location of the fracture initiating crack, the stress at failure, the existence of residual stress or slow crack growth, and knowledge of local processing features that affect the fracture process.[18,19] In addition, it is possible to verify if the failure behavior observed in vitro properly simulates the behavior observed in vivo.

The objective of this study was to evaluate the fracture load of all-ceramic FPDs and to perform a qualitative and quantitative fractographic analysis. The hypothesis of the study was that the framework material influences the fracture load and failure mode of FPDs.

METHODS

Three ceramic systems used as framework material for FPDs, and two veneering porcelains were studied. The ceramic materials are described in Table I.

Stainless steel models simulating prepared abutment teeth were constructed. The prepared die had 4.5 mm height, $6°$ of taper and $120°$ chamfer as finish line.[5] The distance between the centers of the dies was 16 mm, corresponding to the distance between a lower second premolar and a lower second molar (span of 10 mm). Polyvinyl siloxane impressions of the model were taken (Aquasil[TM], Dentsply, Petropolis, RJ, Brazil). A working cast was made using type IV special CAD/CAM stone (CAM-base, Dentona AG, Dortmund, Germany).

Table I. Materials identification

Legend	Material[*]	Composition	Indication
YZ	Vita In-Ceram YZ	yttria partially stabilized zirconia tetragonal polycrystals	framework
IZ	Vita In-Ceram Zirconia	alumina-based zirconia-reinforced glass infiltrated ceramic	framework
AL	Vita In-Ceram AL	alumina polycrystals	framework
VM7	Vita VM7	feldsphatic porcelain	veneer
VM9	Vita VM9	feldsphatic porcelain	veneer

[*] The materials were manufactured by Vita Zahnfabrik, Bad Sackingen, Germany.

The stone cast was digitized by the internal laser scanner component of CEREC inLab unit (Sirona Dental Systems, Charlotte, NC, USA) to generate a tridimensional image that was used to design the FPDs frameworks for YZ, IZ and AL ceramic systems. After the milling process, YZ and AL frameworks were sintered using the Zyrcomat furnace (Vita Zahnfabrik, Bad Sackingen, Germany), and IZ frameworks were glass infiltrated using the Inceramat 3 furnace (Vita Zahnfabrik, Bad Sackingen, Germany). The glass infiltration cycle was performed at 1110°C for 6 hours, according to the manufacturer's instruction. The excess glass (Zirconia Glass Powder, Vita Zahnfabrik, Germany) was removed with air-borne particle abrasion using 50 μm aluminum oxide particles. Only the external surface of the FPDs was air abraded because there was no excess glass on the internal surface.

The frameworks were veneered with the manufacturer recommended porcelain: VM9 for YZ, and VM7 for IZ and AL. A Keramat I furnace (Knebel, Porto Alegre, Brazil) was used to perform the porcelain sintering. The porcelains were sintered according to the following cycle: pre-drying at 500°C for 6 min, heating to 910°C at a rate of 55°C/min under vacuum, heating at 910°C for 1 min and cooling down to room temperature (~6 min). Before veneering, a bonding agent (Effect Bonder, Vita Zahnfabrik, Bad Sackingen, Germany) was applied on the YZ framework and sintered according to the manufacturer instructions. Three porcelain applications were performed and a polishing burr was used to obtain a uniform thickness of, approximately, 1.2 mm around the crowns and pontic and 0.6 mm around the connectors. The uniform porcelain thickness was assured by means of thickness measurement at six different points.[5] Finally, all FPDs were subjected to a glaze cycle using a Keramat I furnace.

The FPDs were cleaned in an ultrasonic bath with distilled water for 15 min and cemented in the metal dies with zinc phosphate cement (SS White, RJ, Brazil), following the manufacturer instructions. A cementation device was used to apply a uniform pressure of 15 N for 10 min on the FPD occlusal surface with the objective to guarantee an adequate fit on the dies.

The FPDs were divided into three groups of ten (n=10): (YZ) YZ/VM9; (AL) AL/VM7; (IZ) IZ/VM7. Fracture load was evaluated using a universal testing machine (Sintech 5G, MTS, Sao Paulo, Brazil) in 37°C distilled water. The load was applied in the center of the pontic with a stainless steel sphere (6 mm diameter) at 0.5 mm/min until failure.

Fracture surfaces were examined using a stereomicroscope (Leica MZ 125, Leica Microsystems, Germany) to determine the mode of failure based on the fracture origin and fractographic principles.[15,17,20] Subsequently, specimens were sputter-coated with gold-palladium and examined using a scanning electron microscope (SEM) to locate the position of the fracture origin and measure the critical flaw dimensions (c). The size of a semi-circular flaw, with equal stress intensity factor, was determined using the following equation:[21,22]

$$c = \sqrt{ab} \qquad (1)$$

where, a is the crack depth and b is half crack width.

In a previous research, the fracture toughness (K_{Ic}) was calculated through fractographic analysis of bar-shaped specimens subjected to a three-point flexural test.[23] Fracture mechanics was used to estimate the fracture stress (σ_f) of FPDs, following Griffith-Irwin equation:[24-26]

$$K_{IC} = Y\sigma_f \sqrt{c}$$

(2)

where Y is a geometric factor, which accounts for the shape of the fracture-initiation crack and loading condition and also depends on the ratio a/b. For semicircular cracks Y is approximately 1.24 and for corner cracks is equal to 1.4.[22]

Fracture load and fracture stress (σ_f) values were statistically analyzed by Kruskal-Wallis and Dunn's Method, since these data showed a non-Gaussian distribution. The c-values were analyzed using ANOVA and Tukey's test. The relation between the material and failure mode was analyzed using Chi-square test, which is a non-parametric test used to analyze qualitative data.

RESULTS

The median values and interquartile range for the FPDs fracture load (N) and estimated fracture stress (σ_f, MPa) are presented in Table II. YZ had the highest median fracture load followed by AL and IZ groups that were statistically similar. The same ranking was observed for the σ_f. The fracture stress for the porcelain was not estimated since this layer was damaged for most FPDs, and it was not possible to measure the critical flaw.

Table II. Median and interquartile range (IQR) of the fracture load (FL) and fracture strength (σ_f) of the experimental groups. Mean and standard deviation (SD) values of the critical flaw size (c) for the framework materials.

	FL (N)		σ_f (MPa)		c (µm)	
	Median**	IQR	Median**	IQR	Mean***	SD
YZ	4540[a]	501	1034[a]	161	25[b]	5
AL	1994[b]	465	457[b]	47	38[a]	4
IZ	2340[b]	187	443[b]	54	38[a]	6

* Values followed by the same letter in the column are statistically similar (p>0.001).
**Kruskal-Wallis and Dunn's test (p<0.001)
*** ANOVA and Tukey's test (p<0.001)

Table II also shows the mean values of the critical flaw (c) for the FPDs framework materials. AL and IZ had higher c-values than YZ. The SEM images of the critical flaw for each framework material are presented in Figure 1.

Macroscopic analysis of the FPDs after the mechanical test showed three different sites of fracture initiation: (1) connector, (2) pontic and (3) crown. In general, the fracture started in the FPD cervical area and propagated obliquely in the direction of the occlusal surface. YZ was the only group that showed crown failure (30%). All fractures from AL group initiated in the pontic area.

It was possible to identify different flaw origins through the analysis of the FPDs fracture surfaces using a stereomicroscope and SEM. In the cervical area of the pontic and connectors, the critical flaw was located either on the porcelain surface or on the interface between porcelain and framework ceramic. When failure of the crown was observed, the critical flaw initiated in the framework material, in the internal surface of the margin adjacent to the connector (Figure 2).

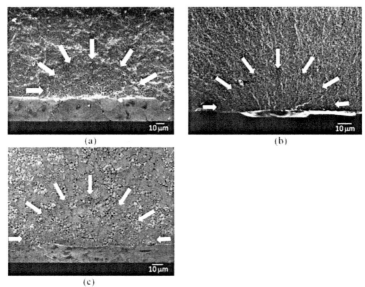

Figure 1. SEM images of the flaw origin for the framework materials:
(a) IZ; (b) YZ; (c) AL. White arrows showing the critical flaw (c).

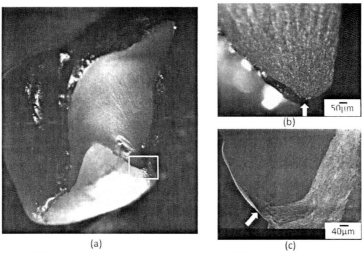

Figure 2. Failure of the crown: (a) stereomicroscope image of the failed crown; (b) and (c) closer view
of the crown margin (white box). White arrow points to the critical flaw.

Table III shows the frequency of each type of flaw origin for each experimental group. The Chi-square test showed that the framework material and the critical flaw location were significantly related ($p=0.017$, $\beta=0.86$). YZ and IZ groups had higher frequency of flaw origin located in the porcelain surface. Interfacial flaws were found for IZ (20%) and AL (60%) groups. It was not possible to locate the critical flaw in all FPDs because in some cases the fracture surface was damaged.

Table III. Frequency of each type of flaw origin (porcelain surface, crown margin and framework-porcelain interface) for each experimental group.

Groups	Porcelain Surface	Crown Margin	Framework-Porcelain Interface	Not Located
YZ	6	3	0	1
IZ	6	0	2	2
AL	4	0	6	0

DISCUSSION

In the present study, YZ group showed the highest fracture load among the experimental groups. This finding is in agreement with the results reported in the literature, in which YZ FPDs showed superior mechanical behavior than other ceramic systems (i.e. IZ, In-Ceram Alumina and IPS Empress II).[4,6,27,28] The fracture stress estimated for YZ framework using fracture mechanics was also superior, which is related to its high fracture toughness (K_{Ic} 6.5 MPa.m$^{1/2,23}$) and small critical flaw dimension (around 25 μm). IZ and AL framework materials showed similar c mean value (38 μm) and K_{Ic} values (3.5 MPa.m$^{1/2,23}$) resulting in similar fracture stress. Only the fracture stress of the framework material was estimated since it was not possible to measure the critical flaw of the porcelain layer which was often extensively damaged after the mechanical test.

Clinical studies reported that the rate of porcelain chipping is higher than the rate of framework failure for all-ceramic FPDs.[29] However, fracture of the framework represents a more critical situation because it requires replacement of the FPD. Ideally, the fracture load of both porcelain and framework should be registered. In the present study, due to the testing methodology, only the fracture load of the framework was obtained.

Macroscopically, the experimental groups showed a similar fracture mode. The fracture initiated in the cervical area of the connector or pontic and propagated obliquely through the pontic occlusal surface, near the area of load application. Only for YZ group the failure initiation site was observed on one of the crowns. In these cases, the critical flaw was located on the internal surface of the margin of the crown in the framework material. The crown margin corresponds to a thin and sharp area in comparison to the pontic and connectors. Therefore, the margin is a site of stress concentration and is susceptible to failure,[30] mainly when a large connector cross-section (16 mm^2) and high toughness material are used, such as YZ.

In the present study the framework material and the critical flaw location were significantly related. For YZ and IZ groups most of the flaws were located on the surface of the porcelain in the cervical area. The flaw propagated through the porcelain layer and deflected in the interface. When the stress magnitude exceeded the framework material toughness another flaw initiated in the framework surface and propagated resulting in catastrophic failure of the FPD. In these cases two critical flaws were identified, one in the porcelain surface and another in the framework surface. Crack arresting was also evidenced by the difference between the fracture surface planes for the porcelain and the framework. A similar behavior was observed in a fractographic study of clinically failed YZ FPDs.[31]

For AL specimens, 60% of the failures started in the interface between the porcelain and the framework material. The critical flaw was located in the framework surface and propagated through the structure as if it was a homogeneous material with no deflection in the interface. This behavior could be related to the great mismatch between the elastic modulus of AL (390 GPa) and VM7 porcelain (67 GPa), inducing higher stress concentration at the interface. A material with higher elastic modulus has less strain and lower stresses are induced in the porcelain layer. In addition, the interface is a source of unique flaws, phase boundaries, and residual stresses induced by thermal mismatches. An investigation with alumina-based (In-Ceram Alumina) FPDs that failed in vitro and in vivo also found that the flaw was located in the interface for most cases (~70%).[32]

CONCLUSION

As the fracture load and failure behavior of the FPDs were influenced by the framework material, the study hypothesis was accepted. YZ showed the highest fracture load and fracture stress values, and the lowest critical flaw dimension among the experimental groups.

ACKNOWLEDGEMENTS

The authors acknowledge the Brazilian agencies FAPESP, CNPq and CAPES for the financial support of this research. This investigation was also supported in part by Research Grant DE013358 and DE017991 from the NIH-NIDCR.

REFERENCES
[1]W.S. Oh, K.J. Anusavice. Effect of connector design on the fracture resistance of all-ceramic fixed partial dentures. *J Prosthet Dent*, 87, 536-42 (2002).
[2]K. Pallis, J.A. Griggs, R.D. Woody, G.E. Guillen, A.W. Miller. Fracture resistance of three all-ceramic restorative systems for posterior applications. *J Prosthet Dent*, 91, 561-9 (2004).
[3]C. Larsson, L. Holm, N. Lovgren, Y. Kokubo, P. Vult von Steyern. Fracture strength of four-unit Y-TZP FPD cores designed with varying connector diameter. An in-vitro study. *J Oral Rehabil*, 34, 702-9 (2007).
[4]J. Tinschert, G. Natt, W. Mautsch, M. Augthun, H. Spiekermann. Fracture resistance of lithium disilicate-, alumina-, and zirconia-based three-unit fixed partial dentures: a laboratory study. *Int J Prosthodont*, 14, 231-8 (2001).
[5]A. Sundh, M. Molin, G. Sjogren. Fracture resistance of yttrium oxide partially-stabilized zirconia all-ceramic bridges after veneering and mechanical fatigue testing. *Dent Mater*, 21, 476-82 (2005).
[6]S. Wolfart, K. Ludwig, A. Uphaus, M. Kern. Fracture strength of all-ceramic posterior inlay-retained fixed partial dentures. *Dent Mater*, 23, 1513-20 (2007).
[7]S. Chitmongkolsuk, G. Heydecke, C. Stappert, J.R. Strub. Fracture strength of all-ceramic lithium disilicate and porcelain-fused-to-metal bridges for molar replacement after dynamic loading. *Eur J Prosthodont Restor Dent*, 10, 15-22 (2002).
[8]M.A. Kilicarslan, P.S. Kedici, H.C. Kucukesmen, B.C. Uludag. In vitro fracture resistance of posterior metal-ceramic and all-ceramic inlay-retained resin-bonded fixed partial dentures. *J Prosthet Dent*, 92, 365-70 (2004).
[9]M.J. Suarez, J.F. Lozano, M. Paz Salido, F. Martinez. Three-year clinical evaluation of In-Ceram Zirconia posterior FPDs. *Int J Prosthodont*, 17, 35-8 (2004).
[10]P. Vult von Steyern, P. Carlson, K. Nilner. All-ceramic fixed partial dentures designed according to the DC-Zirkon technique. A 2-year clinical study. *J Oral Rehabil*, 32, 180-7 (2005).
[11]I. Sailer, A. Feher, F. Filser, H. Luthy, L.J. Gauckler, P. Scharer, et al. Prospective clinical study of zirconia posterior fixed partial dentures: 3-year follow-up. *Quintessence Int*, 37, 685-93 (2006).
[12]M.K. Molin, S.L. Karlsson. Five-year clinical prospective evaluation of zirconia-based Denzir 3-unit FPDs. *Int J Prosthodont*, 21, 223-7 (2008).
[13]A.J. Raigrodski. Contemporary materials and technologies for all-ceramic fixed partial dentures: a review of the literature. *J Prosthet Dent*, 92, 557-62 (2004).

[14]A. Della Bona. Bonding to ceramics: scientific evidences for clinical dentistry. 1° ed. São Paulo: Artes Médicas; 2009.

[15]G.D. Quinn. Fractography of ceramics and glasses. Washington: National Institute of Standards and Technology; 2007.

[16]S.S. Scherrer, G.D. Quinn, J.B. Quinn. Fractographic failure analysis of a Procera AllCeram crown using stereo and scanning electron microscopy. *Dent Mater*, 24, 1107-13 (2008).

[17]J.J. Mecholsky. Fractography: determining the sites of fracture initiation. *Dent Mater*, 11, 113-6 (1995).

[18]A. Della Bona, K.J. Anusavice, P.H. DeHoff. Weibull analysis and flexural strength of hot-pressed core and veneered ceramic structures. *Dent Mater*, 19, 662-9 (2003).

[19]B. Taskonak, J.A. Griggs, J.J. Mecholsky, Jr., J.H. Yan. Analysis of subcritical crack growth in dental ceramics using fracture mechanics and fractography. *Dent Mater*, 24, 700-7 (2008b).

[20]C 1322-02a, Standard Practice for Fractography and Characterization of Fracture Origins in Advanced Ceramics, (2003).

[21]P.N. Randall. Plain strain crack toughness testing of high strength metallic materials. In: Jr BWF, E SJ, editors. ASTM STP 410. Philadelphia: American Society for Testing and Materials; 1966. p. 88-126.

[22]A. Della Bona, J.J. Mecholsky, Jr., K.J. Anusavice. Fracture behavior of lithia disilicate- and leucite-based ceramics. *Dent Mater*, 20, 956-62 (2004).

[23]M. Borba, M.D. de Araujo, K.A. Fukushima, H.N. Yoshimura, P.F. Cesar, J.A. Griggs, et al. Effect of the microstructure on the lifetime of dental ceramics. *Dent Mater*, 27, 710-21 (2011).

[24]A.A. Griffith. The phenomena of rupture and flow in solids. *Philos Trans R Soc*, 221, 163-98 (1920).

[25]G.R. Irwin. Analysis of stresses and strain near the end of crack transversing a plate. *J Appl Mech*, 24, 361-4 (1957).

[26]J.J. Mecholsky. Fracture mechanics principles. *Dent Mater*, 11, 111-2 (1995a).

[27]H. Luthy, F. Filser, O. Loeffel, M. Schumacher, L.J. Gauckler, C.H. Hammerle. Strength and reliability of four-unit all-ceramic posterior bridges. *Dent Mater*, 21, 930-7 (2005).

[28]F. Beuer, B. Steff, M. Naumann, J.A. Sorensen. Load-bearing capacity of all-ceramic three-unit fixed partial dentures with different computer-aided design (CAD)/computer-aided manufacturing (CAM) fabricated framework materials. *Eur J Oral Sci*, 116, 381-6 (2008).

[29]A. Della Bona, J.R. Kelly. The clinical success of all-ceramic restorations. *J Am Dent Assoc*, 139 Suppl, 8S-13S (2008).

[30]J.B. Quinn, G.D. Quinn, J.R. Kelly, S.S. Scherrer. Fractographic analyses of three ceramic whole crown restoration failures. *Dent Mater*, 21, 920-9 (2005).

[31]B. Taskonak, J. Yan, J.J. Mecholsky, Jr., A. Sertgoz, A. Kocak. Fractographic analyses of zirconia-based fixed partial dentures. *Dent Mater*, 24, 1077-82 (2008a).

[32]J.R. Kelly, J.A. Tesk, J.A. Sorensen. Failure of all-ceramic fixed partial dentures in vitro and in vivo: analysis and modeling. *J Dent Res*, 74, 1253-8 (1995).

INTRINSIC STRENGTH, DAMAGE AND FAILURE

C.R. Kurkjian
University of Southern Maine
Gorham, ME 04038

ABSTRACT

This paper is essentially my 'point of view' regarding the state of our understanding of the strength of glass. It was developed as part of preliminary work done at the request of GMIC[1]: The Glass Manufacturers Industrial Council (some of these thoughts were published earlier in the International Journal of Applied Glass Science[2,3]). A few years ago GMIC started an initiative to promote the study of the strength of glass that has resulted in the formation of the "Useable Glass Strength Consortium" - UGSC. More particularly, they saw a need to increase the level of fundamental research on strength in order to take advantage of what they saw as recent experimental and theoretical advances in this area. In this paper I will describe three specific areas. I will review the advances made in the measurement and understanding of the intrinsic (pristine) strengths of oxide glasses. Secondly, as has been known for a very long time, contact damage can reduce this intrinsic strength by three orders of magnitude or more. Thus I will look at the history of the study of this damage and suggest that the application of photoelastic techniques may be of great use and have been mainly overlooked. Finally I will point out that since any failure: failure of a high-strength, flaw-free fiber or a damaged bottle, will occur when the intrinsic failure stress (or more accurately, K, the stress intensity) is applied either to that perfect fiber or to the tip of a crack in the bottle. Clearly, we must understand the behavior of glass at these high stress levels.

INTRODUCTION

Alan Arnold Griffith[4] is commonly thought of as the father both of the theory of so-called Griffith flaws and of fracture mechanics. His work followed that of Inglis[5] who had showed that a geometrical discontinuity in a solid body will lead to an increase in the stress at that point, and would thus lead to an apparent decrease in its strength. While Griffith appeared to show that the strength of a glass fiber would increase as its diameter decreased, this has since been shown to be extrinsic[6]. Understanding of the true intrinsic strength, its composition dependence and the details of the mechanisms of contact damage and thus strength decrease are, at the moment, far from complete. It is my aim in this paper to review some of these issues.

Soon after Griffith's work and the discussion of the so-called Griffith flaws, sustained efforts went into the search for these Griffith flaws: at first thought to be intrinsic, and then to try to understand the reasons for extrinsic flaws as well. On the basis of the Griffiths equation, Irwin[7] formalized the fundamentals of fracture mechanics. Following that, a great deal of work, often using glasses as a convenient solid, went into the development of a framework for describing the behavior of solids with a variety of defects under a variety of mechanical constraints. In what follows I will describe the state of understanding of the time, temperature and composition dependence of intrinsic strength of oxide glasses and will describe the application of photoelastic experimental/theoretical techniques to probe the contact stresses giving rise to mechanical damage/cracking. Finally it is important to realize that in all cases when mechanical failure occurs, at the point of failure, the intrinsic failure stress rather than the lower 'applied stress' is an important parameter and therefore it is necessary to understand the stress dependence of all glass properties.

INTRINSIC STRENGTH

As indicated above, Griffith's data showed that the strength of glass fibers increased very rapidly with a decrease in fiber diameter. Although a completely satisfactory explanation of this behavior cannot be given at the moment because of the lack of detailed data and a satisfactory discussion of his experimental conditions, it now seems that this effect is not fundamental. In fact, in 1967 Neil Cameron[8] at the University of Illinois showed an effect illustrated in Figure 1, which may

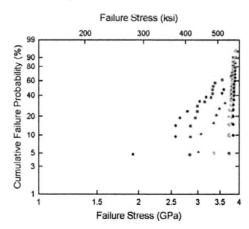

Figure 1. Tensile strengths of E-glass fibers as a function of 'soak' temperature[8]

completely account for Griffith's results. Cameron clearly showed in detail the very important fact that had been suggested earlier by Otto[10] and Thomas.[11] He showed that a certain minimum time and (soak) temperature were required in order to effectively erase the thermal history effects which resulted in a bimodal strength distribution. As this 'soak' temperature and time were increased, even though the draw temperature was maintained at a constant and lower temperature, the low strength tail portion gradually decreased until essentially single mode strength was obtained. Clearly, the *mean or median strengths* increase with increasing 'soak' temperature. These results, together with those of Otto and Thomas, suggest that the use of a lower draw ('soak') temperature for drawing thicker fibers, could lead to a lower median strength. Although a detailed understanding of this has not yet been developed its importance is clear.

In the mid-1970s, many companies around the world were drawing high-strength silica fibers in extremely long lengths for telecommunications use. Testing was often done in 20 or 50 m gauge length and a great deal of data were accumulated. It was noticed that all of these strengths had approximately the same maximum value and all of them had this essentially same Weibull m- value: that is the distribution of strengths appeared to be very similar and had a value of standard deviation (ν) of about 5%. However, when engineers started to employ diameter feedback systems in order to reduce the deviation in fiber diameter, it was noticed that the strength distribution became narrower and the m-value increased to about 150. This latter value corresponds to a variance of 0.8%. It was then argued that this very small distribution really implied that the strength was single-valued[12]. While this might be interpreted to mean that all of the flaws are the same size, we prefer to think of this these fibers as flaw-free, and will describe them as such in what follows.

These long gauge length fibers were studied in order to understand the sources of manufacturing defects. The quality of the glass surface, the quality of the bulk glass, the quality of the furnace atmosphere and the quality of the coating, were all taken into consideration until very high-quality fibers were produced in very long lengths. After that, the interest changed to a study of the fatigue and aging behavior of these fibers. In this case it was found more expedient to use a simpler testing technique and two point bending (TPB)[13] was often employed. This is a very useful technique because it is very simple to use under ambient conditions but perhaps more importantly, it can be simply used under a variety of other conditions: for instance liquid nitrogen, vacuum, wet or dry atmospheres, etc. This two point bend technique has recently been employed with great success at the Missouri University of Science and Technology at Rolla, MO. Lower et al[14] and Tang et al[15] have studied not only composition, time and temperature dependence but also have gone further in attempting to understand the melting temperature and time relations investigated first by Otto, Thomas and Cameron.

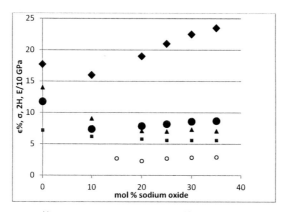

Figure 2. Failure strain[14] (filled diamonds), failure stress[15] (filled circles), 2xH (filled triangles), E/10 (filled squares). Open circles (roughly) represent the data of Pukh[17] and Kennedy.[18]

Figure 2 shows the results of two such point bend measurements under inert conditions, (that is at 77° K), for glasses in the sodium silicate system. The measured failure strains for the system are indicated by the filled diamonds. The failure stresses, calculated from the failure strains using the technique of Gupta et al,[16] are shown by the filled circles. Results of failure stress measured by other investigators[17,18] in three-point bending, are indicated by the small open circles. The reason for the discrepancy between the TPB data and the other two sets is unclear. It is possible that there is a problem with the estimation of the Young's modulus at the failure strain and thus the higher strengths shown by Tang are in error. Neither Pesina et al[17] nor Kennedy[18] showed Weibull plots so that it is possible that the soak temperature problem indicated by Cameron's results may account for the discrepancy. As indicated by the figure caption, the filled diamonds and filled squares correspond to twice the Vickers hardness (2H) and the Young's modulus, E, divided by 10 (E/10). The hardness is multiplied by two in order to convert the room temperature hardness values to those to be expected at liquid nitrogen temperature. Also some qualitative models have indicated that the intrinsic strength would be expected to be of the order 1/10 Young's modulus. It is interesting that 2H approximates the trend shown by the failure stress, while there is relatively little change of Young's modulus with composition: that is, the brittle fracture (strength) and flow behavior (hardness) appear similar.

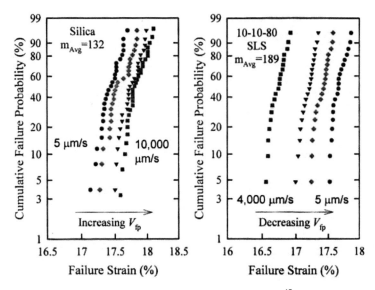

Figure 3. Two point bend failure strain at 77K.[17]

Another interesting thing found in the studies from Rolla is that in testing in the absence of water, e.g. at liquid nitrogen, an 'inert dynamic fatigue affect' (IDFE) appears: see also figure 3. In silica this behavior is similar to that of normal 'stress corrosion' fatigue. That is to say, as testing is carried out more quickly, the failure stress increases. This is what was called 'thermal fluctuation' fatigue by Russian workers[19] many years ago. An interesting and potentially very useful finding was that while this 'normal' behavior was found for silica, an anomalous glass, for normal glasses such as soda lime glass or even in multicomponent glasses in which 20% or more alkali is added, this 'anomalous' IDFE effect was found. For these glasses, the strength *decreases* with increasing testing rate! This is an unexpected anomaly which suggests some kind of relaxation process at these high stresses.

Now that I hope I have convinced you that there is no diameter dependence of strength, I'm afraid I need to show some recent results of Brambilla and Payne[20] on nano fibers (500 nm diameter). These are shown in figure 4. These results seem self-explanatory, but the reasons for this striking deviation from the claimed diameter dependence is unclear!

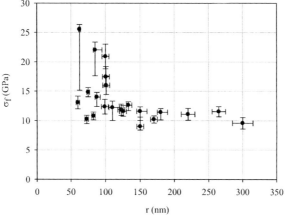

Figure 4. Tensile strength measurements of nano silica fibers (500 nm) at room temperature[20].

CONTACT DAMAGE

In 1880, Hertz[21] analyzed the elastic stresses that result from contact between a ball and a flat plate. Since that time, ball indentations have been used in glass studies primarily to understand the surface flaw distributions in, for instance, plate glass. The elastic analysis by Hertz proved very useful in describing how these stresses will propagate ring/cone cracks in a region near its contact circle and therefore such a technique can be used to map surface flaws distributions. In early studies of indentation using a pointed indenter, it was first argued that cracking due to this type of indenter was similar and thus pre-existing cracks[22] were blamed and no crack nucleation processes were considered. In the 1970s Peter[23] and Hagan[24] however, showed the appearance of what they called 'shear flow lines' under a pointed (Vickers) indenter in a soda lime glass. They argued that the intersection of these shear lines might be the nucleation points for flaws and thus it was not necessary to postulate 'preexisting flaws'. Subsequently Ernsberger[25] measured the index change under a silica glass and suggested that this index (density) change resulted from the extremely high stresses under a pointed indenter in so-called anomalous glasses. While no one has yet seen shear lines under a pointed indenter in silica glass, it has been pointed out by Lathabai et al[26] that there are 'apparent' shear faults in the indentation surface itself (Figure 5).

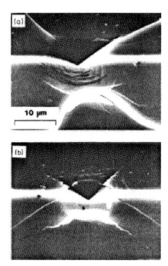

Figure 5. Shear lines *under* an Vickers indent in soda lime glass, and shear lines *on the indentation surface* in the silica glass.[26]

Thus, clearly the indentation behavior in normal and anomalous glasses is somehow quite different. More recently Yoshida et al[27] have estimated the fraction of the deformation under Vickers indentations which is the result of volume and shear deformations. They found that while soda lime glasses have roughly equal contributions from bulk and shear deformation, fused silica showed more than 90% densification.

Over the years workers have shown substantial differences in the load required to produce cracking in different glass compositions. In particular, Sehgal and Ito,[28] have described what they call a 'less brittle' glass. The next figure (6) shows how they found that rather small changes (~4or 5%) in the composition of a normal soda lime silica glass can result in quite a substantial increase in the load necessary to produce cracking under Vickers indenter. Clearly such behavior should be studied in much more detail.

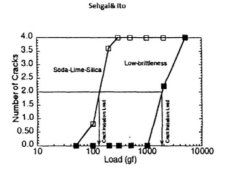

Figure 6. Data of Sehgal and Ito[28] showing the change in load on a Vickers diamond necessary to produce cracks in 'standard' and 'less brittle' soda lime glass.

PHOTOELASTICITY

We have recently applied the technique of micro-photo elasticity to this problem of contact damage in glasses. Although photo elasticity has been used since the late 19th century to look at annealing and tempering of glass objects and even though in 1920 Dalladay and Twyman[29] applied it to the study of stresses that resulted from scratching the glass, very little work has been done since that time. In 1976, Arora et al[30] studied the birefringence surrounding several cracked normal and anomalous glasses and by the application of fracture mechanics similarity analysis and argued that the stresses generated in anomalous glasses were only half that found in normal glasses. They rationalized this by recognizing that the densification that is known to occur in anomalous glasses would tend to reduce the development of stresses because the volume of the material is directly reduced. In the case of normal glasses there is shear flow which will then cause the buildup of stress. The next two images (Figure 7[31]) are photoelastic images of a normal glass (standard soda lime glass composition) and an anomalous glass (silica). In order to simplify the analysis of such an indentation, a cone indenter with the same included angle as Vickers indenter was employed since the cone gives an axially symmetric indent which simplifies the analysis. While these analyses are still being carried out, qualitatively it can be seen that the two glasses are quite markedly different. The black background corresponds to zero retardation and the lighter the color corresponds to increased retardation. The arrows indicate the slow axis of the resultant stress. Analysis of these retardation maps is being done by the so-called onion-peeling technique of Anton et al,[32] which is applicable in the case where there is axial symmetry and these results are being compared to fracture mechanics calculations done both by Yoffe[33] and Chiang et al[34] and described in detail by Cook and Pharr.[35] This photoelastic technique promises to give a great deal of information about the stresses generated both by blunt/elastic indentations as well as by pointed/plastic indentations.

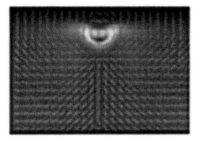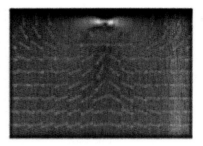

Figure 7. Retardation and slow axis direction in soda lime glass (left) and silica (right) imaged perpendicular to the (vertical) loading direction.[31]

HIGH STRESS EFFECTS

High-strength, flaw-free silica fibers were tested at liquid nitrogen temperature as early as 1965 by Proctor Whitney and Johnson[36] and they found a failure stresses of ~14 GPa. Under ambient conditions this failure stress is reduced by fatigue to ~5.5 GPA. Thus, if 15 pounds of tensile load is applied to a 125 μ fiber at room temperature, this fiber will fail in about 10 seconds. With a perfectly uniform fiber, that 15 pounds will be experienced by every portion of the fiber. On the other hand, if there is a large flaw in that fiber, or if there is a plate of silica glass with a large flaw, they will not fail at 15 pounds but will fail at a lower load as described by the well-known Griffith or fracture mechanics equations. However, the silica oxygen bonds in the silica or the sodium - oxygen - silicon bonds in both cases will fail when these bonds experience that same intrinsic stress of 5.5 GPa, even though a very much lower *applied* load is involved. Thus it would seem obvious that in order to understand and model any of these processes: indentation cracking or tensile strength, we must have knowledge of all of the glass parameters which might conceivably be involved. That is, certainly mechanical parameters such as Young's modulus and Poisson's ratio must be evaluated at high stresses.

There are many illustrations of measurements in the literature at moderate stresses or pressures: e.g., changes in the Young's modulus as a function of strain[37, 38, 39] as well as very interesting work of Schlaap[40] on the nonlinear behavior of creep.

The early work of Charles[41] on the stress dependence of electrical conductivity of a soda lime silica glass is shown in Figure 8 and illustrates quite clearly the importance of studying high stress behavior.

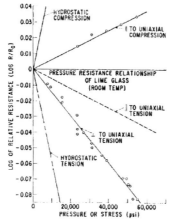

Figure 8. Change of electrical resistivity of a soda lime glass with applied stress.[41]

A particularly spectacular change in physical properties is shown in the next figure (9). This shows results from a shock experiment.[41] Here the Poisson's ratio of a soda lime glass is shown to increase from its normal value of 0.22 a value about 0 .5 at a strain of about 10%!

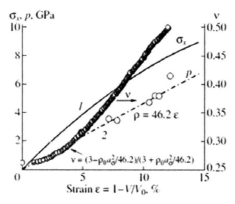

Figure 9. Effect of strain (ε) on Poisson's Ratio of a soda lime glass (open circles).[42]

Clearly knowledge of these high stress parameters is absolutely essential in order to do any kind of modeling: either the finite element or MD simulation.

SUMMARY
In this paper I have tried to illustrate where I believe we stand with respect to our understanding of the intrinsic strength of glass, as well as of our understanding of the processes which reduce this strength by contact damage. Since all processes having to do with the strength of glass involve very high

stresses: stresses of the order of 3 to 14 GPa, glass properties at these high stresses must be investigated.

REFERENCES

[1] http ://www.gmic.org/Strength%20In%20Glass.html

[2] C.R. Kurkjian, P.K. Gupta and R.K. Brow, The Strength of Silicate Glasses: What do we know, what do we need to know? *Int.Journ Appl.Glass Sci.*, **1** (1)27-37 (2010).

[3] C.R. Kurkjian, Toward a Glass with Higher Usable Strength: the need to develop collaborations! *Int.Journ.Appl.Glass Sci.,* **1** (3)216-220 (2010).

[4] A. A. Griffith, The Phenomena of Rupture and Flow in Solids, *Phil. Trans. Roy. Soc.*, **A221**, 163 (1920).

[5] C. E. Inglis, Stresses in a Plate due to the Presence of Cracks and Sharp Corners, *Inst. Naval Architects Trans.*, **55**, 219-241(1913).

[6] C.R Kurkjian, From Griffith Flaws to Perfect Fibers: A History of Glass Research, *J. Non-Cryst. Solids,* **373**, 265-271 (1985).

[7] G.R Irwin, Fracture, in Handbuch der Physik, Springer-Verlag, Berlin, **6** (1958).

[8] N. M. Cameron, Relation between Melt Treatment and Glass Fiber Strength. *J. Am. Ceram. Soc.*, **49**, 144-148 (1966).

[9] C.R. Kurkjian, and W.R., Prindle, Use of Early Maps to Guide Us Along the Road to a Stronger Glass of the Future, in: A Global Roadmap Ceramic and Glass Technology, Stephen Freeman editor Wiley (2007).

[10] W. H. Otto, Relationship of Tensile Strength of Glass Fibers to Diameter, *J. Am. Ceram. Soc.* **38**, 122 (1955).

[11] W. F. Thomas, An Investigation of the Factors Likely to Affect the Strength and Properties of Glass Fibres, *Phys. Chem Glasses*, **1**, 4-18. (1960).

[12] C. R. Kurkjian and U. C. Paek, Single-Valued Strength of 'Perfect' Silica Fibers. *Appl. Phys. Letts.*, **42**, 251 (1983).

[13] M. J. Matthewson, C. R. Kurkjian and S. T. Gulati, 'Strength Measurement of Optical Fibers by Bending,' *J. Am. Ceram. Soc.***69** 815 (1986).

[14] N.L. Lower, R.K. Brow, C.R. Kurkjian, Inert Failure Strain Studies of Sodium, Silicate Glass Fibers*, J. Non-Cryst. Solids*, **349**, 168-172 (2004).

[15] Z. Tang, R.K. Brow, N.P. Lower, P.K. Gupta, and C.R., Kurkjian, Using the two point bend technique to determine failure stress of pristine glass fiber, to be published (2011).

[16] P.K. Gupta and C.R Kurkjian, Intrinsic Failure and Nonlinear Elastic Behavior of Glass, *J. Non Cryst. Solids*, **351**, 2324-8 (2005).

[17] C. R., Kennedy, R. C., Bradt, and G. E. Rindone, Strength of Binary Alkali Silicate Glasses. *Phys. Chem Glasses,* **21**, 99-105 (1980).

[18] Pesina, T.I., Baikova, L.G., Pukh, V.P., Novak, I.I. and Kireenko, M.F., Strength and Structural Features of Glasses of the Na_2O-Al_2O_3-SiO_2 System, *Soviet Physics and Chemistry of Glass*, **12**, 1(1986).

[19] S.N., Zhurkov and B.N. Narzullaev, Time dependence of Strength of solids, *Zhur. Tekh.Fiz.***23**, 1677 (1953).

[20] G. Brambilla and D. N. Payne, The Ultimate Strength of Glass Silica Nanowires, *Nano Lett.* **9**, 831-835 (2009).

[21] H. Hertz,.,On the Contact of Elastic Solids, *J. ReineAngew. Math.* **92**, 156-171(1881).

[22] B.R. Lawn, and A.G Evans, A Model for Crack Initiation in Elastic /Plastic Indentation Fields*, Journal of Materials Science* **12**, 2195 (1977).

[23] K. W. Peter, Densification and Flow Phenomena of Glass in Indentation Experiments, *J. Non-Cryst. Solids* **5**, 103-115. (1970).

[24]J.Hagan, Shear Deformation Under Pyramidal Indentations in Soda-lime Glass. *J. Matls. Sci.* **15**, 1417-1424 (1980).

[25]F. M. Ernsberger, The Role of Densification in Deformation of Glass under Point Loading. *J. Am. Ceram. Soc.* **51**, 545-547 (1968).

[26]S. Lathabai, J. Rodel T. Dabbs and B.R. Lawn, Fracture Mechanics Model for Subthreshold Indentation Flaws, *J. Matls Sci.,* **26**, 2157-2168 (1991).

[27]S. Yoshida, J.-C. Sangleboef and T. Rouxel, Quantitative Evaluation of Indentation-induced Densification in Glass. *J. Mater. Res.*, **20**, 3404-3412 (2005).

[28]J.Sehgal amd S.Ito, A New Low-Brittleness Glass in the Soda-Lime-Silica Glass Family, *J. Am. Ceram. Soc,* **81**, 2485-2488 (1998).

[29]A.J. Dalladay and F. Twyman, The Stress Conditions Surrounding a Diamond Cut in Glass, *Trans. Optical Soc.* **23**, 165-169 (1922).

[30]A. Arora, D.B.Marshall and B.R. Lawn, Indentation Deformation/Fracture of Normal and Anomalous Glasses. *J. Non-Cryst. Solids* **31**, 415-428 (1979).

[31]A.Errapart, H. Aben, R. Oldenbourgh and C.R.Kurkjian, Application of Integrated Photoelasticity for the Measurement of Residual Stress in Locally Plastically Deformed Glass, *Proc.Photomechanics 2008,* Loughborough, U.K.(2008).

[32]J. Anton, A. Errapart, H.Aben, and L.Ainola, A Discrete Algorithm of Integrated Photoelasticity for Axisymmetric Problems, *Exp.Mech.,* **48** (5) 613-620 (2008).

[33]Yoffe, E. H. Elastic Stress Fields Caused by Indenting Brittle Materials. *Phil. Mag.* **A 46**, 617-628. (1982).

[34]S. S Chiang, D.B. Marshall and A.G Evans, The Response of Solids to Elastic/Plastic Indentation. 1. Stresses and Residual Stresses, *Journal of Applied Physics* **53**, 298-311(1982).

[35]R. F. Cook and G. M. Pharr, Direct Observation and Analysis of Indentation Cracking in Glasses and Ceramics, *J. Am. Ceram. Soc* **73** 787-817 (1990).

[36]B.A. Proctor, L. Whitney and J.W.Johnson, The Strength of Fused Silica. *Proc. Roy. Soc.* **A297** 534-557(1967).

[37]F. Mallinder and B.A. Proctor, Elastic Constants of Fused Silica as a Function of Large Tensile Strain*, Phys. Chem. Glasses*, **5**, 91 (1964).

[38]M.H.Manghnani and B. K. Singh, Effect of Composition, Pressure and Temperature on the Elastic, Thermal and Ultrasonic Properties of Sodium Silicate Glasses, *Proc. Xth ICG (1974).*

[39]R.Bruckner and G. Pahler, in: Strength of Inorganic Glass, ed. C.R.Kurkjian, Plenum Press (1985).

[40]D.M.Schlaap, Creep Properties of Glass under High Strain, *Phys. Chem. Glasses* **6**, 168-170 (1965).

[41]R.J.Charles, Effect of Uniaxial Stress on Glass Resistivity, *J.Am.Ceram. Soc.* **52** (6)350 (1969).

[42]A.S.Savinykh, G.V.Garkushin, S.V. Rasorenov and G.I.Kanel, Longitudinal and Bulk Compressibility of Soda Lime Glass at Pressures to 10 GPa, *Russian Journal of Technical Physics*, **52** (3), 328-332 (2007).

THE NOTCHED BALL TEST – CHARACTERISATION OF SURFACE DEFECTS AND THEIR INFLUENCE ON STRENGTH

Tanja Lube,[1,1] Stefan Witschnig,[1,2] Peter Supancic,[1,2] Robert Danzer,[1,2] Oskar Schöppl[3]
[1] Institut für Struktur- und Funktionskeramik (ISFK), Montanuniversitaet Leoben,
Franz-Josef-Straße 18, A-8700 Leoben, Austria
[2] Materials Center Leoben Forschung GmbH, Roseggerstrasse 12, A-8700 Leoben, Austria
[3] SKF Development Cluster Ball Bearings, Development Office Steyr, Seitenstettnerstraße 15,
A-4401 Steyr, Austria

ABSTRACT

The use of the notched ball test to measure the strength of balls is demonstrated. A long narrow notch is ground into the equatorial plane of a ball. To fracture the ball, it is then compressed diametrically perpendicular to the notch. This brings about bending stresses in the ligament in front of the notch with tensile stresses at the balls' surface. These stresses depend on the applied load, the geometry of the notch and – since the stress field is moderately biaxial – on Poisson's ratio.

The strength of silicon nitride balls with 47.625 mm diameter, which were ground by two different procedures, was determined. An optical investigation of the balls showed that different defect populations are present on the balls' surfaces. The influence of these defects on the static strength of the balls was measured by positioning the defects into the zone with the maximum tensile stress during the notched ball fracture test. A significant difference of the surface strength levels could be recognized. Fractography revealed that it was caused by different defect populations like Hertzian cracks, grinding scratches or material inhomogeneities.

INTRODUCTION

In the last decades ceramic balls are used to an increasing extent, for example in valves or in high performance hybrid ball bearings.[1,2] Because of the materials specific properties such as low wear rates, low density and high corrosion resistance, the use of structural ceramics is advantageous for applications requiring high operating speeds, electrical insulation (e.g. in windmills) or those involving severe corrosive attack (e.g. in the chemical or food industry). The load capacity and the lifetime of components are determined by their strength.[3] In the case of balls for bearings, the strength depends to a great extent on the quality of their original surface which can be influenced by machining, handling and operation.

Although information on the strength of balls is essential, testing this material property on balls is not straightforward. It is possible to machine bending specimens out of balls in order to perform bending tests (beams: uniaxial, discs: biaxial[4]), but thus only the strength of the interior (bulk) of the balls is measured. Additionally these methods are not applicable to small balls. In another procedure, called the "crush test", one to three balls are aligned between two punches and squeezed together until fracture of one of the balls occurs[5]. In this test, the original components and not specially produced specimens are tested. The resulting quantity from this test is the fracture load which depends on the ball size. The fracture load does not clearly correlate to material properties like strength or fracture toughness and no fractographical analysis is possible because the balls are usually completely crushed. For this reason, tests like the C-sphere Test[6] or the Notched Ball Test (NBT)[7] were developed. A notch is cut into the equatorial plane of the ball (to approx. 80 % of the diameter) and a compressive load is applied at the poles. In this way the notch is squeezed together and high tensile stresses occur in the surface region of the ball opposite to the notch root.

In analogy with beam bending tests these tests produce two main fractured fragments, which permits a simple detection and characterization of the failure-causing critical flaws and thus a clear correlation of surface defects and the strength of the specimen.

THE NOTCHED BALL TEST

Recently, the "Notched Ball Test" (NBT) was established at the Institut für Struktur- und Funktionskeramik to measure the strength of ceramic balls for structural applications.[7] In this test, a notch is cut along the equatorial plane of the sphere with a length of about 80% of the ball diameter and a width of approximately 10% of the ball diameter. The ball is then squeezed together perpendicular to this plane in order to generate a tensile stress at the ball surface opposite to the notch, Figure 1. As indicated in Figure 1, the stress field in the notched ball is moderately biaxial. In the region with the highest tensile stresses at the surface, the first principal stress σ_θ is approximately 10 times the second principal stress σ_φ. The value of σ_θ at position 1 is defined as the strength of the notched ball specimen, σ_{NBT}, and is calculated using:

$$\sigma_{NBT} = f_N \cdot \frac{6 F}{h^2}$$ (1)

In eq. (1), F is the fracture load and h the ligament thickness.

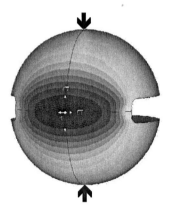

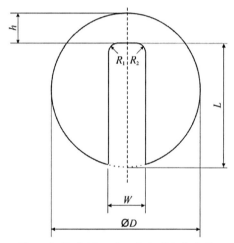

Figure 1. Schematic of the notched ball specimen. Contours of the distribution of the first principal stress σ_θ on the surface of the specimen. Black arrows indicate the loading direction.

Figure 2. Definition of symbols: D is the ball diameter, L and W are the notch length and width respectively, $h = D - L$ is the ligament thickness and R_1 and R_2 are the notch root radii.

The dimensionless factor f_N depends on the relative notch geometry, i.e. the relative notch length $\lambda = L/D$, and to a lesser extent on the relative notch width $\omega = W/D$, the shape of the notch root $\rho = 0.5 (R_1 + R_2)/W$ and Poisson's ratio v. The definition of the symbols is explained in Figure 2. f_N was evaluated numerically by FEM calculations for the parameter range given in Table I.

The notched ball test is a robust and simple test. The notch can be machined into a large number of balls at one time. All quantities that are necessary to determine the strength can be measured easily. Unless very small balls (i.e. $D < 5mm$) with very long and narrow notches are tested the error in the determined strength is dominated by the error in the measurement of the fracture load.[7] Until now the NBT has been used to test balls with diameters ranging from 5 mm to 47 mm from different ceramics like silicon nitride, silicon carbide, alumina and zirconia.[8, 9]

For the testing of the strength of balls, the notched ball test has clear advantages over standard bending tests on specimens machined from the balls. It offers the possibility to test the original machined surface with its typical defects. If these defects are known in advance, they may be aligned with respect to the stress field. If sufficient care is taken during the manufacturing of the notch no further strength limiting damage will be introduced into the specimens.

Table I. Parameter range used for the numerical evaluation of the factor f_N used in eq. (1).

relative notch length λ	$0.75 \leq \lambda \leq 0.92$	relative notch root radius ρ	$0.00 \leq \rho \leq 0.50$
relative notch width ω	$0.05 \leq \omega \leq 0.20$	Poisson's ratio ν	$0.20 \leq \nu \leq 0.30$

EXPERIMENTAL

Experiments were conducted on a total of 65 balls with a nominal diameter of $D = 47.625$ mm made from bearing ball grade silicon nitride. The balls were machined by two different procedures, P1 and P2. The surfaces of all balls were impregnated with fluorescent penetration dye and investigated using a stereo microscope and UV light to classify typical surface defects. The biggest defects were marked. This optical surface investigation procedure resulted in four different specimen sets, each one containing balls with a specific defect population (numbers in parentheses indicate the number of balls in the set): P1-A (15), P1-B (15), P2-C (15), P2-D (20). Specimens of the sets P1-A and P2-C were strength tested using the NBT in random orientation, i.e. no special attention was paid to position the defects in the tensile stress region. For the other sets the notches were positioned in a way to subject the biggest defect to the maximum tensile stress.

To cut the notches, batches of a maximum 6 balls were carefully glued into a V-block. This arrangement is made to ensure that the notches are in the equatorial plane of the balls and that the length of the notch in the batch is consistent. Special care was taken to not damage the surface area of the balls opposite the notch (in the ligament). At one end of the row a glass plate was fixed in the guide rail perpendicular to the row direction, see Figure 3. A commercial diamond grinding wheel with a nominal width of 4 mm was used. The grinding wheel was redressed to a square edge profile before and during machining each notch. The specimen preparation procedure was easy to perform but – due to the large dimensions of the balls – time consuming. It is assumed that the notch in the flat end-plate is representative of the notches in all the balls of the batch. The width and notch root radii R_1 and R_2 of the notches were determined on the notch in the glass plate for each particular cutting batch on a micrograph. The notch length of each individual ball was measured. To check the notch plane position with respect to the ideal equator plane, the offset of the notch was determined. An offset smaller than 0.5 % of the radius is tolerated. The mean geometry of the specimens is given in Table II.

Table II. Range of geometrical data of the notched balls (absolute values ± standard deviations, corresponding relative quantities are also given. The symbols are explained in Figure 2.

ligament thickness h [mm]	7.112 ± 0.084		
notch length L [mm]	40.513 ± 0.084	L / D	0.851
notch width W [mm]	4.210 ± 0.033	W / D	0.088
mean radius of the fillet R_N [mm]	0.685 ± 0.141	R_N / W	0.163
z-offset (% of the ball radius)	$0.328 + 0.235$		

To fracture the notched balls, a universal testing machine (MIDI, Messphysik) was used. The specimens were placed between two parallel punches. The end plates of these punches were made of silicon nitride to minimize any frictional effects. The notch was aligned perpendicular to the direction of load application using a positioning aid. The angular alignment of the equatorial plane was better than $\pm 0.5°$ for all tested balls. A pre-load of about 10 % of the estimated fracture load was applied and then the positioning aid was removed. The load was increased at a crosshead speed of 4 mm/min rate so that fracture occurred within 5 to 15 seconds.

Figure 3. V-block with six balls and a glass plate (arrow) after cutting the notch.

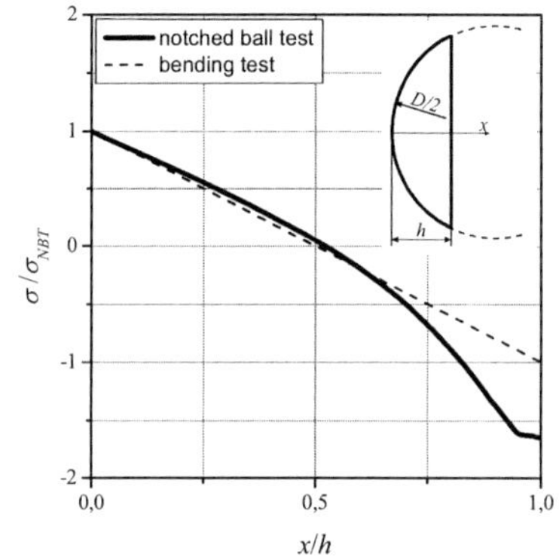

Figure 4. Stress gradient in the ligament of the notched ball (through line) compared to the stress in bending specimens with the same height h (dashed line).

The dimensionless factor f_N was determined for each specimen according to its notch geometry by interpolation of the FEM results from the parametric study. The values in the present investigation were in the range from $f_N = 0.890$ to $f_N = 0.936$. The parameters of the Weibull strength distributions were calculated according to EN 843-5.[10] Biased values are reported for the Weibull modulus.

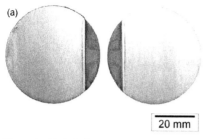 

20 mm

1 mm

Figure 5. (a) Both fragments of a fractured ball. The bright regions correspond to the notch, the dark segments in the centre of the image are the ligament showing the typical features of ceramic fracture surfaces. (b) Detail of the fracture surface shown in (a). The fracture origin is in the center of the image. Typical features – fracture mirror and hackle – are clearly visible.

Fractography was done with a stereo microscope for all specimens and scanning electron microscopy for selected fracture surfaces. An overview of the fracture surface of a fractured notched ball is shown in Figure 5. Due to the nearly uni-axial stress field with its weak gradient over the ligament thickness (see Figure 4) the fracture surfaces are easy to interpret.

RESULTS AND DISCUSSION

Typical Defects on the Balls' Surfaces
The optical inspection of the balls surfaces revealed different defects which allowed a classification of the balls into four sets. Specimens classified as P1-A have a negligible fine dispersed porosity but no other defects. Specimens in the set P1-B have straight surface scratches, Figure 6a. These scratches appear as single scratch, arrays of parallel scratches, or as scratches crossing each other. Scratches are narrow and vary in length. Balls in the set P2-C have C-cracks, i.e. partial Hertzian ring cracks, Figure 6b. P2-D specimens have C-cracks and scuff marks (SM). Scuff marks are several times wider than scratches and appear to be deposits on the surface of the balls, Figure 6c.

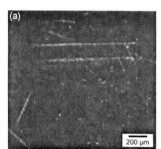 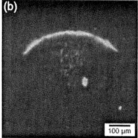 

200 μm

100 μm

500 μm

Figure 6. Defects on the surfaces of balls. (a) crossing scratches, set P1-B, (b) C-Cracks, set P2-C, (c) scuff mark, set P2-D. Note the scales: scratches are usually shorter than 1 mm and narrow while scuff marks may reach a length of several mm and are up to 200 μm wide.

Strength Distributions and Defects

The strength distributions of the sets P1-A and P1-B are plotted in a Weibull plot in Figure 7. The parameters of the Weibull strength distribution are summarized in Table III. It is evident, that the scratches reduce the characteristic strength of the balls. Even though all specimens in P1-B were tested in a way to load the biggest scratch with the highest tensile stress, fractography showed that not all specimens fractured due to the scratches. Specimens that failed due to scratches are marked by "S" symbols in the Weibull plot. All other specimens of set P1-B, as well as the specimens from P1-A, failed due to material inhomogeneities.

Table III. Parameters of the Weibull strength distributions for balls from the machining process P1 after microscopical surface inspection and fractographical classification. Numbers in brackets indicate the 90% confidence intervals.

| | after surface inspection | | after fractography | |
	P1-A	P1-B	P1-agglomerates	P1-scratches
number of specimens	15	15	22	8
characteristic strength [MPa]	741	641	725	591
	(720 – 763)	(611 – 674)	(708 – 743)	(562 –623)
Weibull modulus [-]	17.5	10.4	14.9	10.2
	(11.2 – 22.7)	(6.7 – 13.5)	(10.5 – 18.7)	(5.0 – 14.1)

Details of a fracture surface from a specimen that fractured due to a scratch are shown in Figure 8. Fracture started from a scratch approximately 20μm deep with a slightly different orientation than the rest of the fracture surface. The scratch is associated with serious damage of the microstructure up to approximately 5μm beneath the surface. Specimens from set P1-B which did not fail due to scratches, as well as specimens from set P1-A fractured due to agglomerates of the sintering aids yttria and alumina (Figure 9).

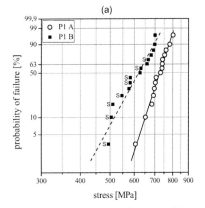
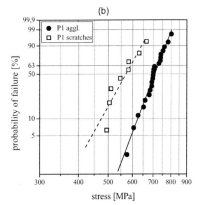

Figure 7. Weibull probability plots of the strength values measured on the specimens from the procedure P1. (a) Strength distributions due to the optical classification, sets P1-A ○ and P1-B ■. Specimens that failed due to scratches are marked by "S" symbols. (b) Strength distributions due to the classification after fractography. Specimens that failed due to scratches are shown as open squares □, specimens that failed due to agglomerates are shown as full circles ●.

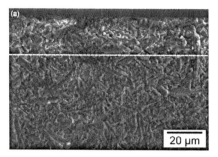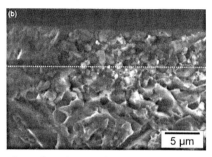

Figure 8. A scratch as typical fracture origin in specimens from the procedure P1-B. (a) Fracture started from a scratch approximately 20μm deep with a slightly different orientation than the rest of the fracture surface, dashed line. (b) The scratch is associated with damage of the microstructure up to approximately 5μm beneath the surface, dotted line.

After fractography, a new classification of the specimens is possible: separate strength distributions can be plotted for specimens that failed due to scratches and for those that fractured due to agglomerates, see Figure 7b. Obviously, the scratches determine the strength only if they are big enough. There is a small overlap with strength values due to material inherent defects. The parameters of these new Weibull strength distributions for balls from the grinding process P1 are also given in Table III.

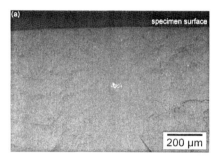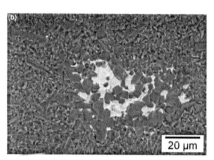

Figure 9. Agglomerate as typical fracture origin in specimens from the procedure P1-A. (a) Fracture is caused by a defect that is located approximately 330 μm below the surface of the ball. (b) The backscattered electron image allows a differentiation of chemical elements: the agglomerate is rich in yttria and alumina.

The strength distributions of the specimen sets P2-C and P2-D are shown in Figure 10a. There is a complete overlap of the two strength distributions. Fractography revealed that all specimens – from set P2-C as well as from P2-D - with a strength lower than 560 MPa failed due to C-cracks. In set P2-C, C-crack fractures also occurred in specimens which were orientated randomly with respect to the maximum stress. Failure due to scuff marks was only found on two specimens. Since the strength values for these specimens were high (636 MPa and 696 MPa) this kind of defect is recognized to be not critical for the static strength of the balls. All other specimens failed due to agglomerates similar to those found in sets P1-A and P1-B, Figure 9.

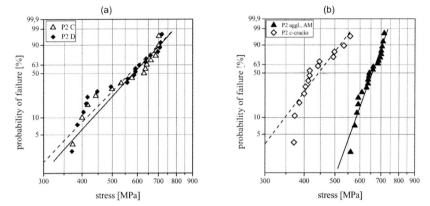

Figure 10. Weibull probability plots of the strength values measured on the specimens from the procedure P2. (a) Strength distributions due to the optical classification, sets P2-C △ and P2-D ◆. (b) Strength distributions due to the classification after fractography. Specimens that failed due to C-cracks are shown as open diamonds◇, specimens that failed due to agglomerates or scuff marks are shown by full triangles▲.

A typical fracture surface of a specimen with a C-crack as fracture origin is shown in Figure 11. C-crack fractures are usually easily detectable at low magnifications, i.e. observation with a stereo microscope is sufficient. The more detailed image made using the SEM reveals that the double c-crack of this example penetrated approximately 100 μm into the ball.

Scuff marks seem to introduce small pre-cracks into the ball without damaging the material's microstructure. This can be concluded from the tilted fracture surface in Figure 12. In contrast to the scratches (Figure 8) the material immediately beneath the surface is not damaged in this case.

Since the strength distributions of the sets P2-C and P2-D overlap completely and since both distributions contain two types of fracture origins, C-cracks on the low strength end and agglomerates at the high strength end, a new classification of the specimens is suggested. All failures due to C-cracks are evaluated in one sample and all other specimens are evaluated in another sample. These strength distributions are shown in Figure 10b. The parameters of the Weibull distributions are summarized in Table IV. It can be recognized that these two strength distributions do not overlap at all. Specimens with C-cracks have a characteristic strength which is more than 200 MPa lower than the strength of balls that fail due to agglomerates. Compared to the original distribution of all P2 specimens, the characteristic strength is increased by more that 100 MPa and the Weibull modulus is significantly increased from around $m_1 \approx 6$ to $m_2 > 16$ if failures due to C-crack can be avoided.

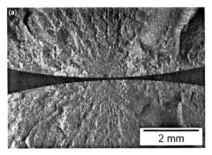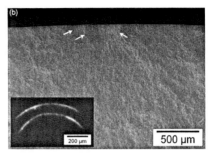

Figure 11. Typical fracture origins in specimens from the procedure P2. (a) Fracture due to a C-crack. The fracture origin is visible at a low magnification. (b) The SEM image indicates the presence of a double C-crack (see insert) that penetrated approximately 100 μm into the material prior to the strength test.

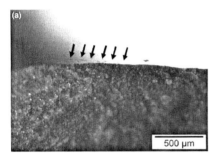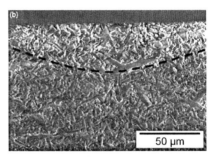

Figure 12. Typical fracture origins in specimens from the procedure P2. (b) Fracture due to a scuff mark. The scuff mark is indicated by arrows. (b) The fracture origin was a shallow semi-elliptical pre-crack of approximately 40 μm depth indicated by the dashed line.

Table IV. Parameters of the Weibull strength distributions for balls from the machining process P2 after optical and fractographical classification. Numbers in brackets indicate the 90% confidence intervals.

| | after surface inspection | | after fractography | |
	P2-C	P2-D	P2-agglom. & SM	P2-C-cracks
number of specimens	15	20	21	14
characteristic strength [MPa]	619	606	671	466
	(571 - 671)	(564 - 562)	(654 – 689)	(435 – 499)
Weibull modulus [-]	6.3	10.4	16.2	7.7
	(4.0 – 8.2)	(4.1 – 7.4)	(11.3 – 20.4)	(4.8 – 10.1)

SUMMARY AND CONCLUSIONS

The strength of balls with a diameter of 47.625 mm was tested using the notched ball test (NBT). On most balls, defects that were identified on the surface prior to notching were positioned in the region with maximum tensile stress during the NBT. The determination of the strength of selected defects was thus possible. The fracture surfaces of the notched ball specimens are expressive and permit an identification of fracture origins.

The investigated specimens were machined by two different procedures. These two grinding procedures produced clearly distinguishable defect types. The balls were classified into four sub-sets by optical selection according to the different defect types. The strength tests were used to validate these selection criteria with respect to their influence on the static strength of the balls. Fractography led to a more detailed description of the different defects.

- Deep narrow scratches on the surface of balls from P1 grinding lead to severe damage of the microstructure and to pre-cracks. The strength of balls with scratches is decreased significantly as compared to balls without scratches. Scratches are detectable on the surface of the balls during optical inspection. The optical classification was capable of sorting out the specimens with the most severe damage.
- Scuff marks that appear on balls from procedure P2 are shallow grooves in the surface that may be accompanied by pre-cracks. Since these pre-cracks are small, the strength of balls that contain such defects is high. Scuff marks are thus not a strength limiting defect type for the balls.
- C-cracks, which are only present on balls from grinding procedure P2, are the most dangerous defects in the balls. The clear separation of strength values determined by C-cracks from those due to other defects indicates that a stringent criterion is needed to sort these balls out.
- Balls that did not fail from any of the surface defects described above, failed due to agglomerates of sintering aids.
- Only procedure P1 produces balls that have no surface defects at all. Balls from procedure P2 always have small surface defects and therefore also a lower strength.

REFERENCES

[1]G.T.Y. Wan, A. Gabelli, and E. Ioannides, Increased Performance of Hybrid Bearings with Silicon Nitride Balls, *Tribology Transactions*, **40**, 701-07 (1997).

[2]L. Wang, R. W. Snidle, and L. Gu, Rolling Contact Silicon Nitride Bearing Technology: A Review of Recent Research, *Wear*, **246**, 159-73 (2000).

[3]D. Munz and T. Fett, Ceramics, pp. 298, Vol. 36, Springer, Berlin, Heidelberg, 1999.

[4]A. Börger, R. Danzer, and P. Supancic, Biaxial Strength Test of Discs of Different Size using the Ball on Three Balls Test, *Ceramic Engineering and Science Proceedings*, **25**, 283-89 (2004).

[5]D.K. Shetty and V. Pujari, Load-Bearing Capacity in Quasi-static Compression and Bearing Toughness of silicon Nitride Balls, *Tribology Transactions*, **47**, 522-26 (2004).

[6]A.A. Wereszczak, T.P. Kirkland, and O.M. Jadaan, Strength Measurement of Ceramic Spheres Using a Diametrally Compressed "C-Sphere" Specimen, *J. Am. Ceram. Soc.*, **90**, 1843–49 (2007).

[7]P. Supancic, et al., A New Test to Determine the Tensile Strength of Brittle Balls - The Notched Ball Test, *Journal of the European Ceramic Society*, **29**, 2447-59 (2009).

[8]P. Supancic, et al., The Notched Ball Test - A New Strength Test for Ceramic Spheres, pp. 327-35 in Ceramic Materials and Components for Energy and Environmental Applications, Vol. 210, *Ceramic Transactions*. Edited by D. Jiang, et al., 2010.

[9]P. Supancic, et al., Strength Tests on Silicon Nitride Balls, *Key Engineering Materials*, **409**, 193-200 (2009).

[10]EN 843-5, Advanced Technical Ceramics - Monolithic Ceramics - Mechanical Properties at Room Temperature: Part 5 - Statistical Evaluation.

ANALYTICAL MODEL FOR IMPROVED RESIDUAL SURFACE STRESS APPROXIMATION

Roberto Dugnani, Ph.D., P.E.
Zixiao Pan, Ph.D., P.E.
Exponent Failure Analysis Associates
Menlo Park, CA 94025

Traditionally the strength of glass has been estimated using empirical relationships between dimensions obtained from the fracture surface's topological features and the magnitude of the failure stress. A typical example of this is Orr's relationship $\sigma_f \cdot R_m^{1/2} = A_m$, where σ_f is the failure stress, R_m the radius of the mirror/mist region, and A_m a fitting constant. In order to determine the value of the constant A_m, the failure stress is determined by performing mechanical testing, whereas R_m is determined by careful fractographic analysis. In some cases, the relationship between the failure stress and the mirror-mist region has been extended to approximate surface residual stresses, i.e., $(\sigma_f + \Delta\sigma_0) \cdot R_m = A_m$, where $\Delta\sigma_0$ is the magnitude of the surface residual stress. This paper will show that the direct application of this empirical relationship to measure the extent of residual stresses might be misleading under certain conditions. A more precise estimation requires careful selection of the loading rate. This problem has been overlooked in previous studies. In this work, an analytical model that takes into consideration the effect of the loading rate will be presented. A non-dimensional constant, $\alpha \equiv \left(\dot{\sigma} \cdot K_{IC}^2\right) / \left(v \cdot \sigma_f^3 \cdot Y^3\right)$ is also introduced to help determine whether the effects of the loading rate are significant. It was found that for low values of α and high values of R_m/H, the mirror radius over the thickness of the sample, H, the effect of the loading rate should be considered.

INTRODUCTION

The strength of glass and other brittle materials is often estimated by the expert fractographers using established empirical relationships between the failure stress and the distance of the topological features from the fracture's origin. For instance, the magnitude of the failure stress has been shown by various authors, including Orr,[1] to correlate with the inverse of the square root of the radius from the origin to the mirror/mist boundary:

$$\sigma_f \cdot R_m^{1/2} = A_m \tag{1}$$

where σ_f denotes the stress at fracture and R_m the radius of the mirror-mist boundary region, and A_m is an empirically determined constant. To establish the value of the fitting constant A_m, mechanical testing such as four point bending tests (4PBT) as prescribed in ASTM C158[2] are carried out. The testing is supplemented by careful measurement of the mirror radius R_m as described in ASTM C1256[3]. Figure 1 shows a schematic example of how the mirror-mist boundary region is normally measured on a typical glass fracture surface. Some authors[1,4] have also proposed extending the relationship in equation 1 to approximate surface residual stresses in samples of interest. In principle, this could be achieved by expressing the stress in equation 1 as the sum of the applied stress, σ_f, and the magnitude of the surface residual stress to be approximated, i.e., $\Delta\sigma_0$:

$$(\sigma_f + \Delta\sigma_0) \cdot R_m^{1/2} = A_m \tag{2}$$

Although the technique appears to be a quick and simple method to estimate the residual stresses in glass samples, in some cases the results will lead to erroneous estimates. This paper shows that as a result of subcritical crack growth and diffusion-limited crack growth during testing, the value of $\Delta\sigma_0$ in equation 2 is expected to be a non-zero, positive number even for the case of fully annealed glasses as documented in the literature.[5]

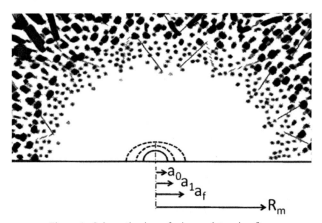

Figure 1. Schematic view of mirror-mist region for a typical glass fracture surface near the failure origin.

CRACK GROWTH IN GLASS

Various research papers have shown[6, 7] that the sub-critical crack in glasses prior to catastrophic failure follows a power-law relation, with the crack velocity, da/dt, scaling with the stress intensity factor to some power, n. Four distinct regions are generally recognizable when describing the growth of a crack under loading. In region 0 no crack extension is observed. In region I, the crack length increases rapidly as a result of stress-corrosion reactions at the crack's tip under the effects of water or other stress-corrosion effective molecules. In region II the slope of the curve decreases as a result of the limited access of the corrosive molecules to the crack's tip (diffusion-limited zone). Finally, in region III, the crack undergoes rapid fracture. Figure 2 shows a schematic view showing the relationship between the crack velocity, v, and the stress intensity factor, K_I, based on the two-region power-law model. It should be noted that instability and micro-branching at the crack's tip in region III introduce the familiar topographical features (i.e., the mirror region) on the fracture surfaces of glasses used in equation 1 and described in Figure 1.

The two-region power law prior to rapid fracture can be described by the relationship:

$$\frac{da}{dt} = v \cdot r^{n2-n1} \left(\frac{K_I}{K_{Ic}} \right)^{n1} \quad for \quad K_I < r \cdot K_{IC} \text{ (Region I)} \tag{3}$$

$$\frac{da}{dt} = v \left(\frac{K_I}{K_{Ic}} \right)^{n2} \quad for \quad K_I \geq r \cdot K_{IC} \text{ (Region II)} \tag{4}$$

K_I is the mode I stress intensity factor, K_{IC} the critical stress intensity factor, v the critical crack velocity, and r, n_1, and n_2 are fitting constants depending on the material tested and strongly dependent on environmental conditions such as temperature and humidity.

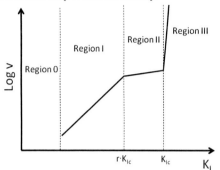

Figure 2. Schematic drawing of the two-region power law model.

MIRROR RADIUS INCLUDING CRACK-GROWTH

Equations 3 and 4 provide a general empirical relationship to describe the rate at which the crack extends prior to catastrophic failure. Some simplifications to these equations can be made for the special case when the loading is applied to the sample at a constant rate i.e., $d\sigma/dt=$constant. Constant loading rate is typically used in any laboratory test setting and is also strictly required for tests complying with ASTM C158.

In this paper we will be denoting a_0 as the crack dimension prior to loading, a_1 as the crack size at the transition between region II and region III, and a_f as the crack size at failure. Also, t_1 refers to the time lapse necessary for the crack to extend to a length a_1 and t_f as the time to failure. To generalize the solution of the posed problem, non-dimensional parameters for time, τ, and crack length ξ, are introduced:

$$\xi_i = \frac{a_i}{a_f} \qquad \tau_i = \frac{t_i}{t_f} \tag{5}$$

At the transition between region I and region II, the following relationship holds true:

$$r \cdot K_{Ic} = r \cdot Y \cdot \sigma_f \sqrt{a_f} = Y \cdot \sigma_1 \sqrt{a_1} \tag{6}$$

Assuming that the value for the geometric factor, Y, in equation 6 does not vary considerably between its value at failure and its value at the transition between region I and II, a relationship between the stress, σ_1, and the non-dimensional crack extension, ξ_1, at the transition between region I and II can be derived:

$$\sigma_1 = r \cdot \sigma_f \left(\frac{a_f}{a_1}\right)^{1/2} = \frac{r \cdot \sigma_f}{\xi_1^{1/2}} \tag{7}$$

Since for constant loading rate $\sigma_1=(d\sigma/dt)\cdot t_1$ and $\sigma_f=(d\sigma/dt)\cdot t_f$ equation 7 can be expressed in its non-dimensional equivalent form:

$$\sigma_1 = \frac{d\sigma}{dt}t_1 = r \cdot \frac{d\sigma}{dt}t_f\left(\frac{1}{\xi_1^{1/2}}\right) \quad \Rightarrow \quad \tau_1 = \frac{r}{\xi_1^{1/2}} \tag{8}$$

Equation 3 describing the power law in region I can also be re-written in a non-dimensional equivalent form:

$$\frac{a_f}{t_f}\left(\frac{t_f}{a_f}\frac{da}{dt}\right) = v \cdot r^{n2-n1}\left(\frac{Y \cdot d\sigma/dt \cdot t \cdot \sqrt{a}}{Y \cdot \sigma_f \cdot \sqrt{a_f}}\right)^{n1} = v \cdot r^{n2-n1}\left(\frac{t}{t_f}\sqrt{\frac{a}{a_f}}\right)^{n1} \quad \Rightarrow$$

$$\frac{\alpha}{r^{n2-n1}}\left(\frac{d\xi}{d\tau}\right) = \tilde{\alpha}\left(\frac{d\xi}{d\tau}\right) = \left(\tau \cdot \xi^{1/2}\right)^{n1} \qquad \text{for } 0 < \xi < \xi_1 \tag{9}$$

where α is a non-dimensional constant defined as $\alpha \equiv \left(d\sigma/dt \cdot K_{IC}^2\right)/\left(v \cdot \sigma_f^3 \cdot Y^3\right)$ and $\tilde{\alpha} \equiv \dfrac{\alpha}{r^{n2-n1}}$

An explicit relationship relating τ_1 to ξ_1 can be obtained by integrating equation 9:

$$\tilde{\alpha}\int_{\xi_0}^{\xi_1}\frac{d\xi}{\xi^{n1/2}} = \int_0^{\tau_1}\frac{d\tau}{\tau} \tag{10}$$

$$\frac{2\tilde{\alpha}}{2-n_1}\left(\xi_1^{1-\frac{n1}{2}} - \xi_0^{1-\frac{n1}{2}}\right) = \frac{\tau_1^{n1+1}}{n_1+1} \tag{11}$$

A relationship analogous to the one described in equation 11 for the crack growth in region I can also be extended to region II. For region II, equation 4 reduces to the form:

$$\alpha\left(\frac{d\xi}{d\tau}\right) = \left(\tau \cdot \xi^{1/2}\right)^{n2} \quad \text{for } \xi_1 < \xi < \xi_f \tag{12}$$

and

$$\alpha\int_{\xi_1}^{1}\frac{d\xi}{\xi^{n2/2}} = \int_{\tau_1}^{1}\frac{d\tau}{\tau} \tag{13}$$

Direct integration of equation 13 will yield the expression:

$$\frac{2\alpha}{2-n_2}\left(1-\xi_1^{1-\frac{n2}{2}}\right) = \frac{1}{n_2+1} - \frac{\tau_1^{n2+1}}{n_2+1} \tag{14}$$

By combining the results from equations 8, 11, and 14, a system of three equations and three unknowns is obtained and the values of the variables ξ_0, ξ_1, and τ_1 can be solved for. Also, to a first degree approximation, it can be shown that the normalized crack extension prior to failure can be written in the form:

$$\xi_0 \approx \left[\frac{n_1 - 2}{2\alpha(n_1 + 1)} + 1 \right]^{\frac{2}{2-n1}} + r - 1 \tag{15}$$

Equation 15 allows us to estimate the crack extension at failure based on the non-dimensional parameter α. Equation 15 indicates that as α increases (e.g., high loading rate, low failure stress) the effects of the subcritical crack growth become less important.

MIRROR-MIST BOUNDARY

Because of the effect of slow crack growth prior to catastrophic failure, the failure stress, σ_f, will have some dependency on the loading rate. Denoting by σ_f the failure stress for a finite loading rate and by σ_∞ the hypothetical failure stress for the instantaneous loading rate case, equation 16 can be obtained:

$$\frac{\sigma_f}{\sigma_\infty} = \frac{Y_0}{Y_\infty} \sqrt{\frac{a_0}{a_f}} \approx \sqrt{\xi_0} \tag{16}$$

Y_0 is the geometric factor for the crack at the end of region II, while Y_∞ is the geometric factor for the initial flaw. Although the ratio Y_0/Y_∞ is likely less than unity, it was approximated as approximately equal to unity in equation 16. The dimension of the mirror radius, R_m, will also be affected by the crack growth in region I and II. In equation 17 we assume that the mirror will manifest itself as soon as a critical fracture intensity factor $K_1|_{mist}$ is reached at the tip of the crack. The ratio between the mirror-radius for the finite loading case over the instantaneous loading can be approximated as:

$$\frac{R_m}{R_\infty} \approx \left(\frac{Y(\overline{R}_\infty)\sigma_\infty}{Y(\overline{R}_0)\sigma_0} \right)^2 = \frac{1}{\xi_0} \left(\frac{Y(\overline{R}_\infty)}{Y(\overline{R}_0)} \right)^2 \tag{17}$$

$Y(\overline{R}_\infty)$ and $Y(\overline{R}_0)$ are the average geometric factors for the crack. To simplify the calculations we assumed that the average value of the geometric factor occurred at the average crack length, $\overline{R}$ in each case.

In general, the aspect ratio between the depth of a crack and its width tend to follow a precise relationship for a four-point bending case. It has been shown[9] that for a sub-critical crack under bending load, the aspect ratio between the depth, d, and the half-width, R, is a function of d/H where H is the thickness of the sample:

$$\frac{d}{R} = 0.84 - 1.33 \cdot (d/H) - 0.05 \cdot (d/H)^2 + 0.72 \cdot (d/H)^3 \tag{18}$$

Although, strictly speaking equation 18 is valid for subcritical crack growth, it can be considered a first degree approximation of the actual aspect ratio of a propagating crack. Using the result in equation 18 and the static stress intensity factor solutions of Newman and Raju for bending stress fields, an approximate solution for Y as a function of R/H can be obtained:

$$Y_{Surf}(R/H) \approx 1.2988 - 0.3952 \cdot \frac{R}{H} \tag{19}$$

Equation 18 approximation was derived for R/H<1 only.

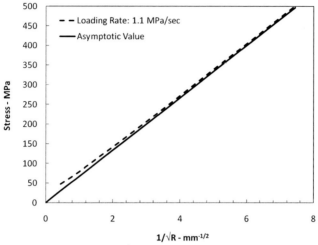

Figure 3. Failure stress vs. $1/\sqrt{R_m}$ for $d\sigma/dt=1.1$ MPa/sec case (RH 65%, RT)

Defining $D^* \equiv a_0/R \approx 1/12$ as the fractal dimension of the material[10], the following relationships can be derived:

$$\frac{R_m}{H} \approx \frac{a_f}{2 \cdot H \cdot D^*} = \frac{a_0}{2H \cdot D^*} \frac{1}{\xi_0} \tag{20}$$

Equation 17 can be re-written combining the results from equation 19 and 20:

$$\sqrt{\frac{R_\infty}{R_m}} \approx \left(\frac{1.2988 - 0.3952 \cdot \dfrac{R_\infty}{H\xi_0}}{1.2988 - 0.3952 \cdot \dfrac{R_\infty}{H}} \right) \sqrt{\xi_0} \approx \left(1 - 0.304 \frac{R_\infty}{H\xi_0} \right) \sqrt{\xi_0} \tag{21}$$

Further simplification of equation 21 yields the form:

$$\sigma_f = \frac{\left(\sigma_\infty \sqrt{R_\infty} \right)}{\left(1 - 0.304 \dfrac{R_\infty}{H\xi_0} \right)} \frac{1}{\sqrt{R_m}} = \frac{A_m\left(\xi_0, \dfrac{a_0}{H} \right)}{\sqrt{R_m}} \tag{22}$$

Equation 22 is an approximation believed to be valid for R/H<1. The equation indicates that the mirror-mist constant, A_m, for the four point bending case is a function of the loading rate as well as the relative thickness of the glass. It also suggests that for small values of α and high values for $a_0/(H \cdot D^*)$, the trend will tend to deviate the most from its asymptotic value (i.e., Orr's formula) shown in equation 1.

Figure 3 shows an example of the expected trend for failure stress versus $1/\sqrt{R_m}$ for $d\sigma/dt=1.1$ MPa/sec as recommended per ASTM C158. The finite loading case curve was obtained solving equations 8, 11, and 14 and substituting ξ_0 in equations 15 and 16. In the calculations, $v=0.2\cdot10^{-3}$ m/sec, $n_1=22$, $n_2=0$, $K_{Ic}=0.75$ MPa·m$^{0.5}$ were assumed based on typical values reported in the literature for 65% relative humidity (RH) at 27°C. [7] The thickness of the glass was assumed equal to H=2.5mm for the example considered. It is apparent that the trend diverges from the ideal Orr's relationship at low stress levels and, in general, the trend will appear to cross the σ-axis at a positive number even in the absence of the residual stresses that are reported in the literature[5] as the cause of this deviation.

RESIDUAL STRESS PREDICTION

Contrary to what is generally assumed, in some cases the mirror constant A_m is a function of the loading rate and the relative glass thickness as shown in the previous section. For most practical application, as for instance to predict the failure stress using equation 1, treating A_m as constant will not introduce any significant errors. Nonetheless, when attempting to infer the residual stress level by extrapolating data obtained at low loading rates (i.e., $R_m \rightarrow \infty$) might lead to erroneous estimations. Figure 4 shows a schematic example of how extrapolating a limited number of data will lead to a miscalculation in the value of the residual stress, $\Delta\sigma_0$. In figure 4, the trend for the failure stress versus $1/\sqrt{R_m}$ based on calculations for data at 65% RH and 27°C is shown in a dotted line. Also the figure shows hypothetical data obtained for relatively low values of σ_f. Since the slope of the curve decreases as the failure stress decreases (for the same loading rate), a positive intercept with the σ-axis is to be expected when extrapolating failure stress data versus $1/\sqrt{R_m}$ especially for a relatively low-stress data set. Figure 4 also allows for a quick estimation of the magnitude of the artificial residual stress $\Delta\sigma_0$ as a function of the average failure strength for a set of glasses based on typical reported data at 65% RH at 27°C. [8] For the given condition, an apparent residual stress equal to $\Delta\sigma_0\approx10$ MPa is to be expected for a set of data with average fracture strength of 200MPa. This result is consistent with data found in the literature. [1,5]

Figure 5 shows data obtained from the literature[11] for biaxial loading of silica disks as well as the trend predicted given the conditions the test was carried out at (H=2.5mm, loading rate 1.1MPa/sec). The trend and the data appear to be in good agreement and in both cases a positive intercept on the σ-axis is predicted although residual stresses might not be present in the samples tested. The trend in this case was computed assuming testing at room temperature and RH=65%.

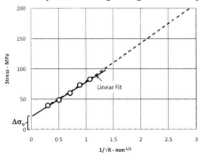

Figure 4. "Apparent" residual stress, $\Delta\sigma_0$, at glass surface based on extrapolation from a hypothetical set of stress-mirror data.

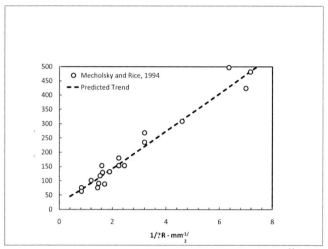

Figure 5. Mirror-mist versus failure stress from Mecholsky and Rice[11] as well as the predicted trend for a loading rate of 1.1MPa/sec, 65% RH at RT.

CONCLUSIONS

This paper described a method to estimate the effect of the loading rate on the mirror constant, A_m. Both the failure stress, σ_f, and the length of the mirror region, R_m, were derived for the general case of finite, constant loading rate. It was found that the value of the mirror constant A_m is a function of a newly proposed non-dimensional parameter $\alpha \equiv \left(\dot{\sigma} \cdot K_{IC}^2 \right) / \left(v \cdot \sigma_f^3 \cdot Y^3 \right)$ as well as the relative mirror region size with respect to the glass's thickness $a_0/(H \cdot D^*)$. Due to the rate dependence of the mirror constant, A_m, the estimation of residual stresses by extrapolation of the stress versus $1/\sqrt{R_m}$ curve (i.e., $R \rightarrow \infty$), will lead to inaccurate results especially for high values of $a_0/(H \cdot D^*)$ and low values of α. The model proposed also predicted that the resulting "apparent" residual stress $\Delta\sigma_0$ will be a positive number (i.e., apparent surface compressive stress). The results from this study, based on empirical data obtained at 65% relative humidity at 27°C, were found to be consistent with data reported in the literature[5]. Further work is necessary in the future to extend the model to different temperature and humidity values. Mirror mist measurements at low stress values on annealed samples should be carried out to confirm the validity of the proposed model. In order to improve the reliability of data for surface stress estimation, higher loading rates, thicker samples, and dry testing conditions should be preferred. Whenever possible, it is recommended to use uniaxial testing rather than four-point bending tests or ring-on-ring tests for surface stress estimations. Finally, it is possible to obtain better estimates of residual stresses by carefully censoring test data with low values for $a_0/(H \cdot D^*)$.

REFERENCES
[1]L. Orr, "Practical Analysis of Fractures in Glass Windows," *Mater. Res. Stand.*, 12 [1] 21-23, 47(1972).
[2]ASTM C158-02 (2007), "Standard Test Methods for Strength of Glass by Flexure (Determination of Modulus of Rupture)", 1995.
[3]ASTM C1256 – 93 (2008) Standard Practice for Interpreting Glass Fracture Surface Features.

[4]J.C. Conway, Mecholsky J. J., "Use of crack branching data for measuring near-surface residual stresses in tempered glass," *J. Am. Ceramic Society*, 72 [9], 1584-1587 (1989).

[5]J.B. Quinn, "Extrapolation of Fracture Mirror and Crack-Branch Sizes to Large Dimensions in Biaxial Strength Tests of Glass," *J. Am. Ceram. Soc.*, 82 [8], 2126-32 (1999).

[6]S.M. Wiederhorn, "Influence of Water Vapor on Crack Propagation in Soda-Lime Glass," *J. Am. Ceram. Soc.*, 50, 407-414 (1970).

[7]G.S. Glaesemann, D.A. Clark, T.A. Hanson, and D.J. Wissucherk, "High-Speed Strength Testing of Optical Fiber," *Reliability of Photonics Materials and Structures*, Edited by E. Suhir, M.Fukuda, and C.R. Kurkijan, 531, 249-260, Proceeding of the Materials Research Society, San Francisco, 1998.

[8]M. Haldimann, "Fracture strength of structural glass elements: analytical and numerical modelling, testing and design", Ph.D. Thesis, École Polytechnique Federale de Lausanne, 2006.

[9]P. Dwivedi, D.J. Green, "Indentation crack-shape evolution during subcritical crack growth", J. Am. Ceram. Soc., 78 [5], 1240-46 (1995).

[10]J.J. Mecholsky, R.W. Rice, S.W. Freiman, "Prediction of Fracture Energy and Flaw Size in Glasses from Measurements of Mirror Size," *J. Am. Ceram. Soc.*, 82, 57 [10], 440-443 (1974).

[11] J.J. Mecholsky, and R.W. Rice, "Fractographic Analysis of Biaxial Failure in Ceramics," *American Soc. for Testing and Materials*, STP 827 (1984).

THERMAL SHOCK BEHAVIOR OF Si₃N₄-SPECIMENS – INFLUENCE OF ANNEALING AND EDGES

THERMAL SHOCK BEHAVIOR OF Si_3N_4-SPECIMENS – INFLUENCE OF ANNEALING AND EDGES

Walter Harrer, Robert Danzer
Institut für Struktur- und Funktionskeramik (ISFK), Montanuniversitaet Leoben
A-8700 Leoben, Austria

Karl Berroth
FCT Ingenieurkeramik
D-96528 Rauenstein, Germany

ABSTRACT

Due to the outstanding mechanical and thermal properties, silicon nitride can be used for roller tools in the metallurgical industry. However, in some applications rapid heating and cooling of tools can lead to high thermal stresses. These thermal stresses may cause thermal fatigue or catastrophic failure of the rollers. Therefore the thermal shock behavior of silicon nitride is of great interest.

During thermal shock (caused by quenching), the complete surface of a component including all edges and corners, is put under tensile load. Therefore damage at the surface (e.g. scratches, contact damage, machining cracks) or at the edges (flaking, cracking) is highly relevant for the thermo shock resistance of each component. Removal of surface damage will be beneficial for the thermo shock resistance of the components.

In this work a heat treatment (annealing) of silicon nitride specimens of different sizes (from standard bending specimens to rings with almost 300 mm diameter and 30 mm wall thickness) is made at 1000°C in atmospheric pressure. Bending and water quench tests were made on annealed specimens and the experimental results were compared to the results of non-annealed components. It can be shown that the annealing caused an increase in strength of approximately 200 MPa and – depending on the specimen size and geometry – an increase in quenching resistance between 50°C and 200°C.

It has been observed that most of the thermal shock cracks started at the edges of the specimens. Therefore the bending strength (where failing from edges is avoided) is not relevant for the description of our thermal shock test. For a high thermal shock resistance proper machining of the edges is of highest relevance.

INTRODUCTION

Because of their brittleness ceramics are sensitive against thermal shock. Through cooling (or heating) thermal strains develop in the material. Inhomogeneous temperature fields lead to localized different thermal expansion. If this expansion is constrained, thermal stresses develop. If these thermal stresses reach the strength of the material, cracks can propagate and can cause failure of the components.[1,2] The basics to describe thermal shock behavior are well known and can be found in the literature, e. g..[3-8] A general theoretical approach has been found by Hasselman.[8]. The analytical description of thermal stresses is based on the well known Biot-concept, where the dimensionless Biot-modulus Bi is given by

$$B_i = \frac{h_f \cdot d}{\lambda} \qquad (1)$$

The Biot-modulus depends on the thermal conductivity λ, a characteristic length d and the heat transfer coefficient h_f. The thermo-mechanical stress at the surface of a cooled body is:

$$\sigma_{th} = \alpha E' \cdot \Delta T \cdot \sigma^*(B_i) \qquad (2)$$

where α is the coefficient of thermal expansion, ΔT is the temperature change during cooling, and E' is the appropriate elastic constant.[9] At an edge, where an uniaxial stress state exists, it holds: E'=E, and in a plane surface, where a biaxial stress state exists, it holds: E'=E/(1-v), where v is the Poisson's ratio. σ*(Bi) is a dimensionless normalized stress which depends only on the Biot-modulus. For very large Biot numbers, say B_i>50, the thermal shock is very hard and σ*→1. This is a worst case scenario.

Failure occurs (i.e. thermal shock cracks come to existence[10,11]) if the thermal stress reaches the strength σ_f of the material, for the worst case scenario ($B_i \rightarrow \infty$, $\sigma^* \rightarrow 1$). The corresponding critical temperature difference ΔT_c is called first thermo shock parameter R, which is an appropriate ranking parameter for brittle materials loaded with a hard thermal shock (here the modulus for a plane stress state is used):

$$R = \frac{\sigma_f(1-v)}{E \cdot \alpha} \tag{3}$$

In general the loading is softer as assumed in eq. 3 and $\sigma^* < 1$. Then the critical quenching temperature difference is:

$$\Delta T_c = \frac{\sigma_f(1-v)}{E \cdot \alpha \cdot \sigma^*(B_i)} \tag{4}$$

In the past it has been shown for several materials (especially Si$_3$N$_4$, SiC and SiC/Si$_3$N$_4$ composites), that an appropriate heat treatment in an air environment can increase the strength of specimens. It is claimed that this heat treatment causes the formation of thin oxide layers (e. g. SiO$_2$ and Y$_2$Si$_2$O$_7$) at the surface of the specimens. These penetrate surface cracks (which are – in general – caused by machining) and cause a sealing of (most of) these cracks. This effect is called "crack healing."[12-16] In previous works it can be shown, that for the Si$_3$N$_4$ material which was used in this study, the glassy layer fills the micro pores and the machining-induced flaws and heals these defects, if they are not too large.[17] This effect resulted in a significant increase in the biaxial tensile strength of specimens made of this material.

In the aluminum and the steel industry, very large components made from silicon nitride ceramics exist, e.g. immersion heater sheaths and riser tubes in aluminum foundries or rolls and wear parts in steel rolling mills. These components are in contact with liquid metal melt or with metals at rolling temperatures. Some of these components are very large having a length of more than one meter and a diameter of some ten centimeters. These are conditions where thermal shock may occur.

The aim of this study is to investigate the thermal shock behavior of a commercial silicon nitride material, which is used for immersion heater sheaths and rollers in the metal producing industry. Tests will be made on small laboratory specimens but also on some large specimens cut out of large tubes. A second aspect investigated is the improvement of the thermal shock behavior of this material. Laboratory specimens as well as parts of real tubes used in the metallurgical industry will be heat treated and water-quench tested to see if the critical temperature difference can be improved.

INVESTIGATED MATERIAL AND SPECIMEN PREPARATION

A commercial silicon nitride with Alumina and Yttria additives from FCT Ingenieurkeramik, Germany, is investigated. Discs with a diameter of 250 mm and a thickness of 5 mm were hot pressed. Large tubes were produced by cold isostatic pressing of cylindrical preforms, green machined to near net shape by turning, then gas pressure sintered and finally ground to requested diameters, length and tolerances. The microstructure of the sintered silicon nitride material is shown in Fig. 1.

Almost one hundred standard bending test specimens (sample ID: BT, $\geq 45 \times 4 \times 3$ mm^3) were machined out of the discs following EN 843-1.[18] It should be noted, that according to the standards (e.g. EN 843-1), only the edges at the tensile side of the specimens were chamfered. An overview on some important materials properties is given in Table I.

Table I: Some properties of FSNI ([a] measurements at ISFK, [b] data from FCT)

Young's modulus E [GPa]	Poisson's ratio ν [-]	coefficient of thermal expansion α [10⁻⁶·K⁻¹] at 1.000 °C	Fracture toughness K$_{Ic}$ (SEVNB-method) [MPa√m]
297[a]	0.28[b]	3.2[b]	5.0[a]

The other specimens were rings which were cut out of gas-pressure-sintered and ground large tubes. The dimensions of the rings cut from the tubes are: diameter 270 mm, wall thickness 20 mm, height 47 mm. The dimensions of the second type of rings are: diameter 201 mm, wall thickness 15.5 mm, height 50 mm. All surfaces and chamfers of the rings (with the exception of the inner surface) were finished with a grain size of 15 μm (D15). The inner surface was honed and had therefore an average surface roughness that was twice as high as the roughness of the other sides. Since both type of ring specimens showed very similar behavior it will not be distinguished between them in the following (sample ID: R).

One of the smaller rings was cut into segments (length 50 mm, sample ID: S). The edges of the segments were chamfered with a grain size of 15 μm (D15). Some of the segments were cut into almost cuboid-shaped specimens (edge length of about 15 mm, sample ID: C). The edges of some of the cuboid-shaped specimens were chamfered by hand. The edges of the other specimens remained as they were after cutting. Since chamfered and non-chamfered cuboid specimens showed very similar behavior, both types of specimens will not be discerned in the following. Ring, segment and cuboid specimens are shown in Fig. 2.

To improve the surface quality of the specimens and to heal out surface cracks, a fraction of each type of specimen was heat treated after machining at 1000°C in atmospheric pressure in an electric furnace with a heating and cooling ramp rate of 5°C/min. The holding time was 10 hours. To distinguish the non-annealed from the annealed samples, the sample identifiers were given prefixes (n) and (a), respectively. An overview on the prepared specimens is given in Table II.

Table II. Overview on used specimens and on completed tests

Sample identifier.	Number of specimens		Comments
	Non-annealed (n)	annealed (a)	
BT 1	30	15	Used for bending tests at room temperature (no quenching).
BT 2	20	30	Used for quenching tests followed by bending tests.
SR	3	7	Used for quenching tests; the tests results on both sets are similar, therefore both samples are evaluated commonly (sample identifier: R).
LR	1	1	
S	3	5	Used for quenching tests.
C1	6	4	Used for quenching tests. A common evaluation of chamfered and non-chamfered specimens was made.
C2	4	4	Vickers intends were made into these specimens (on the a-specimens before annealing). Then they were quenched.

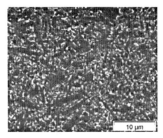

Fig. 1. Microstructure of the investigated silicon nitride); SEM picture (plasma etched).

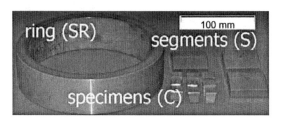

Fig. 2. Gas-pressure-sintered silicon nitride ring (SR), segments (S) and cuboid specimens (C) cut out of the ring.

EXPERIMENTAL PROCEDURE AND RESULTS

Conventional bending tests according to EN 843-1[18] were made on 30 non-annealed (n-BT 1) and 15 annealed (a-BT 1) specimens. The results are printed in Table III.

Table III. Bending strength test results, (results of hot pressed silicon nitride)

Sample identifier	Characteristic strength σ_0 [MPa]	Weibull-modulus m [-]
n-BT 1	839 [815-864]*	11.5 [8.6-14.0]*
a-BT 1	1012 [981-1045]*	15.9 [10.9-20.7]*

*the values in the squared brackets refer to the 90% confidence intervals

Water quench tests[11,19] followed by bending tests were performed on samples n-BT 2 and a-BT 2 at ISFK (five specimens for each quenching temperature). The specimens were heated at a rate of 5°C/min to the quenching temperature in a vertical tubular air furnace and held for one hour. Afterwards the specimens were dropped by free fall into cold water (20°C). To have a constant heat transfer from the sample to the water, it was continuously agitated so that no vapor layer could develop on the specimens' surface.[11,19] The lowest quenching temperature used was 570°C (giving a temperature difference ΔT=550°C). For the following subsets of the sample, the quenching temperature was increased by increments of 50°C. After quenching, the samples were checked for a possible crack formation by use of a fluorescent dye penetration. Then they were strength tested in four-point-bending test.[18]

Fig. 3 shows a summary of the water-quench experiments on bending specimens. The annealing caused an increase in strength (about 200 MPa without quenching) as well as in thermal shock resistance, as a result of some crack healing. For the non-annealed specimens the critical quenching temperature difference is about ΔT=650°C, and for the annealed specimens it is increased to almost ΔT= 800°C.

After strength testing, an extensive fractographic analysis[20] was performed. The fracture surface of specimens that failed from material inherent flaws (Fig. 4) looks quite different from that of specimens that failed from thermal shock damage (Fig. 5 and 6). The number of failures caused by inherent flaws and by thermal shock cracks is also indicated for each subsample shown in Fig. 3.

In more than 10 % of the specimens of a-BT2, failure occurred at relatively low strength values. The fracture origins were large material flaws inside the specimens. An example is shown in Fig. 4. The fracture origin is a very large pore. Although all bending specimens were cut out of a single large plate, and the classification into samples and subsamples was random, unfortunately most of specimens containing such volume flaws were found in the annealed sample (and were used for tests with ΔT=650°C to ΔT=750°C). This fact slightly confuses the clear definition of a critical thermal shock temperature.

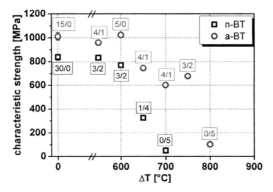

Fig. 3. Strength vs. temperature difference after water-quench test for non-annealed (n-BT,) and annealed (a-BT) specimens (plotted is the characteristic strength, i.e. the stress for a probability of failure of 63 %). Also given is the number of specimens on which no thermal shock cracks could be found and on which thermal shock cracks were detected.

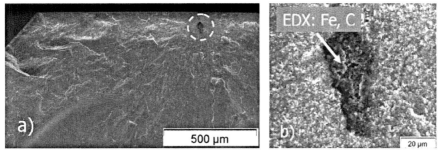

Fig. 4. Heat treated specimen a-BT19 after the water-quenching (ΔT=700°C) and bending testing (σ_f = 406 MPa): (a) SEM-micrograph of the fracture surface. The mirror is marked by a circle. (b) The fracture origin could be identified as a large pore near the surface.

Bending test specimens have a tensile and a compressive loaded surface. Since brittle fracture (in a bending test) is triggered by tensile stresses, the tensile loaded surface is machined carefully to avoid machining damage and the tensile loaded edges are carefully chamfered. But since – in bending testing – the compressive side is not relevant for fracture, chamfers at this side are not requested in standards (e.g. EN 843-1). But in a quenching test all surfaces and all edges of the specimens are tensile stressed. Therefore machining damage at edges may become relevant. In the actual case badly machined edges at the compressive side and the end faces of the specimens caused thermal shock cracks to be initiated and detected at relatively low quenching temperatures. An example is shown in Fig. 5. The origin of thermo shock cracks is a very large edge defect caused by inadequate machining. It is obvious that this damage is too severe to be healed by the annealing procedure (where a surface film of glass with a thickness of 1-2 μm is built, which grows into cracks). Five of the 30 a-BT2 specimens had thermal shock cracks which were caused by edge defects (above all starting from the end face of the specimen), which all were formed at quenching temperature differences ΔT<800°C. The same happened in most of the non-annealed specimens at quenching temperature differences ΔT<700°C.

The fracture surface of a non-annealed specimen, where the fracture origin was a thermo shock crack (ΔT=650°C) is shown in Fig. 6.

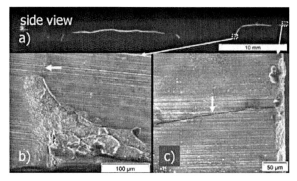

Fig. 5. Heat-treated specimen a-BT3 after the water quenching (ΔT=550°C). The specimen had strength of σ_f= 585 MPa: (a) Side view of the specimen after the treatment with a dye penetrant. (b) One crack (arrow) starts at a large defect on the edge of the lateral face and (c) propagates to the end face of the specimen.

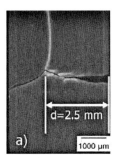 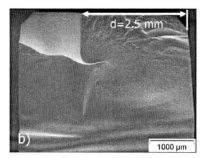

Fig. 6. Not-heat-treated specimen n-BT12 after water quenching (ΔT= 650°C) and bending testing (σ_f = 65 MPa): (a) View on tensile side (stereo microscope) and (b) fractured surface (SEM).

Water-quench tests were also made in the case of the other samples C1, S, and R (for non-annealed as well as annealed samples). The heating was made as described for the bending specimens, but for S and R a conventional electric furnace was used. The specimens were quenched and then checked for a possible crack formation by use of a fluorescent dye penetrant. Starting from the lowest quenching temperature (ΔT=300°C), the procedure was repeated with the same specimens with increased quenching temperature (in steps of +50°C), until thermal shock cracks could be found in all investigated specimens. The results are summarized in Fig. 7, where the fraction of specimens having thermal shock cracks are plotted for the cuboid, segment, and ring specimens respectively.

It should be noted that most thermal shock cracks started at the edges of the specimens. Examples (for R specimens) are shown in Fig. 8. Although the number of tested specimens was small and the statistics is therefore poor, it can clearly be recognized, that the thermal shock resistance decreases with the size of the specimens. The length of the edges for the C, S and R specimens is approximately 120 mm, 480 mm, and 3000 mm, respectively. Again an improvement of thermal shock resistance can be recognized by annealing, but the effect is less pronounced as recognized on bending specimens. For example in the case of C-specimens the occurrence of the first thermal shock damage is shifted from ΔT = 450°C to ΔT = 550°C due to annealing. In the case of S and R specimens a shift of approximately 50°C can be recognized.

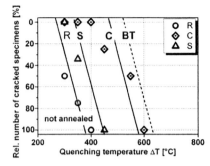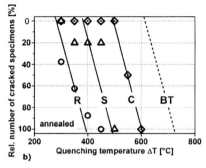

Fig. 7. Fraction of quenched specimens having thermal shock cracks vs. quenching temperature difference for bending (BT), cuboid (C), segment (S) and ring (R) specimens respectively (a) for non-annealed and (b) for annealed specimens. The dashed line of the BT specimens is shown for comparison additional.

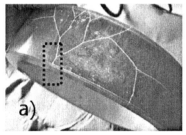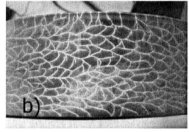

Fig. 8. Ring specimens after thermal quenching. Cracks are made visible by a dye penetrant. (a) Specimen a-SR5 after ΔT =350°C and (b) specimen a-LR1 afterΔT =400°C. The thermal shock crack in (a) started at the edge (rectangle).

Finally – in order to produce well defined starter cracks for the quenching experiment - eight C-specimens were Vickers indented before annealing and quenching. Four different indentation loads of 9.8 N (HV1), 49.1 N (HV5), 98.1 N (HV10) and 294.3 N (HV30) were used. The length of the cracks, which were found at the edges of the indents, was measured by use of a light microscope. Then half of the specimens were annealed and the crack length was determined again. Quenching tests (following the procedure described above) were made and the change of the crack length was measured with increasing quenching temperature.

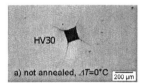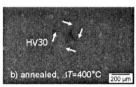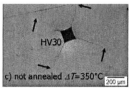

Figs. 9. Influence of annealing and water-quenching on HV30 indents: (a) indentation cracks before annealing and quenching. (b) the same indent after annealing and a quenching with ΔT=350°C and (c) non-annealed cracks after quenching at ΔT=350°C.

First crack growth in n-C specimens with Vickers indents started at quenching temperature differences of ΔT=300°C (at HV30 indents), at ΔT=350°C (at HV10 indents) and at ΔT=400°C (HV5 indents), respectively. On the HV1 indents no crack growth occurred in the investigated temperature range. In the a-C specimens (which were annealed after indentation) no growth of cracks at the indentations could be observed until a ΔT=600°C.

Fig. 9 shows cracks at HV30 indents. Before annealing (a) cracks (length approximately 160 μm) exists at the edges of the indent. As result of annealing the cracks healed partially and even after the water quenching at ΔT=400°C the cracks did not grow (b). However in the non-annealed specimen (c) crack growth (at a length of about 300 μm) occurred even after water quenching with ΔT=350 K.

In Fig. 10 a comparison of the crack growth in non-annealed and annealed specimens is shown, which were indented with HV10. In the non-annealed specimens first (slightly) crack growth could be observed at ΔT=350°C (a). For experiments with ΔT≥450°C severe crack growth starting from the edges of the HV10-indentation can be observed (e).

In the case of the annealed specimens no thermal shock damage could be observed at ΔT=400°C and ΔT=450°C (b and d). Even at ΔT=600°C no crack growth or development of thermal shock cracks starting at the indentations could be observed (f), but in this case some large cracks started from edge defects.

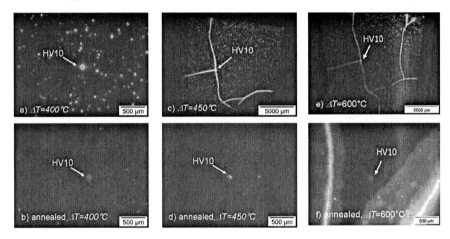

Fig. 10. Behavior of cracks at HV10 indents in non-annealed (a, c and e) and annealed specimens (b, d and f). At ΔT=400°C no crack growth occurs neither in the n- (a) nor in the a-specimens (b). At ΔT=450°C crack growth starts in the n- (c) but not in the a-specimen (d). At ΔT=600°C the cracks in the n-specimen are further grown but in the a-specimen no crack growth occurs at the indent. But some large cracks, which started at the edges, have been formed.

DISCUSSION

It is obvious that the annealing of the specimens has a beneficial effect on the bending strength and the thermal shock behavior of the investigated material (Table III., Figs. 3, 7, 9 and 10). This increase in strength is a consequence of the healing of small surface cracks, which – in most cases – are introduced during machining. Although in conventional bending testing the specimen preparation is optimized in that way that only little machining damage should be significant in the bending test, the RT- strength is increased by annealing for almost 200 MPa. This clearly indicates that machining cracks exist in non-annealed specimens and that they can be healed out – at least partially by the applied heat treatment.

On the basis of simple bending tests an analysis of thermal shock loading would predict that critical quenching temperature differences are much higher as determined by experiments. Remember that the maximum tensile stress in a water quench test is $\sigma_{th}= \alpha E' \cdot \sigma^*(Bi)$, where the factor $\alpha E' \cdot \Delta T$ is the tensile stress for extremely hard quenching and the factor $\sigma^*(Bi)$ describes the reduction of constraint for the actual situation. In the case of the actual experiments the Biot number is between 2 and 5 and the factor $\sigma^*(Bi)$ is between 0.3 and 0.4 (for details see [6,7,9,11]). Using eq. 4, the strength values determined on bending specimens at room temperature (Table III) and these values for the factor $\sigma^*(Bi)$, no thermal shock cracks should occur up to quenching temperature differences of about 1589°C to 2119°C in the case of non-annealed specimens. In the case of the annealed specimens, the critical temperature differences should be even higher (about 1917°C to 2555°C). But in reality cracks in BT- specimens occurred at $\Delta T=550$°C and in R-specimens even at $\Delta T=300$°C. The reason for this surprising behavior is that the strength of the thermally shocked components is much lower than the strength of the bending specimens. Under quenching conditions the whole surface of the components is tensile stressed (in the case of the rings about 10^5 mm^2; in comparison the effective surface of bending specimens is more than 1000 times smaller) and the length of the tensile stressed edges in the rings is about 2000 mm (in bending specimens the tensile stressed edges are perfectly machined and their length is about 80 mm). Therefore, the probability of finding surface or edge defects in the "large" rings is much higher as than in "small" bending specimens. In this context, it should also be noted that, in the case of the bending specimens, only the edges at the "tensile stressed surface" were done properly. In fact, most of the thermal shock cracks started at edge defects. Therefore the strength value used for the above analysis is not relevant for our tests.

In our quenching experiments almost all thermo shock cracks started at the edges. Healing of cracks by annealing worked nicely for short cracks, which have tightly closed borders. But in the case of edge damage, where flaking occurs and very deep cracks can develop, only partial crack healing occurs. This shows that the quality of the edge finish is of utmost relevance for the thermal shock resistance of specimens and components.

On the non-annealed specimens crack growth starting at HV5 indentations (length about 30 µm, measured from the edge of the indentation) could be observed at a $\Delta T= 400$°C. On the other hand the cracks starting at HV1 indentations of the non-annealed specimens had a length of less than 10 µm. Here no crack growth could be observed, which means that such small cracks are not sensitive to thermal shock up to quenching temperatures of $\Delta T=600$°C. The thermal shock experiments on specimens with Vickers indents show that the healing of indentation cracks works very effectively. In the case of annealed HV30 cracks (crack length before annealing about 160 µm) no crack extension could be observed up to quenching temperature differences of $\Delta T=600$°C. This value almost reaches the quenching temperature difference of nicely machined bending specimens. In the case of HV10 cracks on annealed specimens (crack length before annealing about 120 µm) no crack extension occurred at $\Delta T=600$°C, but other (edge) cracks have extensively grown. In other words the healing has shrunk the indentation cracks to a size smaller than the edge cracks.

CONCLUSIONS

There are three main conclusions which can be made on the basis of the experiments described above:

i) For designing against thermal shock, the use of conventional bending strength data is – in general – not adequate. During quenching the complete surface of the component (including all edges and corners) is tensile stressed, and the strength of the edges may become strength limiting. It should be noted that the defect-free machining of edges is very demanding.

ii) The applied heat treatment is a simple and effective action to reduce surface and edge damage and to heal - at least partially - surface and edge cracks. If cracks are too deep or if their opening is too wide the healing is less effective. In any case, the applied annealing improves the thermo shock resistance of specimens and components.

iii) It can be shown that there exists a size effect on the thermo shock resistance of the investigated specimens.

REFERENCES

[1]R. Danzer, T. Lube, P. Supancic and R. Damani, Fracture of Ceramics, *Adv. Eng. Mat.*,**10**, 275–298 (2010).

[2]R. Danzer, Mechanical Failure of Advanced Ceramics: The Value of Fractography, *Key Eng. Mat.*, **223**, 1-18 (2002).

[3]T. K. Gupta, Strength Degradation and Crack Propagation in Thermally Shocked Al₂O₃, *J. Am. Ceram. Soc.*, **55**, 249-253 (1972).

[4]P. F. Becher, D. Lewis III, K. R. Carman and A. C. Gonzalez, Thermal Shock Resistance of Ceramics: Size and Geometry Effects in Quench Tests, *Am. Ceram. Soc. Bull*, **59**,, 542-545 (1980).

[5]G. A. Schneider and G. Petzow, Thermal Shock and Fatigue Behavior of Advanced Ceramics, Kluwer Academic Publishers NATO ASI Series, Series: Applied Sciences **241** (1993).

[6]B. Boley and J. Weiner, Theory of Thermal Stresses, Krieger Publishing Company, Malabar, Florida, (1985).

[7]H. Parkus, Thermoelasticity, 2nd revised and enlarged edition, Springer-Verlag, Wien, New York (1976).

[8]D. P. H. Hasselman, Unified Theory of Thermal Shock Fracture Initiation and Crack Propagation in Brittle Ceramics, *J. Am. Ceram. Soc.*, **52**, 600-604 (1969).

[9]J. D. Achenbach and R. B. Hetnarski: Thermal Stresses I, Mechanics and Mathematical Methods, Elsevier Science Publishers B. V., (1986).

[10]M. Fellner and P. Supancic, Thermal Shock Failure of Brittle Materials, *Key Eng. Mat.*, **223**, 97-106 (2002).

[11]D. Munz and T. Fett., Ceramics: Mechanical properties, failure behaviour, materials selection. Berlin: Springer-Verlag. (1999).

[12]Y. H. Zhang, L. Edwards and W. J. Plumbridge, Crack Healing in a Silicon Nitride Ceramic. *J. Am. Ceram. Soc.*, **81**, 1861-1868 (1998).

[13]M. Nakatani, K. Ando and K. Houjou, Oxidation behaviour of Si₃N₄/Y₂O₃ system ceramics and effect of crack-healing treatment on oxidation, *J. Eur. Ceram. Soc.*, **28**, 1251-1257 (2008).

[14]K. Ando, K. Takahashi, K. Nakayama and S. Saito, Crack Healing Behaviour of Si3N4/SiC Ceramics under Cyclic Stress and Resultant Fatigue Strength at the Healing Temperature. *J. Am. Ceram. Soc.*, **85**, 2268-2272 (2002).

[15]K. Houjou, K. Ando, SP. Liu, S. Sato, Crack-healing and oxidation behaviour of silicon nitride ceramics. *J. Eur. Ceram. Soc.*, **24**, 2329-2348 (2004).

[16]D. R. Clarke and F. F. Lange, Strengthening of a Sintered Silicon Nitride by a Post-Fabrication Heat Treatment. *J. Am. Ceram. Soc.*, **65**, C51-C52 (1982).

[17]W. Harrer, R. Danzer, R. Morrell, Influence of surface defects on the biaxial strength of a silicon nitride ceramic-increase of strength by crack healing, submitted to *J. Eur. Ceram. Soc.* (May 2010).

[18]EN 843-1: Advanced technical ceramics, monolithic ceramics-Mechanical properties at room temperature, part 1: determination of flexural strength, 1995.

[19]ENV 820-3: Advanced technical ceramics, monolithic ceramics-Thermomechanical properties, part 3: Determination of resistance to thermal shock by water quenching, 1994.

[20]G. D. Quinn, Fractography of Ceramics and Glasses. NIST Special Publication 960-16. Washington (2007).

A FRACTURE ANALYSIS OF GADOLINIUM-DOPED CERIA CERAMICS TO IMPROVE STRENGTH AND RELIABILITY

Kouichi Yasuda, Kazuhiro Uemura, and Tadashi Shiota
Department of Metallurgy and Ceramics Science, Faculty of Science and Engineering
Graduate School, Tokyo Institute of Technology
2-12-1-S7-14, Ookayama, Meguro-ku, Tokyo 152-8552, JAPAN

ABSTRACT
    Strength-limiting flaws were identified in gadolinium-doped ceria (GDC) ceramics in order to improve their reliability for use as a solid oxide fuel cell. GDC ceramics were prepared by dry pressing and sintering at various temperatures. The modulus of rupture was determined by 4-point bending tests and the Weibull distributions were plotted. The fracture origins were identified to determine the cause of fracture at each sintering condition.

INTRODUCTION
    Solid oxide fuel cells (SOFCs) are one of the most promising power generators for high efficiency and multi-fuel compatibility.[1] The working temperature of SOFCs is >1000°C when yttria-stabilized zirconia is used as electrolyte. To attain long-term durability, a lower working temperature is needed for next-generation SOFCs. Gadolinium-doped ceria (GDC) ceramics is one of the most promising candidates as the new electrolyte.[2,3]
    GDC ceramics exhibit good electrochemical properties as low as 800°C, however, the mechanical properties have not been fully investigated. Sammes et al.[4] showed that the fracture strength of GDC ceramics (sintered at 1600°C) was around 140 MPa at room temperature and it was decreased to above 500°C. They did not state the reason. Sameshima et al.[5] reported that samarium-doped ceria ceramics (sintered at 1600°C) had fracture strength of around 70 MPa. They found a very large defect on the fracture surface, and the size was considerably larger than a pore size. Little data on fracture strength of GDC ceramics is present in the literature, but the reported values are not consistent.
    In this paper, GDC ceramics were prepared by sintering $Ce_{0.9}Gd_{0.1}O_2$ powder compacts at 1200, 1300, and 1400°C. The GDC ceramics were cut into bending specimens ($4 \times 3 \times 36$ mm^3), and 4-point bending tests were conducted. After strength testing, the fracture surface of each specimen was examined by optical microscopy to find the location of fracture origin. Secondary electron microscopy was used to identify the fracture origin size, shape and chemical composition. From statistical analysis, the causes of fracture were determined and classified to improve the strength and reliability of GDC ceramics.

EXPERIMENTAL PROCEDURE
Sample Preparation
    10 mol% gadolinium-doped ceria powder (Anan Kasei) was used as raw material. The GDC powder was compressed in uniaxial pressing under 7 MPa for 2 minutes, and then compressed again by cold isostatic pressing under 100 MPa for 2 minutes. The compacts were sintered in air at 1200, 1300, and 1400°C, respectively. The holding time was 2 hours, and the

heating and cooling rates were 5°C/min. Rectangular-sectioned specimens were cut from the GDC ceramics, followed by mirror polishing with diamond slurry. The dimensions of the specimens were 4 mm (breadth) ×3 mm (thickness) ×36 mm (length). Finally, the edges were chamfered.

Characterization and Mechanical Properties of the GDC Ceramics

The true density was measured by pycnometry, and the apparent density was measured by the Archimedes method. The relative density was calculated from these values.

A 4-point bending test[6,7] was conducted with a Universal Materials Testing Machine (Shimadzu, Autograph DSC-R-10TS). The upper and lower spans were 10 mm and 30 mm, respectively. The crosshead speed was 0.5 mm/min. Young's modulus (E) and fracture strength (σ_f) were calculated from the following equations:[6,7]

$$E = \frac{\Delta W (L-l)^2 (2L+l)}{4bt^3 \Delta u} \tag{1}$$

$$\sigma_f = \frac{3W_f (L-l)}{2bt^2} \tag{2}$$

L is the lower span, l is the upper span, b is the breadth, and t is the thickness. Δu is the net displacement of the specimen after correction by displacement of loading train, Δw is the load change against Δu, and W_f is the load at fracture. The number of strength tested specimens was ~ 17 for each sintering condition.

The half size specimen (18 mm length) was prepared and a straight through notch was introduced into the specimen by cutting with a diamond blade (0.2 mm thickness). The notch root was sharpened with a razor blade and diamond paste. The final notch length (a) was ~ 2 mm and the a/W ratio was 0.5. The single edge V-notched beam specimen[8] was tested by 3-point bending with a lower span of 16 mm under the crosshead speed of 0.5 mm/min. The fracture toughness (K_{IC}) was calculated by the following equation:[9]

$$K_{IC} = Y \frac{3W_f L}{2BW^2} \sqrt{a} \tag{3}$$

Y is the shape parameter, W_f is the load at fracture, L is the lower span, B is the breadth (3 mm), W is the width (4 mm), and a is the notch length.

Fracture Analysis Methodology

After the 4-point bending test, fracture surfaces of all the specimens were observed by optical microscopy to find the location of the fracture origin. After gold coating, the fracture surface was observed by secondary electron microscopy to identify the fracture origin size, shape and chemical composition. The accelerating voltage was 30kV, and the magnification was used from 50 to 2000 (JOEL: JSM-5300LV equipped with EDX).

RESULTS AND DISCUSSION

Characterization and Mechanical Properties of Specimens

Figure 1 shows that the relative density of the GDC ceramics decreases with increasing sintering temperature. This is due to the formation of nonstoichiometric phases of ceria resulting in an oxygen gas release.[10,11] In the previous papers,[4,5] both the samples were sintered at 1600°C, and therefore the relative density is about 93% when Figure 1 is extrapolated. This explains the low strength in the literature. The grain size was also measured by an intercept method. As shown in Figure 2, the grain size increases from 0.6 to 1.8 μm with an increasing sintering temperature.

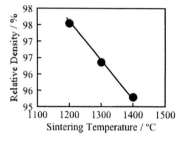 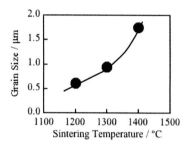

Figure 1. Relative density of the specimens.   Figure 2. Grain size of the specimens.

Figures 3 and 4 show Young's modulus (E) and fracture strength (σ_f) of the GDC ceramics. The decrease in Young's modulus is attributable to pore formation at higher sintering temperatures. The change in the fracture strength will be discussed after identifying fracture origins. It must be noted that the fracture strength values in this paper are relatively larger than those in the previous papers.[4,5] This results from the sintering conditions in this study.

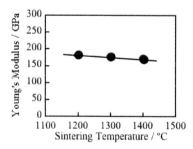 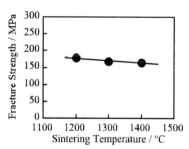

Figure 3. Young's modulus of the specimens.   Figure 4. Fracture strength of the specimens.

Figure 5 shows Weibull plots of the fracture strength. As stated later, the GDC ceramics have multiple causes of fracture, however, we will confront the loss of accuracy to estimate shape and scale parameters to apply multi-modal Weibull distribution. Therefore, it is assumed that the strength distribution can be approximated by the following single-mode Weibull distribution:

$$F(\sigma) = 1 - \exp\left\{-\left(\frac{\sigma}{\sigma_0}\right)^m\right\} \tag{5}$$

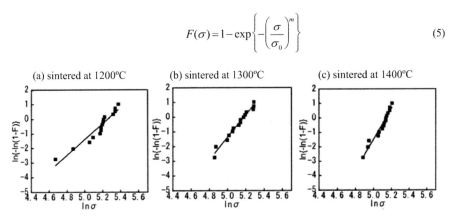

Figure 5. Weibull plots of fracture strength of the specimens.

F is the fracture probability, m is the shape parameter, and σ_0 is the scale parameter. Linearity is present in these plots, so the shape and scale parameters in single mode Weibull distribution are estimated by linear regression. Table I. summarizes its estimation. With increasing sintering temperature, the shape parameter (m) is increased and the scale parameter (σ_0) decreases.

Table I. Shape and scale parameters of Weibull distributions.

Specimen	Shape parameter m / 1	Scale Parameter σ_0 / MPa
Sintered at 1200°C	5.4	192
Sintered at 1300°C	7.6	178
Sintered at 1400°C	10.6	171

Identification of Fracture Origins by Fractography

Figure 6 shows representative optical microscope views of the fracture surfaces. In most specimens, the fracture origin is at the surface. However, in some specimens the fracture origin is at the edge. In Figure 6, the white triangles indicate the location of fracture origins.

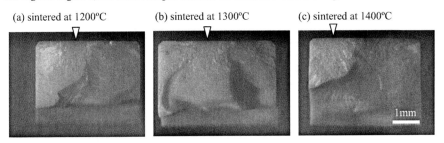

Figure 6. Fracture surfaces by optical microscopy.

Figure 7 shows typical fracture origins in the GDC ceramics which were sintered at 1200°C. The defects at the fracture origins are classified into (a) pores, (b) coarse grains, (c) impurities, and (d) unsintered regions. By EDX analysis, an impurity was found to be a silicon particle although this contaminant source is not known. The frequency of each fracture-causing defect is shown in Table II. Pores and impurities are the two major strength-reducing factors in these samples.

Figure 8 shows fracture origins in the GDC ceramics sintered at 1300°C. These fracture-causing defects are either (a) pores or (b) unsintered regions. However, most of the specimens were fractured by pores as shown in Table II. Figure 9 also shows fracture origins in the GDC ceramics sintered at 1400°C. In this case, the fracture-causing defects are (a) pores, (b) coarse grains, and (c) unsintered regions. With increasing sintering temperature, pores become the major fracture origin and cause a strength-reduction of GDC ceramics.

In Figure 10, the strength values are plotted against the cause of fracture. The fracture strength values are almost the same in the specimens fractured by the different defects (pores, coarse grains, and unsintered regions). However, the fracture strength of the samples with an impurity at the fracture origin is slightly lower than the other defects.

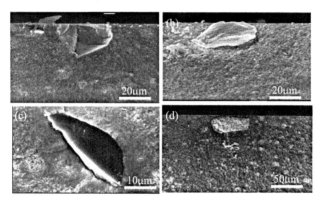

Figure 7. Fracture origins of the specimen sintered at 1200°C,
(a) pore, (b) coarse grain, (c) impurity, (d) unsintered region.

Table II. Frequency of each strength-reducing defect.

Specimen	Pore	Coarse Grain	Impurity	Unsintered Region
Sintered at 1200°C	5	2	8	2
Sintered at 1300°C	13	0	0	3
Sintered at 1400°C	12	3	0	2

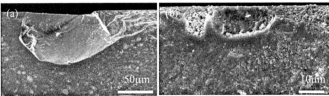

Figure 8.   Fracture origins of the specimen sintered at 1300°C,
(a) pore, (b) unsintered region.

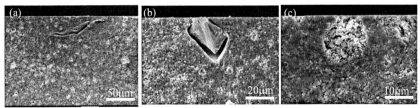

Figure 9.   Fracture origins of the specimen sintered at 1400°C,
(a) pore, (b) coarse grain, (c) unsintered region.

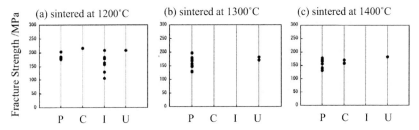

Figure 10.   Fracture cause dependence of fracture strength,
P: pore, C: coarse grain, I: impurity, U: unsintered region.

Verification of Fracture Origins

In order to verify the identification of fracture origins, the size of the fracture origin was measured and the apparent fracture toughness (K_{IC}^{app}) was calculated from the fracture strength (σ_f) and fracture origin size (b) by using the following relations:[12,13]

$$\sigma_f = K_{IC}^{app} \frac{1}{Y_1 \sqrt{\pi b}} \quad (5)$$

$$\sigma_f = K_{IC}^{app} \frac{1}{2Y_2 \sqrt{\frac{b}{\pi}}} \quad (6)$$

Y_1 is the shape parameter for semi-ellipsoidal cracks, Y_2 is the shape parameter for ellipsoidal cracks, and b is either of depth for semi-ellipsoidal surface cracks or the minor axis for ellipsoidal inner cracks. It is noted that some specimens fractured from the inner defects near the surfaces after precise observation by SEM. The size of the fracture origin includes the halo region,[14,15] as shown in Figure 12.

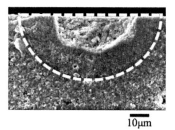

10μm

Figure 12.   Measurement of fracture origin size including halo region.

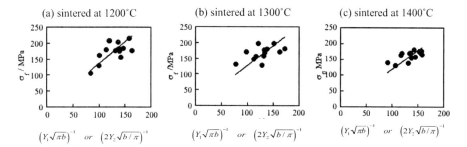

Figure 13.   Relation between fracture strength and reciprocal square root of origin size.

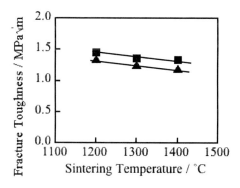

Figure 14.   Fracture toughness of the GDC ceramics,
■: K_{IC} by SEVNB method, ▲: apparent fracture toughness (K_{IC}^{app}).

Figure 13 shows the relationship between fracture strength and reciprocal square root of the fracture origin size. Regression analysis gives the slope as the apparent fracture toughness

(K_{IC}^{app}). The values are plotted with the fracture toughness (K_{IC}) by SEVNB method, as shown in Figure 14. From this figure, the apparent fracture toughness (K_{IC}^{app}) is approximately the same as the fracture toughness (K_{IC}) by SEVNB method. Therefore, the fracture origins are correctly identified in this paper.

Improvement of Strength Reliability of GDC Ceramics

Among the strength-limiting defects in GDC ceramics, impurities, coarse grains and unsintered regions can be eliminated with an improvement in processing. However, pores can not be removed since pores are formed by the nonstoichiometry of ceria at high temperatures. Therefore, the strength reliability of the GDC ceramics is estimated with the removal of impurities, coarse grains and unsintered regions from the GDC ceramics. In this analysis, a bi-modal Weibull distribution function is used on an assumption that GDC ceramics contain two types of strength-limiting defects (pore and the others). The result is summarized in Table III. For the specimens sintered at 1300°C and 1400°C, there is no significant change in the shape and scale parameters since these specimens were originally fractured by pores. However, in the specimen sintered at 1200°C, the shape parameter is increased by a factor of two. These specimens contain several types of defects at the fracture origins, especially impurities. As shown in Figure 10, the specimens containing impurities show a lower and broader strength distribution, and therefore an increased shape parameter is obtained when impurities are removed from the fracture causes in the specimen sintered at 1200°C. Although the number of specimens should be > 30 when the shape parameter is discussed, this estimation (by using 17 specimens) is consistent with the results in Figure 10. From this analysis, the strength reliability of the GDC ceramics can be significantly improved when impurities are eliminated from the GDC ceramics.

Table III. Improvement of strength reliability of GDC ceramics.

Specimen	Shape parameter m / 1	Scale Parameter σ_0 / MPa
Sintered at 1200°C	12.3	208
Sintered at 1300°C	7.1	181
Sintered at 1400°C	9.6	174

CONCLUSION

The causes of the strength-reducing factors in GDC ceramics were classified as pores, coarse grains, impurities, and unsintered regions. With increasing the sintering temperature, pores became the major strength-reducing defects. There was no remarkable change in the fracture strength for all defects except for impurities. From Weibull distribution analysis, it was shown that strength and reliability can be improved by eliminating impurities from GDC ceramics.

ACKNOWLEDGEMENT

The authors greatly express our gratitude to Kenneth. J. Krause (Kohler Co.) for his comments.

REFERENCES
[1]B.C.H Steele, A. Heinzel, "Materials for Fuel-Cell Technologies," *Nature,* 414: 345e52 (2001).

[2]Y.J. Leng, S.H. Chan, S.P. Jiang, K.A. Khor, "Low-Temperature SOFC with Thin Film GDC Electrolyte Prepared in Situ by Solid-State Reaction," *Solid State Ionics*, 170, 9-15 (2004).

[3]G. Laukaitis, J. Dudonis, D. Milcius, "Formation of Gadolinium Doped Ceria Oxide Thin Films by Electron Beam Deposition," *Materials Science* (*Mendziagotyra*), 13, 23-26 (2007).

[4]N. Sammes, G. Tomsett, Y. Zhang, "The Structural and Mechanical Properties of $(CeO_2)_{1-x}(GdO_{1.5})_x$," *Denki Kagaku*, 64, 674-680 (1996).

[5]S. Sameshima, T. Ichikawa, M. Kawaminami, Y. Hirata, "Thermal and Mechanical Properties of Rare Earth-Doped Ceria Ceramics," *Materials Chemistry and Physics*, 61 31-35 (1999).

[6]JIS R1602, "Testing methods for elastic modulus of fine ceramics".

[7]ISO 14704:2008, "Fine ceramics (advanced ceramics, advanced technical ceramics) – Test method for flexural strength of monolithic ceramics at room temperature".

[8]H. Awaji and Y. Sakaida, "V-Notch Technique for Single-Edge Notched Beam and Chevron Notch Methods," *J. Am. Ceram. Soc.*, 73, 3522-3523 (1990).

[9]W.F. Brown, Jr. and J.E. Srawley, "Plane Strain Crack Toughness Testing of High Strength Metallic Materials", *ASTM STP410*, ASTM, Philadelphia (1966).

[10]N.N. Greenwood, "Ionic Crystals, Lattice Defects and Nonstoichiometry", Chapter 5, Butterworth & Co. Ltd., London (1968).

[11]D.J.M. Bevan and J. Kordis, "Mixed Oxides of the Type MO_2 (fluorite)—M_2O_3—I Oxygen Dissociation Pressures and Phase Relationships in the System CeO_2-Ce_2O_3 at High Temperatures" *J.Inorg. Nucl. Chem.*, 26, 1509-1523 (1964).

[12]M. Ishida, H. Noguchi and T. Yoshida, "Tension and Bending of Finite Thickness Plate with a Semi-Elliptical Surface Crack," *Int. J. Fracture*, 26, 157-188 (1984).

[13]M. Ishida, H. Noguchi, "Tension of a Plate Containing an Embedded Elliptical Crack," *Eng. Fract. Mech.*, 20, 387-408 (1984).

[14]J.J. Swab and G. Quinn, " Effect of Precrack "Halos" on Fracture Toughness determined by the Surface Crack in Flexure Method," *J. Am. Ceram. Soc.*, 81, 2261-2268 (1998).

[15]J-k. Park, K. Yasuda, Y. Matsuo, "Effect of Crosshead Speed on the Fracture Toughness of Soda-lime Glass, Al_2O_3 and Si_3N_4 Ceramics Determined by the Surface Crack in Flexure (SCF) Method," *J. Mater. Sci.*, 36, 2335-2342 (2001).

TRIPLE BLIND FRACTOGRAPHY STUDY OF YTTRIA-STABILIZED ZIRCONIA THIN-FILM-MODIFIED PORCELAIN

Robert L. Smith and Jeffrey Y. Thompson
Section of Prosthodontics, Nova Southeastern University
Fort Lauderdale, FL USA

Ryan N. Chan
Department of Materials Science and Engineering, North Carolina State University
Raleigh, NC USA

Brian R. Stoner and Jeffrey R. Piascik
RTI International Center for Materials and Electronic Technologies
Research Triangle Park, NC USA

John J. Mecholsky Jr.
Department of Materials Science and Engineering, University of Florida
Gainesville, FL USA

ABSTRACT
Objective
    To determine if experience and fractography methodology of examiners results in differences in fracture toughness estimation of thin-film modified glass-ceramic.

Methods
    Glass-ceramic blocks (Empress CAD, Ivoclar Vivadent, Schaan, Liechtenstein) were cut into bars (2x2x15mm) and polished through 1200-grit using abrasive paper. One side of each bar was air-abraded (50μm alumina, 0.28MPa, 20s) and ultrasonically cleaned (acetone, 5min) before yttria-stabilized zirconia (YSZ) films were deposited using radio frequency magnetron sputtering. Bars were separated into five groups (n=12/group): Polished (Control-1), Polished and air-abraded (Control-2), YSZ film applied with no bias, YSZ film applied with 100W bias, and YSZ film applied with 150W bias. Bars were fractured in three-point flexure using an electromechanical testing machine in de-ionized water at 37°C. Four specimens per group were randomly selected and flaw size and fracture toughness of specimens was determined by three independent examiners using flexure strength data and fractography techniques. Differences in flaw size between examiners and groups were statistically analyzed using ANOVA (Tukey's test for multiple comparisons).

Results
    There was a significant difference in flaw size between groups for one examiner. There was a significant difference in flaw size in air-abraded and 150W bias groups between examiners.

Conclusion
    Statistically significant differences in fractographically-calculated flaw size values were observed between examiners. Experience and fractographic methods were shown to affect estimation of fracture toughness. Results indicate that strict fractography methodology protocols are important for accurately estimating fracture toughness.

INTRODUCTION

Fractography is defined as the method for characterizing a fractured specimen or component[1]. The use of fractography is commonly recognized as a technique for determining the cause of failure in structures and materials. In dentistry, fractographic analysis of fractured dental components can give quantitative and qualitative information about the failed material. For example, fractography can help determine the initiation site of critical fractures in dental crowns (occlusal surface vs. intaglio surface; veneer vs. core). Knowledge of failure location can help in optimizing the design of dental prostheses to limit areas of high stress concentration. Fractography can also be used to determine the mechanical properties of a material in order to aid in the improvement of strength and fracture resistance of dental materials.

Two important material properties that can be ascertained from fractographic analysis of dental materials are fracture strength, the stress at which material failure occurs from fracture, and fracture toughness, the measure of a material's resistance to fracture. Fractography can be used to determine the fracture toughness of a material, e.g., a new composition of porcelain to be used in a posterior crown, or to determine the fracture strength of an all-ceramic crown that failed in-vivo. This is accomplished by locating and determining the size of the fracture-inducing critical flaw. With this information, along with knowledge of the fracture strength of a material, the fracture toughness can be calculated using the equation[1]:

$$K_{IC} = Y\sigma_f a^{\frac{1}{2}} \tag{1}$$

where K_{IC} is the mode I critical stress intensity factor, i.e., fracture toughness (MPa·m$^{1/2}$), Y is the stress intensity shape factor, a constant that is based on loading geometry as well as the location and geometry of the flaw, σ_f is the fracture strength (MPa), and a is the length of the semi-minor axis, i.e., (usually) the depth of the surface flaw (μm) [Fig. 1]. Using the same equation, the fracture strength of a material can be determined if the fracture toughness of that material, along with the critical flaw size, is known, i.e.:

$$\sigma_f = K_{IC}/Ya^{\frac{1}{2}} \tag{2}$$

Although information about fractography and its methodology is available for review in technical literature and short courses and workshops can be attended for training, fractography is not broadly employed in dentistry. There are several reasons as to the absence of fractographic analysis in dentistry. One reason is that fractography is a subject that is not normally taught in most schools, dental or otherwise. Therefore, most dentists do not even know about the existence of fractography. Another reason for the absence of fractography in dentistry is that fractography is seen as subjective in nature[2]. It is thought that different observers will obtain different results from examining the same fracture surface and therefore the technique is not considered reliable. Experience is thought to be the main reason for the subjective view of fractography; novices are seen as not having enough experience to properly analyze fracture surfaces. However, experience should not limit one from obtaining useful information from a fracture surface. Novices, with access to fractographic analysis equipment and a general knowledge of the methodology from guides or technical literature, should be able to accurately locate the site of a crack-inducing flaw and determine fracture toughness or strength. Therefore, in this study, the fracture toughness of dental porcelain modified with a thin surface film of zirconia as a method to increase fracture strength and fracture toughness, was determined using mechanical testing and fractographic analysis. Fracture surfaces were evaluated by three examiners, with varying levels of experience, to determine if differences in fractography methodology and experience affect determination of flaw size and fracture toughness.

MATERIALS AND METHODS

For this study, leucite-based dental porcelain blocks (IPS Empress CAD, Ivoclar Vivadent, Schaan, Liechtenstein) were used as the substrate for all groups. Blocks were cut into bars (2x2x15mm) using a diamond wafering blade (15 LC IsoMet Diamond Wafering Blade, Buehler, Lake Bluff, IL, USA) with a low-speed diamond saw (IsoMet 1000, Buehler, Lake Bluff, IL, USA). After sectioning, each bar was wet-polished on all sides through 1200-grit using silicon carbide (SiC) abrasive paper (CarbiMet 2, Buehler, Lake Bluff, IL) and a grinding-polishing wheel (MetaServ 2000, Buehler, Lake Bluff, IL, USA) to decrease the size and uniformity of surface flaws present from the cutting process. The edges of the bars were rounded through 1200-grit using SiC paper to decrease the size of edges flaws.

After polishing, bars were separated into five groups (n = 12/group) and processed following previously implemented procedures:[3,4]

1. Polished (Control 1): Porcelain with no additional surface treatment and ultrasonically cleaned (acetone, 5min).
2. Air-abraded (Control 2): Porcelain, air-abraded (50μm alumina, 0.28MPa, 20s) and ultrasonically cleaned.
3. No Bias: Porcelain, air-abraded, ultrasonically cleaned, followed by deposition of yttria-stabilized zirconia (YSZ) film deposited using radio frequency magnetron sputtering (5mT, 25°C, 30:1 Ar/O$_2$ gas ratio) with a thickness of 3μm with no substrate bias.
4. 100 W Bias: Porcelain, air-abraded, ultrasonically cleaned, modified with a 3μm YSZ film deposited at 100W substrate bias.
5. 150 W Bias: Porcelain, air-abraded, ultrasonically cleaned, modified with a 3μm YSZ film deposited at 150W substrate bias.

All bars were fractured according to ASTM C1161[5]. A three-point flexure fixture (support span = 10mm) was used to fracture bars with a electromechanical testing machine (Instron 5542, Canton, MA, USA) at a crosshead speed of 0.5mm/min. in de-ionized water at 37°C to stimulate biological conditions.

After testing, four specimens from each group were randomly selected and analyzed by three different examiners to estimate flaw size and fracture toughness. All specimen fracture surfaces were gold sputter coated (Cressington 108 Auto, Watford, England, UK) to improve optical examination of the specimens. Additional examination of fracture surfaces was conducted using a scanning electron microscope (SEM - Quanta 200, FEI, Hillsboro, OR).

Examiners were selected with varying levels of experience: Examiner 1 – intermediate level of fractography experience of ceramics, Examiner 2 - little to no experience in fractography and Examiner 3 - extensive experience in fractography of ceramics. Examiners selected their preferred method for determining flaw size and fracture toughness. All examiners used the methodology outlined in ASTM Standard C1322 as a template for conducting their examinations. However, Examiner 2 calculated the fracture toughness using Eq. 1 while Examiners 1 and 3 calculated fracture toughness using the following equation:[6]:

$$K_{IC} = Y\sigma_f c^{1/2} \qquad (3)$$

where c is the flaw size calculated from the equation:

$$c = (ab)^{1/2} \qquad (4)$$

where a is the length of the semi-minor axis of the flaw, i.e., depth of surface flaw (μm), and b is the length of the semi-major axis of the flaw, one-half width of flaw (μm) [Fig. 1]. Using Eq. 4 normalizes a non-uniform flaw, allowing for the measurement of asymmetrical shapes and removing the need to calculate a different Y for each flaw[6]. All fracture toughness values were determined under the assumption that all fractures were Mode I (tensile load normal to the plane of the crack) and that all flaws were surface-based with a semi-elliptical geometry.

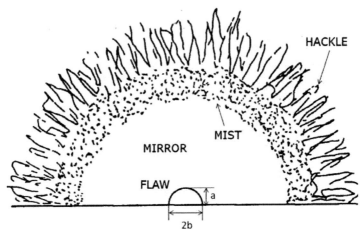

Figure 1. Schematic illustrating ideal fracture surface of a brittle material, where "a" is the semi-minor axis and "b" is the semi-major axis of the flaw (After Mecholsky[7]).

The effect of fractography technique on determination of flaw size values between groups and examiners was evaluated using a one-way analysis of variance (ANOVA) with a level of significance set at 0.05 (p = 0.05). A post-hoc Tukey's test was performed if comparison of flaw size values were determined to be significantly different (p < 0.05).

RESULTS
The average flaw size (Fig. 2) and fracture toughness (Fig. 3) values were determined for the five specimen groups. When comparing flaw size between specimen groups for an examiner, there was only a statistical difference in flaw sizes for Examiner 3 between the polished group and all other groups. Statistical analysis showed that there was a statistical difference in flaw size for the air-abraded and 150 W bias groups when comparing groups between examiners.

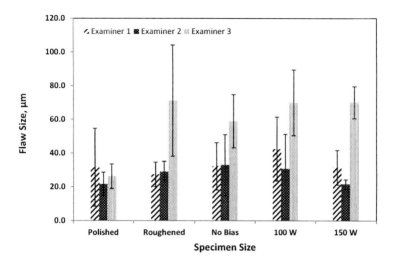

Figure 2. Mean flaw size values for specimen groups.
Error bars represent standard deviation.

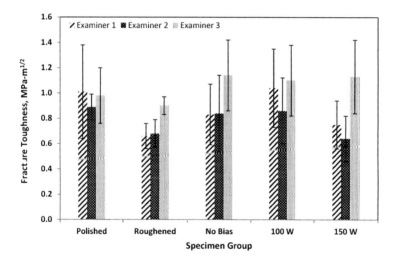

Figure 3. Mean fracture toughness values for specimen groups.
Errors bars represent standard deviations.

When examining the site of flaw location, the majority of specimens (85 %) fractured from flaws located centrally on the surface (Fig. 4A). Only 3 specimens (15 %) failed from edge-located, surface flaws (Fig. 4B).

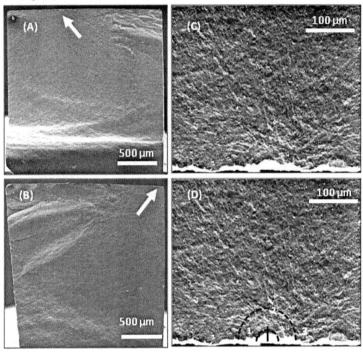

Figure 4: Fracture surface of zirconia thin-film modified porcelain. A) Arrow indicates location of centrally-located surface flaw. B) Arrow indicates location of edge flaw.
C) Magnified image of edge flaw in image (A). D) Example of flaw size measurement by examiners. Numbers denote flaw size identified by examiner.

DISCUSSION

Although the methodology of fractographic analysis is outlined in literature, fractography is not normally applied in dentistry. It is seen as a skill that requires years of experience to obtain important information from fracture surfaces. However, with little training, dentists can ascertain important information from the fracture surface of failed dental materials. To illustrate this point, the fracture toughness of leucite-based dental porcelain, with a zirconia thin-film applied to increase fracture strength and fracture toughness, was analyzed by examiners of varying levels of experience to evaluate if experience affects fractographic analysis.

The material under examination can determine the difficulty in identifying key features of the fracture surface, i.e., mirror, mist, or hackle, and identifying and measuring the crack-inducing flaw. Single crystal and amorphous materials such as glass are easier to analyze than glass ceramics, such as lithia-disilicate, or polycrystalline materials, such as alumina or zirconia, that produce a coarse fracture surface as the crack propagates through the material[2,8]. This makes it difficult to recognize the features of the fracture surface. Although it is possible to locate most initial flaws in brittle materials by identifying surface features such as river marks and Wallner lines, accurately determining flaw size can be a difficult task. It is thought that the coarse fracture surface of the porcelain used in this study is the main reason for the difference in flaw size and fracture toughness values reported by examiners. Experience with analyzing coarse fracture surfaces improves the accuracy of flaw size measurements. Additionally, the porcelain could have been subject to some degree of stress corrosion, producing critical flaws larger than the initial flaw from which fracture occurred. It is thought that Examiner 3, who is more knowledgeable about conducting fractography on a wide range of ceramic fracture surfaces than the other examiners, was able to determine more accurate mean flaw sizes and fracture toughness values for the specimen groups (Fig. 4D).

Determining the correct Y value is also important for properly calculating fracture toughness or fracture strength. Location and shape of, along with the type of loading on, the flaw are all used to select the Y value. In our study, differences in choice of Y value (1.29 – Examiner 1 and 2 vs. 1.26 – Examiner 3)[1,9] did not significantly affect the calculation of fracture toughness for most specimen groups. For the air-abraded and 150 W bias groups, it is thought that differences in flaw size and fracture toughness values could be attributed to an incorrect Y value. Examiners 1 and 2 neglected to recognize that because of the presence of an edge flaw in some of the specimens, a different Y value of 0.8 needed to be used in calculating fracture toughness[10]. This, along with the fact that Examiner 3 determined specimens had a significantly greater flaw size than that measured by Examiners 1 and 2, could explain the difference in fracture toughness estimation amongst examiners for these groups.

Surprisingly, there was no statistically significant difference in flaw size between Examiner 1 and 2, even though there was a difference in how the flaw size was measured by each examiner. Determination of the flaw size by measuring only the depth of the flaw (Eq. 1) or calculating the flaw size (Eq. 3) was shown to have no significant affect on calculating fracture toughness. This is due to the semi-circular shape of the flaws. If the shape of the flaws were more semi-elliptical, i.e., the semi-major axis is significantly greater than the semi-minor axis, then it is possible that a difference in flaw sizes could have been observed between the two examiners.

Difference in flaw size between groups was only exhibited by Examiner 3. The mean flaw size of the polished group was significantly less than all other groups however; there was no difference in fracture toughness values between the polished group and other groups. The polished group is expected to have a smaller average flaw size since the surface was not air-abraded after polishing, resulting in smaller failure inducing critical flaws. This difference in mean flaw size was not observed by either Examiner 1 or 2. Experience may again have been the reason for the difference measured by Examiner 3.

CONCLUSION

The fracture toughness of zirconia thin-film modified dental porcelain was determined using fractographic analysis. Experience was shown to affect determination of flaw size for some specimens. It is thought that difficulty in measuring flaw sizes on the coarse fracture surface of the dental porcelain used in this study was the main reason for the difference in flaw size values. Although experience is important in accurately analyzing a fracture surface, those with a limited knowledge of fractography can obtain useful information from analyzing and comparing fracture surfaces of dental materials by following proper fractography methodology.

ACKNOWLEDGEMENTS

This work was supported by the National Institute of Health-National Institute of Dental and Craniofacial Research, under Grant No. R01-DE013511. The authors would like to thank Jim

Rothrock of the Bioscience Research Center at Nova Southeastern University for his help in conducting SEM of the fracture surfaces.

REFERENCES
[1]ASTM Standard C1322-05b "Standard Practice for Fractography and Characterization of Fracture Origins in Advanced Ceramics," ASTM International, West Conshohocken, PA, 2010.
[2]G.D. Quinn, "Special Publication 960-16: Fractography of Ceramics and Glasses," 1st ed. National Institute of Standards and Technology: Washington D.C., (2007).
[3]E.C. Teixeira, J.R. Piascik, B.R. Stoner, and J.Y. Thompson, "Effect of YSZ thin film coating thickness on the strength of a ceramic substrate," *J. Biomed. Mater. Res. Part B,* **83B**[2] 459-63 (2007).
[4]J.R. Piascik, J.Y. Thompson, C.A. Bower, and B.R. Stoner, "Evaluation of crystallinity and film stress in yttria-stabilized zirconia thin films," *J. Vac. Sci. Technol., A,* **23**[5] 1419-24 (2005).
[5]ASTM Standard C1161 - 02c(2008)e1, "Standard Test Method for Flexural Strength of Advanced Ceramics at Ambient Temperature," ASTM International, West Conshohocken, PA, 2008.
[6]P.N. Randall, pp. 88-126. in Plane Strain Crack Toughness Testing of High-Strength Metallic Material. Edited by W.F. Brown and J.E. Srawley. American Society of Testing and Materials, Philadelphia, PA, 1967.
[7]J.J. Mecholsky, "Fracture Analysis of Glass Surfaces" in Strength of Inorganic Glass. Edited by C. R. Kurkjian. Plenum Publishing Corporation, New York, NY, 1986.
[8]J.J. Mecholsky, "Fractography of Brittle Materials," pp. 3257-65, in Encyclopedia of Materials: Science and Technology. Edited by K.H.J. Buschow, W.C. Robert, C.F. Merton, I. Bernard, J.K. Edward, M. Subhash, and V. Patrick, Elsevier, Oxford, 2001.
[9]J.C. Newman and I.S. Raju, "An empirical stress intensity factor for the equation for the surface crack," *Eng. Fract. Mech.,* **15** 185-92 (1981).
[10]R. Hertzberg, "Deformation and Fracture Mechanics of Engineering Materials," 4th ed. Wiley: Hoboken, NJ, (1995).

RESIDUAL FAILURE DUE TO INCOMPLETE SINTERING OF VITREOUS CHINA PLUMBING FIXTURES

David L. Ahearn, M.S., P.E.
Fred Schmidt, Ph.D., P.E.
Engineering Systems Inc., 6320 Regency Parkway, Norcross, Georgia 30071

ABSTRACT

Analysis of a slip casted vitreous china toilet tank that fractured while no one was home. The crack resulted in release of continuous running water causing water damage to the building. The general fractography suggested a slow crack growth phenomenon. There was no evidence of surface contact damage or external abuse as a fracture initiation. Analysis showed that incomplete sintering resulted in no vitrification of the center material. It is discussed that fracture initiated during manufacturing at the interface of the vitrification front and pores in the unfinished material resulting in critical fracture growth years later while the unit was in service.

INTRODUCTION

Water is running down the stairs to welcome you home from a weeklong family vacation. The culprit is a plumbing failure that occurred in the upstairs bathroom. It is not a busted pipe, water heater, or clogged drain though. It was a toilet that seemed fine before you left. Even upon first sight the toilet does not seem highly damaged, just a small crack down the side of the tank allowing a steady stream of water to fall to the floor.

This paper will examine a ceramic toilet tank that fractured as the result of a slow growing crack in material that was not sufficiently processed. First is a brief description of the manufacturing processes utilized in toilet tank manufacturing. Then, an analysis of the failed tank is presented.

LITERATURE REVIEW

Porcelain vitreous china water closet tanks are created by casting a slurry of carefully selected and mixed components of clay, feldspar, and flint.[1,2] These components are mixed with water to create a workable material similar to the consistency of a milkshake. This slurry is cast and dried, and at this point the material is referred to as "green".[1-3] The dried casting is then fired in a kiln at controlled temperatures for set lengths of time to create the final material desired. The final product's mechanical properties, such as strength, are dependent on the original materials used, the moisture content of the slurry, the compaction of the particles in the cast, the drying process, and the firing process. [2,4,5]

To create a toilet tank, the slurry is pumped into a mold and pressed by slip casting. Moisture is drawn out of the slurry while in the cast to produce a moist but semi-rigid tank. This partially-dried "green" tank is removed from the cast and moved to the drying process.[1] Too much moisture in a green material can result in rupture during the firing process.[4] It is critical to the final properties that drying is performed at a rate and to a point that relieves the material from moisture without inducing too much pressure in the voids created during evaporation. The drying process for the ceramic body is where most of the shrinkage occurs as the body loses moisture content.[2] The shrinkage from drying decreases the ability of fluid to evaporate, requiring increased pressure in pores in order for the gas to escape the material. This pressure, and the strain of shrinkage during drying, can create cracks in the green material. Cracks in a green material can be present prior to complete drying, but unnoticed due to moisture in the crack creating a negative pressure to hold the crack together. These cracks become mechanically active after drying.[2,4] Active cracks can take years to grow into catastrophic failures during normal use.

After adequate drying, the material is ready to be processed (fired) in a kiln. During the firing process, known as sintering, small particles of the ceramic matrix become vitreous enough to produce larger grains and a glassy phase around the larger particles, binding the material together and lowering the energy of the system.[1] Material density is increased during the vitreous phase of the processing. In a dried green material, porosity is approximately 50% by volume, created by the release of moisture from the material. The removal of this porosity from the material depends on the firing process.[2,6] Some residual porosity is expected, but the majority of pores are relieved during the sintering processes to create a stronger final product.[4,5] The strength of the final product is directly related to the density and the presence of pores. The denser the material, and the less pores in the material, the greater the strength of the porcelain.[3-5] The vitreous process also relieves residual stresses imparted into the material during casting and drying.

Transforming the clay into porcelain whitewares, like toilet tanks, requires soaking the components in temperatures between 1100-1600 degrees Celsius.[1,2,6] The time required for the material to obtain the correct final form depends on the temperature used and the physical dimensions of the component. In plant management terms, higher temperatures cost more money to produce while lower temperatures require longer soak times in kilns resulting in fewer units produced per day. To produce ceramic products, time and temperature are balanced for effective but efficient material processing. Batch materials are selected that maximize this efficiency.[7]

ANALYSIS

The tank inspected was not the first to fail at the subject's residence. There were multiple bathrooms with the same model toilet. The first tank failed while they were home allowing them to prevent water damages. The remaining toilets of same make and model in the home were collected and securely stored. During the time of this investigation, one of the tanks placed into storage cracked in a similar manner as the tank being inspected.

The subject fracture reported as the cause of water loss is located down the side of the tank and remained open after initial fracture (Figure 1). The opening is widest at the top of the tank at approximately 1/32nd of an inch. This fracture is long and smooth without bifurcations along the full height of the tank.

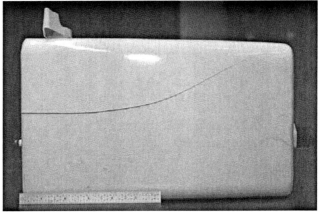

Figure 1. Photograph of the crack down the side of the toilet tank.

Separate hairline shrinkage cracks were noted at the front interior corner of the tank, opposite of the subject crack. These cracks did not penetrate the glaze and did not appear to penetrate the full thickness of the tank wall. No cracks were noted on the bottom of the tank containing the mounting screw holes or the spud hole.

In order to inspect the fracture surface and material the fracture had to be opened. The standard ballcock anti-siphon fill valve, the gravity flush spud, and the flush handle were removed from the tank. Green water based dye was added to the crack tip prior to opening the fracture in order to mark the existing extent of the fracture. The fracture was opened using a neoprene padded screw inclined hand jack mounted on the interior of the tank. The tank was set in a padded box to limit the movement of the portions during separation.

The water closet subject fracture surface revealed a non-homogeneous cross-section of material through the thickness of the tank wall (Figure 2). Inspection of other pieces from the toilet tank revealed this non-homogeneous material present throughout most of the tank.

Figure 2. Subject toilet tank fracture surface shows the non-homogeneous material through cross-section as viewed by stereomicroscope.

Fracture Morphology

The fracture surface was inspected for fractographic relics using a stereomicroscope. The fracture surface of the glaze was inspected using a metallographic microscope with a polarizing lens and an optical microscope. No rapid brittle fracture markings were witnessed along the subject fracture surface. The single smooth fracture surface is indicative of slow, low energy, fracture growth along the side of the toilet tank. The pores present along the fracture surface in the glaze did not have wake hackle markings that would represent rapid fracture growth and direction of fracture growth. The subject fracture did bifurcate as the crack grew through the bottom left front corner of the tank. Bifurcations at areas of sharp dimensional change can be expected due to stress concentration changes at the crack tips.[8,9]

Microstructure Evaluation

Sample material selected for evaluation was sectioned from near the back corner of the tank. The appearance of the sample cross-section was similar to the fracture surface. The sample was mounted in Struers Durofix-2 acrylic and polished. The polished section was viewed on the metallographic microscope. The polished section was then etched with 40% hydrofluoric acid (HF) for 5 seconds and viewed on the metallographic microscope again. The subject sample was then imaged using a scanning electron microscope (SEM) as shown in Figures 3 through 5.

A dense region of relatively low porosity surrounds the porous center region in the subject material and marks the penetration limits of the vitrification, seen as horizontal stripes along the cross-section in Figure 3. The center region contains large sharply shaped voids and small particles that have not been bound by a glassy phase (Figure 4). The bulk of the material outside of these two regions is poorly sintered porcelain material noted by large irregular pores with limited wetting of the quartz particles (Figure 5). Examination at higher magnifications revealed that the particles within the material do not have liquid rims or a high degree of rounding expected of a vitreous ceramics.

Figure 3. Subject toilet tank non-homogeneous materials through the cross-section appear as stripes (SEM micrograph 16x).

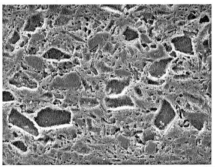

Figure 4. Subject porous material at the center of the cross-section (SEM micrograph 500x).

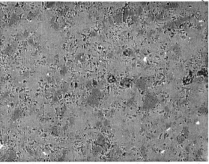

Figure 5. Subject material away from the porous center of the cross-section (SEM micrograph 500x)

Exemplar toilet tanks were purchased for comparative examinations. Figure 6 is a stereomicroscope image of a cross-section taken from an exemplar tank. There is a slight line down the center of the cross-section, typical of slip cast materials. This line is much smaller and less defined than the subject tank. Examination in the SEM provided comparative microstructural views of the porcelain materials (Figures 7 through 9). [It should be noted that in the SEM images of the exemplar material the center line of the cross-section is vertical, turned 90 degrees relative to the other SEM images in this paper.] The voids along the center of the cross-section in the exemplar are rounded and surrounded by a glassy matrix phase. The pores throughout the exemplar material are of consistent size and shape; the pores along the center are merely more densely coalesced.

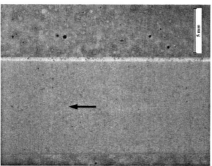

Figure 6. Exemplar toilet tank cross-section stereomicroscope image.

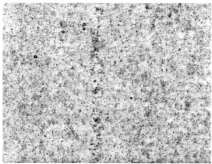

Figure 7. Exemplar tank cross-section revealing typical slip cast voids along the center of the cross-section (SEM micrograph100x).

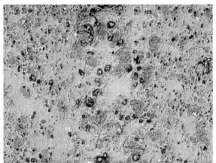

Figure 8. Exemplar material at the center of the cross-section (SEM micrograph 500x).

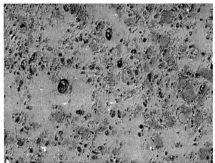

Figure 9. Exemplar material away from the center of the cross-section (SEM micrograph 500x).

Absorption Testing

After opening the fracture surface, portions of the subject water closet tank were utilized for absorption testing. ASME A112.19.2 Section 6.1 requires absorption of water by the ceramic material of the tank not to exceed 0.5% by weight.[10] The average absorption of the subject ceramic tank was 0.8% by weight. The material did not comply with the absorption requirements of the ASME A112.19.2. This test uses absorption properties to verify that the whiteware material has been processed correctly. If the material is not processed correctly, or is made of substandard raw materials, then final properties like absorption or strength will be affected. Improper raw materials could result in a ceramic that has constituents that dissolve in water, quickly degrading a plumbing whiteware in use. Improper processing leaves a material insufficiently sintered or full of pores that can take on moisture. In the subject tank, the material has high porosity and an immaturely developed glassy phase. It has been noted that water is the primary cause of sub-critical, slow, crack growth in ceramics.[3] Failure of this tank material to prevent moisture migration only exacerbates conditions noted above, contributing to fracture growth.

CONCLUSIONS

A slip casted vitreous china toilet tank was analyzed to determine cause of failure. Analysis of the microstructures revealed greater densities of porosity along the center of slip cast toilet tank cross-sections. This is a relic of shear stresses during the removal of the mold and the lack of pore transport during firing.

Small particles, rough edges, and no glassy phase indicated that the center material of the subject casting was not exposed to temperatures sufficient to produce the material transformation to porcelain. The dense surrounding material is evidence of a vitreous liquid sintering transformation occurring that did not reach the very center of the cross-section. Insufficient soak times resulted in incomplete material processing of the subject toilet tank.

Sharply shaped pores noted in the subject material were larger than the grains within the original clay and subsequent porcelain material. Strains developed during cooling of the material from 1000°C during manufacturing created concentrated stresses along the boundaries of these pores. The dissimilar material properties assisted in initiation of fracture sites. In the presence of imperfections, residual stress resulted in fracture upon critical length.

REFERENCES
[1]W.D, Kingery, H.K. Bowen, D.R. Uhlmann, *Introduction to Ceramics*, 2nd Ed., John Wiley & Sons, Inc., 1976.
[2]V.E. Henkes, G.Y. Onoda, W.M. Carty, *Science of Whitewares*, John Wiley & Sons, Inc., 1996.
[3]ASM International, *Engineered Materials Handbook: Ceramics and Glasses*, Vol. 4, 1991.
[4]D.J. Green, *An introduction to mechanical properties of ceramics*, Cambridge University Press, 1998.
[5]J.B. Wachtman, *Mechanical Properties of Ceramics*, John Wiley & Sons, Inc., 1996.
[6]H. Insley, V.D. Fréchette, *Microscopy of Ceramics and Cements*, Academic Press, 1955.
[7]F.H. Norton, Fine Ceramics, McGraw-Hill, 1970.
[8]G.D. Quinn, *Fractography of Ceramics and Glasses*, NIST Special Publication 960-16, 2007.
[9]V.D. Fréchette, Failure Analysis of Brittle Materials: Advances in Ceramics, Wiley-American Ceramic Society, 1990.
[10]ASME A112.19.2-2008/CSA B45.1-2008, "Ceramic Plumbing Fixtures".

LAMINATE DESIGN EFFECTS ON THE FRACTURE PATTERNS OF IMPACT RESISTANCE GLASS PANELS

J. L. Ladner[a], D. L. Ahearn[b] and R.C. Bradt
Dept. of Metallurgical and Materials Engineering
The University of Alabama
Tuscaloosa, AL 35487 – 0202

(a) Huie, Fernambucq & Stewart, Birmingham, AL 35223
(b) Engineering Systems Inc. (ESI), Norcross, GA 30024

ABSTRACT
Impact resistant glass panels are laminate composites with a polymer central layer between glass sheets that serves to maintain integrity of the glass when fracture occurs. This paper compares the crack patterns of two different thickness panel designs, one essentially twice the thickness of the other. Impact is by a 30 caliber copper projectile at velocities from ~100-300 m/s. The number of radial cracks emanating from the impact crush zone and the crack arresting performance of the two panel designs are considered with the variation of the kinetic energy (velocity) of the projectile. The number of radial cracks increases linearly with the kinetic energy of the projectiles. Finally the projectile damage (strain) is also examined as it varies with the kinetic energy of the projectile. The thicker panels are less radial crack prone, have a greater crack arrest capability and create greater damage (strain) of the impacting projectiles.

INTRODUCTION
The application of impact resistant glass panels is continually increasing along with the interest in homeland security and also for applications requiring greater margins of safety. They exhibit a distinctive fracture pattern as discussed by McMaster, et al[1] and illustrated in Figure (1) from their review article of annealed and tempered glass.

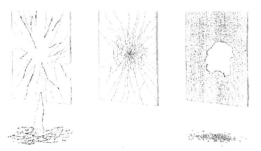

Figure 1. Impact fracture patterns of an annealed glass plate (left), tempered glass plate (right) and a laminated impact resistant glass panel (center).[1]

In Figure 1 it is evident that the annealed glass plate has a readily identifiable central impact point and exhibits the classical radial crack pattern originating from that point. It produces large sharp shards that fall from the broken plate. The tempered glass plate on the right reveals the typical dicing fracture pattern of tempered glass which occurs because of the high level of stored elastic strain energy in the glass plate as discussed by Warren.[2] However, the center impact resistant glass panel, while exhibiting the classical star like radial crack pattern accompanied by circumferential cracks about the impact point and crush zone remains intact as has been previously described by Bradt, et al.[3] The reason for its integrity is its laminate design of two external glass plates with a sandwiched central polymer layer.

Depending on the intensity of the impact it must defeat, the thickness of the glass sheets and their numbers will vary. It is not unusual to utilize multiple sheets of glass and several polymer layers as well for intense impacts. Characteristics of several panel designs relative to their impact resistance, weight and transparency are specified by Underwriters Laboratories.[4] Although it appears obvious that panels with thicker layers should be more impact resistant, there is not a standard design criterion to specify those laminate parameters. This paper addresses one aspect of the design of the panels, namely the thickness of the glass layers, and the central polymer layer in a three layer panel.

PANEL DESIGN AND PROJECTILE IMPACT TESTING

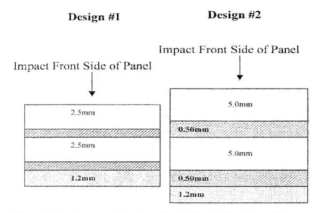

Figure 2. Laminate panel designs #1 and #2 with float glass thicknesses of 2.5mm and 5.0mm, and polycarbonate thicknesses of 0.25 mm and 0.50 mm with a 1.2 mm polyurethane backup layer opposite to the impact side on each.

The two glass panel designs tested in this study are depicted in Figure 2 above with the thickness dimensions of the glass sheets and polymer layers noted in the diagrams. The laminate panels were 12 inches square, a restriction of the ballistic laboratory test stand. This limited size may have effects on the crack arrest capabilities of the panel designs. Five panels of each were prepared and then tested at different projectile velocities on the ballistic test stand in Figure 3 for a single projectile impact at the specimen center.

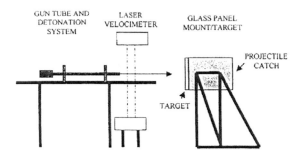

Figure 3. The experimental design was a 30 caliber gun tube without any rifling, propels the flat 30 caliber Cu cylinder. The velocity is controlled by the magnitude of charge loaded into the casings and measured by a dual laser velocimeter and recorded on an oscilloscope. The target panel laminates are mounted just ahead of a projectile catch containing sand to recover the fully penetrating projectiles.

EXPERIMENTAL PROCEDURES

Five individual panels of each of the two designs were impacted at projectile velocities over the range of ~100-300 m/s, but not at precise intervals for the individual grains of the gun powder were large enough to create variability in the charges. For calibration, 100 m/s is ~225 mph and 300 m/s is ~675 mph. The numbers of radial cracks were counted both at the edge of the central spall zone and again at the edges of the panels. These two counts were used to estimate a crack arrest ratio of the panel design from the simple formula:

$$R_{arrest} = \text{\# radial cracks at panel edge} / \text{\# radial cracks at spall edge} \qquad (1)$$

Although it is recognized that this parameter probably will not scale to other sizes, it none the less is useful for a direct comparison of panel designs when the dimensions of the two panel area sizes are the same. The smaller the R_{arrest} value, the greater is the crack arrest capability of the particular panel. As previously noted, the panel area dimensions may be expected to affect the numerator of this panel.

From the laser velocimeter the velocity of the projectile was determined and then applied to calculate the kinetic energy, KE, of the projectile from:

$$KE = \tfrac{1}{2} mv^2 \qquad (2)$$

where the mass of the Cu cylinders was always constant. The OFHC Cu projectiles were 25.4 mm in length, 7.8 mm in diameter and weighed 11.0 g each. Applying Equation (2) to estimate the kinetic energies of the projectiles yields ~55 J for 100 m/s and ~495 J for 300 m/s, so that the kinetic energies varied over just less than an order of magnitude.

The other measurement of interest is the damage parameter applied to the projectile after impact. A simple estimate of strain was applied to the mushroomed end of the Cu cylinder as the back end as illustrated below in Figure 4 appeared undeformed. The deformation strain, ε, was estimated from:

$$\varepsilon = (d_{if} - d_{io} / d_{io}) \times 100\% \qquad (3)$$

Where d_{if} is the diameter of the projectile on the impacting face, the strains are thus those the projectile experienced at its impacting end. The other end remained the original diameter. Strains varied from ~40% to ~80% at the highest velocities, so substantial mushrooming of the Cu cylinders occurred during impact.

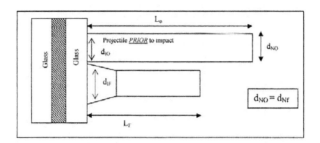

Figure 4. Dimensions of the projectile before and after impact applied to estimate the strain of the impact end during impact.

These parameters, the number of radial cracks, the arrest parameter of the panel design and the strain incurred by the projectile impact end were then plotted versus the kinetic energy of the projectile for basis of assessment and comparison.

RESULTS AND DISCUSSION

The cracking pattern from the impact of design #1 at a projectile impact velocity of 128 m/s is shown in Figure 5. It is presented because in some cases it is relatively easy to distinguish the radial cracks that extend to the edge of the panel and those which are arrested before they reach the panel edge. Both types of cracks were present in every panel impact, but they were not always 100% visible from the same lighting angle. Often the panel had to be exposed to oblique light to see many of the cracks, so actually counting them can be a challenging task, always subject to minor errors, but from the results it is evident that it is possible to make representative meaningful counts of the cracks nonetheless.

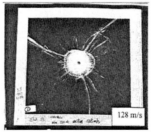

Figure 5. The panel design #1 impacted at a velocity of 128 m/s. Note the three large cracks extending to the edge of the panel and the numerous arrested cracks. Of course there are many other cracks not visible in this lighting.

The number of radial cracks nucleating from the central spall zone is presented in Figure 6 versus the kinetic energy of the projectiles. The numbers of radial cracks are large from ~40 at the lowest projectile kinetic energies to ~140 at the highest projectile kinetic energies. Both of the panel designs appear to follow a linear relationship between the number of radial cracks and the kinetic energy of the projectile. This linear trend has also been observed by Yoshimura, et al.[5] Design #1, the thinner of the glass plates consistently has a greater number of radial cracks and also exhibits a greater rate of increasing of radial cracking with increasing projectile kinetic energy as shown in Figure 6. The greater the impacting kinetic energy of the projectiles, the more susceptible to radial cracking Design #1 becomes relative to Design #2. It is evident that the thicker glass plate design is the more resistant to radial cracking on impact in all respects.

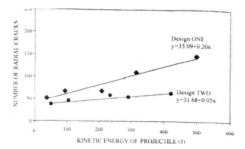

Figure 6. The number of radial cracks emanating from the spall zones of the two panel designs. Design #2 is more crack resistant.

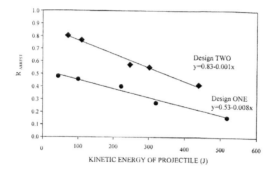

Figure 7. The arrest of the radial cracks in the two panel designs. Again the thicker glass plate design, #2, seems to be superior in arresting radial cracks.

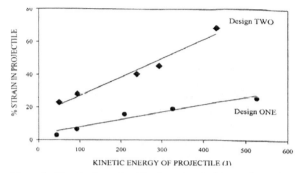

Figure 8. The plastic strain of the impact end of the mushrooming projectile as it increases with the kinetic energy of the projectile.

Figure 7 presents the crack arrest parameter, R_{arrest}, for the two glass panels with increasing kinetic energy of the projectile, which reflects an increasing velocity of the projectile. Both of the two panel designs follow the same trend, as they did for the increasing number of radial cracks with increasing kinetic energy. The trend in the case of crack arrest is that as the kinetic energy of the projectile increases, then the fraction of radial cracks arrested decreases. The higher energy projectiles seem to drive more of the radial cracks all of the way to the panel edge. However, one must be appraised that this parameter is probably test specimen size dependent, for if the panel were smaller then more of the cracks would be driven to the boundary. Were the panel larger, in all likelihood, fewer of the cracks would be able to reach the edge of the panel. This is a good example of test conditions affecting the results, but it is known for this case.

The above results have focused on the damage response of the glass panels to the impacting projectiles. It is a viewpoint of sorts that the results are being considered from the perspective that the projectile is defeating the target for it does accent the radial cracking of the target. Of course, the opposite perspective is that the impact resistant panels are defeating the projectiles, or at least severely damaging the projectiles. The Cu cylindrical projectiles are severely plastically deformed during the impact event. Figure 8 presents the increasing strain in the projectiles after impacting the glass panel targets at increasing velocities. This strain is measured at the mushrooming impact end of the Cu cylinder for the back end of the cylindrical projectiles was not affected by the impact for any of the velocities measured. Similar to the radial cracking of the panels the deformation strain of the projectile also follows a linear trend with the increasing kinetic energy of the projectile. Again, design #2 seems to be preferred as it imparts the greatest plastic flow to the projectiles.

SUMMARY AND CONCLUSIONS

Two panel designs of impact resistant glass panels were tested, the second just twice the thickness of the first. They were impacted with 30 caliber Cu cylinders at velocities from ~100-300 m/s. The two were compared relative to the numbers of radial cracks nucleated during impact, a crack arrest parameter and the plastic flow damage that the glass panel imparted to the Cu projectiles. In every instance, the thicker impact resistant panel was superior to the one that was only half the thickness. It seems like a classic case of bigger is better.

ACKNOWLEDGEMENTS

The authors are grateful to H. Nakai of Nippon Sheet Glass for preparation of the impact resistant panels and to S.E. Jones for his assistance with the ballistic impact testing of the panels. Provision of the drawings by Jeff Motz was also appreciated.

REFERENCES

[1] R.A. McMaster, D.M. Shetterly and A.G. Bueno, "Annealed and Tempered Glass" 453-459 in the *Engineered Materials Handbook*, v 4, ASM Int., Materials Park, Ohio (1991).
[2] P.D. Warren, "Fragmentation of Thermally Tempered Glass", 389-402 in Adv. in Ceramics, V 122, edit by J.R. Varner and G.D. Quinn, Amer. Cer. Soc., Columbus OH (2000).
[3] R.C. Bradt, M.E. Barkey, S.E. Jones and M.E. Stevenson, "Projectile Impact – A Major Cause for Fracture of Flat Glass", 20-23 in The Glass Researcher, v 11, n 2, R.K. Brow editor, Alfred University, Alfred, NY (2002).
[4] Underwriters Laboratories, Inc., ULC 752, "Bullet Resistant Equipment",
http:www.ul.com./info/standard.htm.
[5] H.N. Yoshimura, "Analysis of Projectile Impact Damage in Bulletproof Glass" Provided by the author, to be published, (2005).

APPLICATION OF FRACTOGRAPHY IN PHARMACEUTICAL INDUSTRY

Florian Maurer
SCHOTT AG
Research and Technology Development
Hattenbergstrasse 10, 55122 Mainz, Germany

ABSTRACT

Fractographic investigations were conducted on a glass syringe complaint specimen with breakage at its cone end. The aim of the investigation was to determine whether failure occurred due to mechanical overload or due to unusual damage which led to breakage at moderate mechanical loadings. For this purpose, controlled cone-strength investigations and fracture-mirror size analysis were conducted on the fracture surfaces of reference syringe specimens. From this, an empirical constant of the cone-strength force versus the fracture-mirror size ("syringe mirror constant") was deduced. With these results a comparison of the mirror-mist radii of the reference specimens with the complaint specimen was possible. From Orr's equation it is known that low mechanical loadings create large mirror-mist radii, and vice versa. As the mirror-mist radii of the complaint sample are much greater than for the reference samples, it is deduced that the breakage force for the complaint sample must have been much smaller than the specification level for this syringe. This conclusion is supported by the fact that severe damage is observed in the vicinity of the fracture origin of the complaint sample. It is thus concluded that failure occurred due to damage which was introduced somewhere during hot forming of the glass, syringe assembly, filling, or packaging. This result acquits the syringe operator (e.g., patient or health care staff) from responsibility and shifts it to contributors of the value added chain (e.g., syringe manufacturer, pharmaceutical company, contract fillers or labelers).

INTRODUCTION

Syringes made of glass are widely used for the application of injectable drugs. Several polymeric adaption systems have been designed to be mounted onto the so called syringe Luer* lock cone geometry on which canulas of different sizes can easily be attached (Fig. 1).

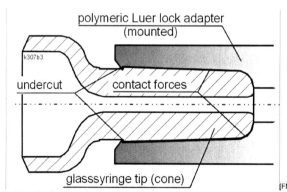

Figure 1. Sketch of a Luer lock syringe cone with an attached polymeric adaption system; the fixation of the adaption system is assured by an undercut at the syringe cone and high contact forces between the syringe cone and the adaption system.

Such Luer lock adaption systems also fulfill the functionality of integrity seal and anti-counterfeiting purposes. They are fixed to the glass syringe Luer cone geometry via an undercut. Additionally, high contact forces between the syringe glass cone and the adaption system are required to prevent rotation around the syringe axis. The complete syringe system is prepared for drug application by manual breakage of small polymeric bars of the Luer lock adapter in torsional, flexural or tensile mechanical load. After breakage of the polymeric bars, common injection canulas can easily be attached to the Luer lock adaption system.

In the paper presented here, a cone-broken glass syringe complaint sample (Fig. 2) with a Luer lock adaption system still attached to the cone was presented to SCHOTT Pharma Services** to be investigated by means of fractography to assess the circumstances for failure. The information was provided that breakage of the syringe occurred due to a flexural mechanical loading (i.e., bending) when the Luer lock adaption system was intended to be opened manually by the operator right before attachment of an injection canula. The specimen of interest represents a complaint specimen, and the aim of the investigations was to conclude whether breakage occurred due to an exaggerated mechanical overload (i.e., abuse) by the operator or due to a reduced strength of the glass syringe cone which led to failure at moderate mechanical loadings. For this purpose, a comprehensive fractographic examination of the complaint specimen was conducted. Additionally, controlled cone strength experiments were accomplished on representative glass syringes that were used as reference specimens. Empirical fracture-mirror analysis[1,2] was conducted on 14 syringes of the reference lot after the controlled breakage. The results of the fracture-mirror analysis were used to correlate the sizes of the fracture mirrors of the reference specimens with their respective force value data of the cone strength. This correlation then was transferred and used for an estimation of the breakage force of the complaint specimen.

Figure 2. Cone-broken glass syringe complaint specimen
(the Luer lock adaption system has been removed).

EXPERIMENTAL
Cone Strength Investigations
Cone strength investigations were conducted on 50 reference glass syringes to simulate the mechanical loadings (i.e., flexural loading) that led to failure of the complaint specimen. The syringes were horizontally fixed in a rotating specimen support while a mechanical force F was vertically applied at the syringe cone (Fig. 3). The force F was introduced with an increasing load rate dF/dt of 50 N/min until breakage. The rotation speed of the specimen support was approximately 22 s/revolution. The distance d from the syringe tip to the point of application of the force F was

chosen to reproduce a mechanical flexural loading with a Luer lock adapter attached to the syringe cone as realistic as possible.

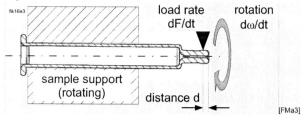

Figure 3. Sketch of the syringe cone strength testing setup.

Fractographic Examination
Fractographic examinations of the cone-broken complaint specimen and fracture-mirror analysis (measurement of mirror-mist radii) on 14 of the reference specimens were accomplished by means of stereo-optical microscopy in combination with enhanced-depth-of-focus software. The measurements of the fracture-mirror analysis on the 14 reference specimens were conducted by two different operators at three different magnifications (32×, 50×, 66×) and using the mean of the measured values.

RESULTS
Complaint Sample Fractography
Fractographic examination of the broken cone of the complaint specimen reveals typical surface topography ("cantilever curl") and fracture surface markings for failure under flexural loading (Fig. 4). The fracture origin is located within a broad region at the outside surface of the cone (Fig. 4, left). Twist hackle[1,3] is observed on both sides opposite to the vicinity of the fracture origin indicating the local propagation direction of the crack (Fig. 4, from the left side to right). As the channel of the syringe cone represents a singularity at which the propagating crack splits and passes around on both sides, a hackle step[1,3] is observed on the cone channel opposite to the vicinity of the fracture origin again indicating the propagation direction of the crack (Fig. 4, right). Additionally, chipping is observed on both sides of the crack paths (Fig. 4, top and bottom). No distinct or pronounced fracture mirror surface markings (e.g., Wallner lines,[1,3] mist hackle,[1,3] velocity hackle[1]) are observed in the vicinity of the fracture origin.

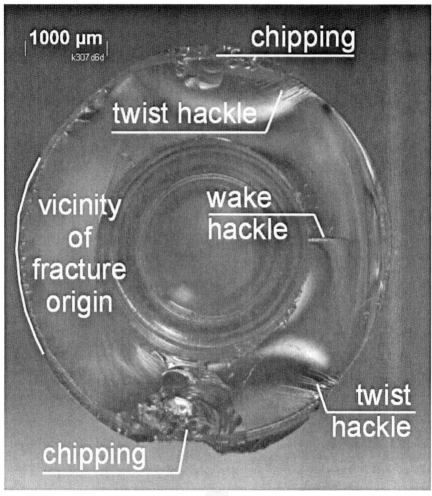

[FMa6]
Figure 4. Image of the cone fracture surface and fracture
surface markings of the complaint specimen.

Reference Specimens Fractography

In contrast to the complaint specimen, typical fracture mirror surface markings are observed for all 14 specimens of the reference lot (representative images are given in Figs. 5 and 6). Breakage occurred under flexural mechanical loading from origins located on the outer surface.

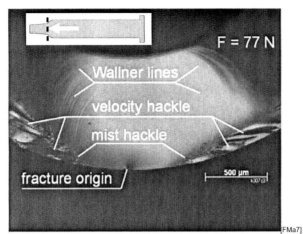

Figure 5. Image of the cone fracture surface and fracture mirror surface markings of one of the reference specimens (breakage force F=77 N); the inset picture shows the viewing direction (white arrow) to the specimen (grey contour, black dashed line: fracture surface).

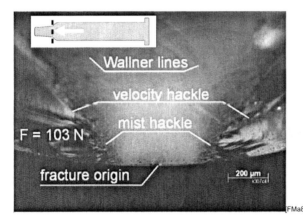

Figure 6. Image of the cone fracture surface and fracture mirror surface markings of one of the reference specimens (breakage force F=103 N); the inset picture shows the viewing direction (white arrow) to the specimen (grey contour, black dashed line: fracture surface).

Cone-Strength Investigations

The distribution of mechanical stresses in a glass syringe under flexural loading is complex and strongly depends on the geometry of the specimen (e.g., length, radii of curvature, wall thicknesses). Thus, an estimation of the strength (i.e., mechanical tensile stress σ at the position of breakage) for every single specimen under investigation is difficult to achieve. Orr's equation[1,2] for the relationship between stress σ and fracture-mirror (mirror-mist) radius $R_{m/m}$ in its original form (where $A_{m/m}$ is the mirror-mist fracture mirror constant) is:

$$\sigma\sqrt{R_{m/m}} = A_{m/m} \qquad (1)$$

For the investigations presented here the fracture-mirror analysis was not correlated to the strength σ of the syringes, but to the breakage force F. Since in this case the maximum bending stress is almost certainly linearly related to the applied bending force, F, then:

$$F\sqrt{R_{m/m(F)}} = A_{m/m(F)} \qquad (2)$$

The subscript (F) refers to the fact that the breakage *force* F is correlated to the mirror-mist radius R instead of the strength σ. $A_{m/m(F)}$, hereafter termed the "syringe mirror constant" has units of force times square root of length, or $Nm^{1/2}$ and it is specific to shape and component. All radii were measured along the surface on either side and not into the interior (since the stress gradient prevented the mirror markings from forming into the depth). The syringe mirror constant is determined by linear fitting of F versus $R_{m/m(F)}^{-1/2}$ which gives a value of $A_{m/m(F)} \approx 1.24\ Nm^{1/2}$ (Fig. 7).

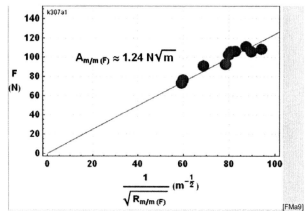

[FMa9]

Figure 7. Plot of force F versus inverse square root of mirror-mist radius $R_{m/m(F)}^{-1/2}$. The syringe mirror constant $A_{m/m(F)}$ is found by linear fitting the data.

For the force F values of the cone-strength of the reference specimens a Weibull distribution is found by means of goodness-of-fit testing to appropriately describe the data (Fig. 8). The specification level of the cone strength of the syringe (i.e., the minimum force required to pass product specification) is defined at F=55 N by this particular test. None of the 50 reference syringe specimens shows force breakage values below this specification level. Thus, all reference syringes would have achieved the required minimum force.

From Eq. (2) and the results of the fracture-mirror analysis, a mirror-mist radius $R_{m/m(F)}$ can be ascribed to every force value F (see upper axis of diagram in Fig. 8).

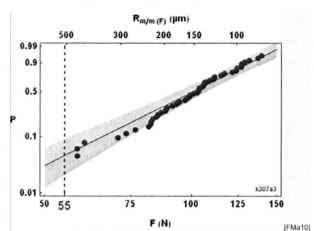

Figure 8. Weibull plot (failure probability P versus breakage force F) of the cone strength of the glass syringe reference specimens. The specification level of the cone strength is F = 55 N (dashed vertical line); solid line: Weibull distribution. The shaded area represents the 95 % confidence interval.

Consequently, the value of $R_{m/m(F)}$ at the specification level force of F=55 N is deduced to $R_{m/m(F)} \approx 553$ µm (Fig. 9, dashed lines). Good accordance is observed for the correlation of the mirror-mist radii $R_{m/m(F)}$ and the cone breakage forces F for the reference specimens.

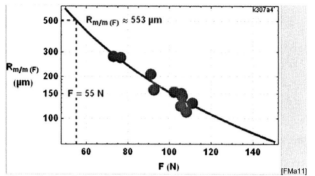

Figure 9. Correlation of mirror-mist radius $R_{m/m(F)}$ and cone breakage force F.

DISCUSSION

The presence, position and orientation of the fracture surface markings observed on the complaint specimen are typical for failure under flexural loading.[1] It is thus supposed that breakage of the syringe occurred under a bending moment applied while the operator intended to open the Luer lock adaption system. The chipping observed on the fracture surface of the complaint specimen (Fig. 4) is supposed not to originate from the primary breakage event but rather from subsequent damaging after breakage. Chipping is not observed on the fracture surface of the complementary syringe body, so it is not ascribed to the primary breakage event.

Fracture-mirror markings like mist hackle and velocity hackle appear when a crack is close to its terminal breakage velocity[1]. The fact that no distinct or pronounced mist hackle or velocity hackle is observed for the complaint specimen is a hint that the acceleration and thus the velocity of the crack were low. Consequently, as low crack acceleration is related to a low amount of stored elastic energy (low forces or stresses) at the occasion of fracture, low forces F were required for breakage. Additionally, from the fracture-mirror analysis it is deduced that for a breakage force F higher than the specification level (F ≥ 55 N) a mirror-mist radius of maximum 553 μm (i.e., $R_{m/m(F)} \leq 553$ μm) must be observable (Figs. 8 and 9). As no mirror-mist radii are observed in this range for the complaint specimen, the force F required for breakage of the complaint specimen is supposed to be much lower than the specification level (i.e. F ≪ 55 N). Furthermore, to fulfill the Griffith equation[4] for brittle fracture and assuming the strength controlling flaw is in a tension or bending stress field:

$$K_{IC} \propto F_{failure}\sqrt{c_{failure}} \qquad (3)$$

Large defect sizes $c_{failure}$ are required when low breakage forces $F_{failure}$ are observed to reach the failure criterion K_{IC} (fracture toughness).

For the complaint specimen, not only one distinct defect but numerous deep cracks are observed along the breakage edge where the fracture origin is located (Fig. 10).

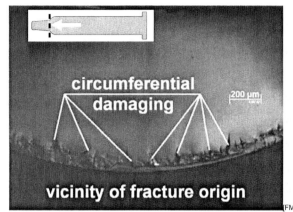

[FMa12]

Figure 10. Image of the vicinity of the fracture origin of the complaint specimen showing severe circumferential cracking damage; the inset picture shows the viewing direction (white arrow to the specimen (grey contour, black dashed line: fracture surface).

These cracks extend as far as 70 to 90 µm below the surface. This type of damage is similar to Case 4 in chapter 10 of ref. 1. It is supposed that breakage of the complaint specimen occurred due to a severe circumferential damaging in the syringe cone region. Due to this severe damaging, mechanical loadings (i.e., forces) even far below the specification level led to cone-breakage failure of this particular specimen. In other words, from the results of the investigations it can be verified that the cone strength of this particular specimen did not fulfill its requirements. However, no proof can be given whether exaggerated forces (i.e., abuse) were applied or not, but the results are sufficient to show that the responsibility for this failure is not at the customer/health care staff rather than somewhere in the process chain of the syringe (e.g., syringe manufacturer, pharmaceutical company, contract filler, contract labeler).

It has to be emphasized that the syringe mirror constant $A_{m/m(F)}$ used for investigations described here was determined empirically only for the purposes of this particular problem; i.e., the value is only valid for the syringe specimens of this lot and the testing setup described above. Usage of this value for other syringe specimens and/or other testing setups is not-appropriate. It is thus strongly recommended to determine the syringe mirror constant $A_{m/m(F)}$ for every new problem, syringe lot, and testing setup.

CONCLUSION

In conclusion, fractographic investigations in combination with mechanical strength measurements have been applied successfully to a problem related to the pharmaceutical industry. A cone-broken syringe complaint specimen was fractographically examined with the aim to determine the circumstances for failure. Controlled strength experiments and subsequent fracture surface analysis were conducted on representative reference specimens. The results then were compared to the complaint specimen. From this, it was deduced from three facts that severe damaging is likely to be responsible for the failure rather than a mechanical overload or abuse by the operator of the syringe:

1. No distinct or pronounced fracture mirror markings are observed for the complaint specimen.

2. The mirror mist radius of the complaint specimen is much greater than predicted for the specification level breakage force ($R_{m/m(F)}$)(F = 55 N) $\approx$ 553 μm.
3. Severe damaging is required to fulfill the Griffith equation and is also observed in the vicinity of the fracture origin of the complaint specimen.

As it is unlikely that the observed failure-inducing severe circumferential damaging has been introduced by the syringe operator (i.e., patient or health care staff) the responsibility for failure is not seen at the syringe operator, but on the side of the syringe manufacturer or pharmaceutical company (also including contract filling and labeling companies). The reasons for the creation of the severe circumferential damaging are manifold (e.g., during processing at the syringe manufacturer), and thus further investigations (e.g., root cause analysis, process improvements) are required.

FOOTNOTES
*Named after German instrument maker Hermann Wülfing Luer († 1883).
**Visit http://www.schott.com/pharmaceutical_systems/pharma_services for further information.

REFERENCES
[1] G. Quinn, Fractography of Ceramics and Glasses, NIST Special Publication 960-16 (2007).
[2] L. Orr, Practical Analysis of Fractures in Glass Windows, *Materials Research and Standards,* **12**, 21-23 (1972).
[3] J. Mecholsky, Jr., S. Powell, Jr., Fractography of Ceramic and Metal Failures, ASTM Special Technical Publication 827 (1982).
[4] D. Munz, T. Fett, Ceramics – Mechanical Properties, Failure Behaviour, Materials Selection, Springer Verlag (1999).

USING REPLICAS IN FRACTOGRAPHY OF GLASSES AND CERAMICS – ADVANTAGES AND PITFALLS

James R. Varner
Kazuo Inamori School of Engineering
NYS College of Ceramics, Alfred University
Alfred, NY 14802, USA

ABSTRACT
This paper discusses reasons for using replicas of fracture surfaces of ceramics, describes several replicating methods and materials, and presents examples that compare replicas with actual fracture surfaces. Advantages of using replicas include being able (1) to examine large pieces without having to cut them down to size, (2) to provide convenient archiving of fracture surfaces, and (3) to eliminate sub-surface scattering of light. Many times, replicas provide clearer views of untreated (uncoated) fracture surfaces than can be obtained by direct observation. Replication using cellulose acetate tape, polyvinylsiloxane, two types of silicone rubber (filled and unfilled), and epoxy are described. Images of replicas of fracture surfaces of glasses and ceramic materials are used to illustrate the power of replicas and also to show pitfalls, such as artifacts.

INTRODUCTION
Ideally, a replica is an exact, but reversed, impression (copy) of a surface. All surface details are present in the replica, but grooves on the real surface are ridges on the replica, and vice versa. Replicas made by a fractographer must be as close to this ideal as possible, but artifacts (imperfections) can occur, and the fractographer must recognize these artifacts. This paper discusses reasons for using replicas, presents several replicating materials and procedures, provides examples of replicas of fracture surfaces of glasses and several ceramics made using these materials, and compares replicas with the real fracture surfaces. It builds on a paper that was presented at Fractography of Advanced Ceramics III.[1]

REASONS FOR USING REPLICAS
Replicas are not difficult to make, but this extra step that adds time to a fractographic examination needs to be justified on a case-by-case basis. The reasons for making replicas are as follows: (1) look at selected areas of fracture surfaces of large pieces. Some fragments are simply too big to get into the chamber of a scanning electron microscope (SEM) or under the lens of an optical microscope. Cutting the fragment is time consuming and runs a certain risk of damaging the fracture surfaces. In some cases, cutting (or any other destructive methods) may not be permitted. Replicas are a convenient way to examine small areas of large pieces without resorting to cutting. (2) Provide a record of a fracture surface. Even if something is done to a fracture fragment that alters or destroys fracture markings, a replica provides a permanent record of that surface. Images provide a record too, but images are selective, especially when made at higher magnifications. (3) Eliminate sub-surface scattering of light. This is especially important for white ceramics, such as porcelain, but it is useful for all ceramics. Sub-surface light scattering makes it very difficult to see fracture markings. This is an especially important reason for using replicas, and the examples will provide evidence to back up this claim. (4) Use transmitted light to examine fracture surfaces. This assumes that the replicas are transparent, and some replicas are not. Having the option of using either reflected or transmitted light in the examination is an important bonus in many examinations. (5) Provide identical specimens for participants in a course on fractography.

REPLICATING MATERIALS AND PROCEDURES

Cellulose Acetate
Cellulose acetate is available in rolls or sheets. One example is Ernest F. Fullam, Thick Replicating Tape, Part No. 11340. It makes a transparent replica. A small amount of acetone is applied to one side of a piece of cellulose acetate tape or sheet. The cellulose acetate softens in about 15 s. After shaking off any excess acetone, the softened side of the tape or sheet is pressed onto the fracture surface, with the pressure being applied for 5-10 s. The cellulose acetate dries in about 15 minutes. Peel it off carefully and tape the replica to a glass slide to help keep it flat.
Cellulose acetate replicas are quick and easy to make, and they are transparent. The tapes and sheets are thin, and the replicas are prone to artifacts (bubbles, tears, etc.).

Filled Silicone Rubber
Filled silicone rubber comes as two liquid components having different colors. Appropriate amounts of each are measured and mixed together by stirring. The homogeneity of the resulting material is judged by the uniformity of its color. One example is Small Parts, Inc., Reprorubber Thin Pour, part # REP-130TP. The blue and yellow components mix to green, and the goal is to have uniform color with no streaks. The mixed liquid is either poured onto the fracture surface, or the surface is dipped into the liquid. This material sets in about 15 minutes at room temperature. Peel it off carefully and place the replica on a glass slide. Filled silicone rubber is also available in a double-cartridge system (e.g., Struers Repliset). The double cartridge (with the two components) is placed in a "gun," and a mixing tip is placed on the end of the cartridge assembly. Piston action of the "gun" forces the two components into the mixing tip, and the mixed material is applied to the piece being replicated. The cartridge system ensures proper mixing and cuts down on waste.
Filled silicone rubber replicas are quick, relatively inexpensive, easy to make, and set quickly at room temperature. The replicas are not transparent, and this can be a significant drawback. The replicas are also prone to bubbles, but vacuum de-airing prior to pouring can eliminate this problem.

Unfilled (Transparent) Silicone Rubber
Unfilled silicone rubber comes as two liquid components. An example is Dow-Corning Sylgard® 184. Appropriate amounts of each are measured and mixed together by stirring. Special mixing "guns" are available to assure correct proportioning and to minimize introduction of air during stirring. The mixed liquid is less viscous than filled silicon rubber, so it may be necessary to contain the liquid until it polymerizes. Unfilled silicon rubber sets in about 24 h at room temperature. It sets much faster if warmed.
Unfilled silicone rubber flows readily into pores and other tight spaces. Replicas are transparent, relatively simple to make, and not especially prone to artifacts (except bubbles). Unfilled silicone rubber is relatively expensive, has a low viscosity, and sets slowly at room temperature.

Polyvinylsiloxane
Polyvinylsiloxane (PVS or VPS) is a family of materials that comes in a variety of viscosities and setting times. In any case, this is a two-component system that is available in double cartridges for use in dispensing "guns" that assure proper mixing and ease of delivery. These materials are used in dentistry to make impressions in patients' mouths, so they are safe and have fast setting times (just a few minutes). Viscosities are usually relatively high, so even vertical fracture surfaces can be replicated without having to build a form to contain the liquid.

PVS replicas are easy to make, and replication of surface details is excellent. The cost per replica is low. These replicas are not transparent, and contact with sulfur must be avoided, since sulfur poisons polymerization.

Epoxy

Epoxy is used to make positive replicas from negative replicas. Epoxy applied directly to a glass or ceramic fracture surface will almost certainly stick to the surface. All of the materials described above are applied directly to fracture surfaces, can be removed easily from these surfaces after setting up, but provide negative copies of the original surfaces. The best combination is to make a PVS replica, which sets up in just a few minutes, and then use this as a mold that gets filled with liquid epoxy. A variety of epoxy materials are available from suppliers such as Struers and Buehler, but all are two-component systems that need to be mixed thoroughly. The viscosities are usually fairly low, so the epoxy needs to be contained until it sets (which takes about 24 h).

Once set, the epoxy replicas are water clear, strong, and hard. They can be viewed using any type of illumination in an optical microscope, or they can be viewed using an SEM.

EXAMPLES

The following figures show images that illustrate success with replicas or which show some of the pitfalls encountered when making and using replicas. Figure 1 shows a glass specimen with an internal fracture origin. The internal origin, a refractory "stone," is clearly seen. Figures 2, 3 and 4 show this origin as seen in the three replicas. Each of the replicas captured surface details, but the best result was obtained with the clear silicone rubber. The filled silicone rubber replica (Fig. 2) does not work with transmitted light, and the reflected-light image (Fig. 3) has uneven brightness. The cellulose acetate replica is thin, so ridges on the other side of the replica disturb the image, and there are artifacts (regions where there was inadequate contact with the specimen surface). Still, the essential details of the origin and fracture markings are seen clearly.

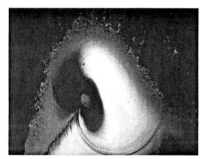

Figure 1 (on left). Reflected-light image of a glass specimen with a refractory "stone" as an internal origin; image width ~ 12 mm. Figure 2 (on right). Filled silicone rubber replica; reflected-light image showing origin and fracture markings of the glass specimen shown in Fig. 1.

Illumination is always a key issue when using a light microscope to observe fracture surfaces. Reflected light works well with both glass and ceramic fracture surfaces. Near-normal incidence usually works best for glass, while oblique incidence usually works best for ceramics. Transmitted light often works well with glass specimens, especially with rough surfaces. Ceramic specimens

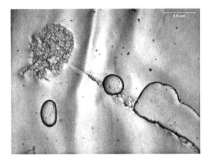

Figure 3 (left). Cellulose acetate replica; transmitted-light image of this same specimen; notice the out-of-focus ridges (on the bottom side of the replica) and the regions where there was no contact between the cellulose acetate film and the fracture surface. Figure 4 (right). Clear silicone rubber replica; transmitted-light image of this same specimen; excellent detail with no artifacts.

usually cannot be observed using transmitted light, but transparent replicas of ceramic specimens work very well in transmitted light. The next images show the usefulness of being able to use transmitted light with replicas to see details of fracture surfaces.

Figure 5 (on left) and Figure 6 (on right). These images show an edge chip on a glass rod. Both were taken using transmitted light. Figure 5 (the original specimen) has significant sub-surface reflections at the edge, but these are of course missing from Figure 6 (a clear silicone rubber replica). (Images courtesy of Kimberly Polishchuk.)

With glass specimens, transmitted light will reveal sub-surface cracks, which is useful information. However, sub-surface reflections also obscure surface details. Figures 5 and 6 are good examples of these points. Figures 7 and 8 provide strong evidence for how effective transmitted light can be when observing water-clear replicas of ceramic fracture surfaces. Details, such as wake hackle at pores and bubbles, get lost in the "noise" of scattered sub-surface light when looking at ceramic specimens using reflected light.

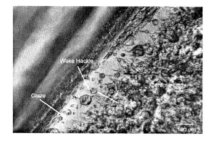

Figure 7 (left). Cellulose acetate replica of glazed electrical porcelain; transmitted light; note the wake hackle at bubbles in the glaze; image courtesy of WonBum Lee. Figure 8 (right). Clear silicone rubber replica of glazed electrical porcelain; transmitted light; composite of two images; image courtesy of Tim Nedimyer.

Making replicas of selected regions of very large specimens is an excellent way to deal with problems associated with doing microscopy of very large pieces. The SiC piece shown in Fig. 9 and 10 is obviously much too large to fit on the stage on an optical microscope or in the chamber on an SEM. Rather than cutting the specimen, which is undesirable for several reasons (cost, time, partial destruction of the specimen), one or more replicas can be made of key regions. In this case, the origin is at one edge, as seen in Fig. 11. A filled silicone rubber replica was made at this location, and an epoxy replica was made from the silicone rubber replica, producing a positive replica of the fracture surface. The epoxy replica was sputtered coated with Au/Pd, and a reflected-light image of this replica is shown in Fig. 12. No images are shown in this article, but this replica was also examined in an SEM.

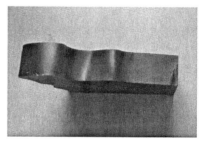

Figure 9 (left) and 10 (right). Two views of a SiC specimen that is clearly too large to put on the stage of an optical microscope or in the chamber of an SEM.

 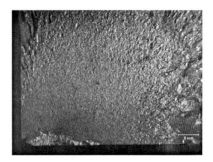

Figure 11 (left). Camera image of the original SiC specimen shown in Fig. 9 and 10; the fracture origin is at the lower left corner. Figure 12 (right). Epoxy replica of this specimen that has been sputter coated with a thin Au/Pd layer; reflected-light image of the origin (along the lower edge) and fracture mirror.

A very effective use of replicas is in teaching fractography. People who are learning about fractography of glasses and ceramics need to look at specimens in order to get used to recognizing and interpreting fracture markings. People also need to recognize features, such as edge chips, that are post-fracture damage, and which have nothing to do with the main event. Usually every class participant has a different specimen. In a set of broken glass rods, although all specimens will have a mirror, Wallner lines, and mist/velocity hackle, individual features will be different. Some specimens will have edge chips, some may have a missing origin, some may have partial mirrors, some may have asymmetric mirrors, etc. For a first look at a broken glass rod, replicas provide the means for every class participant to see exactly the same fracture markings, mirror shape, origin location, etc. Any peculiarities (such as edge chips) can be pointed out and discussed. The next step is to have the students look at real specimens, and they are better prepared to recognize non-essential features and distinctive features.

Figures 13-16 show images of a broken glass rod and of an epoxy replica of this same rod. Multiple PVS replicas were made of the glass rod, and an epoxy replica was then made of each PVS replica. As discussed earlier, this provides a positive replica of the original specimen. In fact, since the epoxy is water clear, it is easy to mistake a replica for the real thing. This is most evident in Fig. 13 and 14, which were taken using transmitted light. Figures 15 and 16 show that the two-step replication process produces a positive replica that duplicates even subtle markings like Wallner lines exactly.

Figures 17 and 18 provide another example of replicas used in teaching. This example also provides evidence for another reason for making replicas; namely, to preserve the appearance of the original specimen. Notice in Fig. 17 that there is missing glass next to the origin, but the replica (Fig. 18) shows the fracture surface with an intact fracture surface. This mystery can be is easily explained. The replicas were made while the original specimen was, in fact, completely intact. However, no images were made of the original specimen before this was done. Before images were taken of the original specimen, it was accidently damaged. Thus, images of the original specimen have the missing glass, but the replicas show the intact specimen.

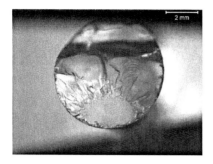 

Figure 13 (left). Broken glass rod; transmitted light. Figure 14 (right).
Epoxy replica of the rod shown in Fig. 13; transmitted light.

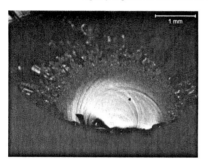 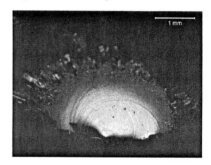

Figure 15 (left) and 16 (right). Same glass rod as shown in Fig. 13 and 14;
original specimen is shown in Fig. 15; epoxy replica is shown in Fig. 16; transmitted light.

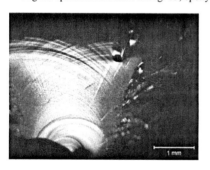 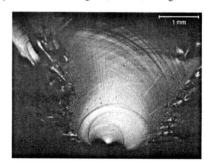

Figure 17 (left). Broken glass rod; notice missing glass to the next to the origin; reflected light. Figure
18 (right). Epoxy replica of the glass rod shown in Fig. 17; replica was made before the original
specimen was damaged; the fracture surface is intact; reflected light.

Replicas exactly duplicate fracture surfaces, when everything goes right. However, artifacts and other pitfalls can happen, and the fractographer needs to recognize these. Most replicating materials are liquids when they are applied to fracture surfaces; therefore, bubbles are the most common artifact (see Fig. 20 and 22). Cellulose acetate is soft, and the most common artifact regions where there was no contact between the soft cellulose acetate and the fracture surface (see Fig. 3 and 19), and tears in the replica (more likely to occur, if it was removed before it set up).

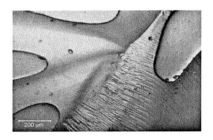 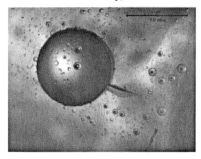

Figure 19 (left). Cellulose acetate replica of a glass fracture surface; notice the excellent duplication of fine twist hackle, and the regions where there was no contact between the cellulose acetate and the fracture surface during replication; transmitted light. Figure 20 (right). Epoxy replica of a glass fracture surface showing a bubble and wake hackle; notice the small bubbles on the surface of the replica; the reason for these bubbles is not known; transmitted light.

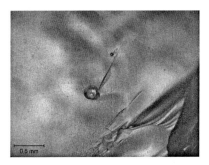 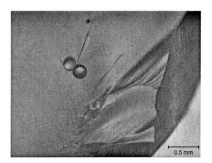

Figure 21 (left). Image of a glass fracture surface showing two bubbles (one is very small) with wake hackle; transmitted light. Figure 22 (right). Image of an epoxy replica of the specimen shown in Fig. 21; the two bubbles and wake hackle have been duplicated exactly, but there is a bubble (an artifact) in the replica right next to the larger, real bubble; transmitted light.

In addition to artifacts, other problems can occur in the replication process. When using PVS, be aware that sulfur poisons the polymerization reaction, and the PVS never sets up. Sulfur is sometimes found in latex gloves, and sulfur cement is sometimes used to attach metal hardware to electrical porcelain parts. Small spheres sometimes appear on the surface of epoxy replicas. These may simply be annoying, or they may render the replica useless. Epoxy seems to react with filled silicone rubber, evidenced by a change in color of the silicone rubber where contact with the epoxy had occurred. The replica, however, may still be fine. Finally, the replica may pick up loose material on fracture surfaces. This may not be a serious issue, but it should be disclosed when discussing replication with clients. It can be a serious issue, when there are deposits on a fracture surface, since the deposits may be evidence, and replication may partially or complete remove them. If all parties agree to proceed with replication, samples of surface deposits should be preserved prior to replication. If similar deposits occur on both halves of a fracture surface, replicating only one side is an intermediate step in an analysis. Where possible, SEM observations with EDS analysis should be done prior to replication.

CONCLUSION

Replicas of glass and ceramic fracture surfaces are relatively quick and easy to make. The procedures are nondestructive. (Some loose fragments or deposits may be pulled from the surface.) Replicas provide the only way to examine fracture surfaces of very large objects without cutting the pieces down to an acceptable size. Transparent replicas allow a much better view of the fracture-surface markings (especially of ceramics) without having to coat the surface. Sub-surface light scattering is eliminated, and transmitted light can be used. Both cellulose acetate and clear silicone rubber provide transparent replicas in one step. Cellulose acetate replicas are easier and faster to make, but they are more prone to artifacts, and they are thin (the other side of the replica may interfere with the view of the fracture markings). Clear silicone rubber is more difficult to apply, since it has a low viscosity, and it takes about 24 hours to set at room temperature. However, it yields the highest-quality replicas. Epoxy replicas of first-step replicas provide hard, durable, transparent replicas that also work in an SEM. Artifacts can occur, especially bubbles and areas of inadequate contact between the replicating material and the fracture surface. Fractographers need to recognize artifacts in order to avoid misinterpreting them as fracture markings.

REFERENCES

[1]J.R. Varner, Replicas as a Technique for Examining the Fracture Surfaces of Ceramics," *Fractography of Advanced Ceramics III*, J. Dusza, et al., Ed., Trans Tech Publications, Zurich (2009).

Author Index

Author Index

Thompson, J. Y., 265
Tsirk, A., 123

Uemura, K., 255

Varner, J. R., 299

Witschnig, S., 225

Yasuda, K., 255
Yoshimura, Y. N., 205